Advanced Coatings and Films for Food Packing and Storage

Advanced Coatings and Films for Food Packing and Storage

Guest Editors

Jun Mei
Jing Xie
Biao Zhang

Basel • Beijing • Wuhan • Barcelona • Belgrade • Novi Sad • Cluj • Manchester

Guest Editors

Jun Mei	Jing Xie	Biao Zhang
Shanghai Ocean University	Shanghai Ocean University	China Jiliang University
Shanghai	Shanghai	Hangzhou
China	China	China

Editorial Office
MDPI AG
Grosspeteranlage 5
4052 Basel, Switzerland

This is a reprint of the Special Issue, published open access by the journal *Coatings* (ISSN 2079-6412), freely accessible at: https://www.mdpi.com/journal/coatings/special_issues/coatings_films_food.

For citation purposes, cite each article independently as indicated on the article page online and as indicated below:

Lastname, A.A.; Lastname, B.B. Article Title. *Journal Name* **Year**, *Volume Number*, Page Range.

ISBN 978-3-7258-2707-7 (Hbk)
ISBN 978-3-7258-2708-4 (PDF)
https://doi.org/10.3390/books978-3-7258-2708-4

Contents

About the Editors

Jun Mei

Jun Mei (1986) graduated from Shanghai Jiaotong University, China, with an M.Sc. in Food technology (2011) and obtained his PhD in Biomedical Engineering (Biotechnology) in 2015. From 2015 to 2017, he worked as a post-doctorate in the postdoctoral research center for horticulture in SJTU. He is currently an associate professor with the Laboratory of Food Cold Chain Logistics and Quality Control at Shanghai Ocean University, China. His research focuses on exploring the mechanism associated with the deterioration of quality in the process of food cold chain preservation, the development of key technologies for food cold chain preservation, and the development of key technologies for aquatic products during keep-live transport. The aim of his research is to understand the main causes of food spoilage during the cold chain preservation process, to develop preservatives from spices and combine physical bacteria-reducing and bacteriostatic techniques to maintain the quality of food in the cold chain process, and to exploit techniques for the transport of fish, such as temporary rearing before transportation, gradient cooling and cold acclimation, and the use of anti-stress agents. He is the Assistant Editor of Food Quality & Safety and a Guest Editor for Coatings and Frontiers in Nutrition. MEI is undertaking or has undertaken three subsidiary subjects from the SKRD Program and one project from NSFC, and he also teaches undergraduate and postgraduate courses at Shanghai Ocean University. He is a (co-)advisor of 18 MS students and currently supervises 15 MS students. He has co-authored over 80 publications and holds two patents.

Jing Xie

Jing Xie (1968) obtained her PhD in Power Engineering and Engineering Thermophysics from the University of Shanghai for Science and Technology in China (2000). From 2001 to 2002, she worked as a post-doctorate in the postdoctoral research center for Food Science at Oregon State University. She has worked at Shanghai Ocean University since 1993 and is the dean of the School of Food Science and Technology. Currently, she is the principal investigator of the Laboratory of Food Cold Chain Logistics and Quality Control at Shanghai Ocean University. Her research focuses on exploring the cold adaptation mechanism of spoilage microorganisms in the cold chain logistics process of aquatic products and developing new cold chain preservation technology and equipment. The purpose of her research is to understand the mechanism associated with the adaptation of spoilage microorganisms to cold chain low temperatures from the aspects of microbial morphological structure and cell membrane composition and metabolism; she also aims to develop new cold chain preservation technology and equipment to inhibit microbial growth, thereby extending the shelf life of food. XIE is undertaking or has undertaken more than 10 national research projects. At Shanghai Ocean University, she teaches undergraduate and postgraduate courses. She is a (co-)advisor of 30 PhD students and currently supervises nine PhD students. She has co-authored over 200 publications and holds 39 patents.

Biao Zhang

Dr. Biao Zhang is a Doctor of Engineering and lecturer at China Jiliang University. He received his PhD in Engineering from the Tianjin University of Science and Technology in 2021. He is committed to the development of highly sensitive and rapid immune detection technology and express products for small molecular hazards such as pesticide and veterinary drug residues, biotoxins, and heterocyclic amines produced during food processing. He has studied the rapid detection methods of biogenic amines, heterocyclic amines, fluoroquinolones antibiotics, oxomethoate, and other pesticides in food, obtaining a series of research results. So far, he has worked in the fields of biosensors, bioelectronics, sensors, and actuators, as well as in the Journal of Agricultural and Food Chemistry. He has published 14 SCI papers in international journals such as Chemical and Food Chemistry; he has also applied for 11 invention patents, an authorized invention patent, and a utility model.

Preface

Consumers have high requirements for food quality and safety, and there is an urgent need to develop new types of packaging to extend the shelf life of food. Active packaging systems can use natural biopolymers such as polysaccharides, proteins and lipids as film-forming materials, and are prepared by adding active ingredients such as antibacterial agents, antioxidants and deoxidizers. The active ingredients can be released or absorbed into the packaged food or its surroundings to act as a bacteriostatic and antioxidant. It is necessary to consider the film-forming characteristics of the material, the added active ingredients, and the physical and chemical properties of the food. At the same time, with consumers concerned about the safety of chemical synthetic substances, researchers are endeavoring to find natural active ingredients that can extend the shelf life of food. This Special Issue summarizes the latest research on active packaging from the perspective of natural biopolymer film-forming materials and the development of natural active ingredients.

Modifying film-forming substrates or adding active ingredients to change the film-forming properties is crucial in active packaging when aiming to extend the shelf life of food. Pashova introduced edible coatings based on lipids and applied them in food preservation. An effective method for extracting/separating plant wax from the cuticle and waste peel of fresh vegetables and fruits before processing (high alcoholic beverages, wine, fructose preserves, fruit juices, vegetable preserves, etc.) was proposed and the wax was prepared into an edible film. Gil and Rudy systematically summarized the composition of the packaging materials and bioactive substances used in the packaging of meat and meat products. A packaging containing active compounds was found to extend the shelf life of foods by reducing the occurrence of adverse effects during storage, such as microbial growth, oxidation and water loss, which also reduce the use of preservatives. Song prepared a double layer of edible iron bean starch containing a waterproof and oilproof layer of corn zein using the casting method, and studied its physical and chemical properties, microstructure and mechanical properties. Sabbah et al. added olive leaf extract (OLE) or guava leaf extract (GLE) to edible pectin films to improve their film-forming performance. The results showed that OLE or GLE could significantly improve the opacity, greenness, antioxidant activity, moisture absorption rate of the film and reduce its water vapor permeability. Treviño-Garza et al. added cardamom (MCA) and officinalis (MCO) to flaxseed slime as the substrate, and found that the addition of MCA and MCO changed the color, increased the total phenol content and antioxidant activity (DPPH and ABTS), and reduced the hardness, deformation and fracturability. The structure of MCO films is more uniform (heterogeneous structure with protrusions). Weng et al. prepared a sodium alginate-based repellent and polymethyl dimethacrylate-based water repellent via crosslinking and modification, mixing these in different proportions to obtain a composite water repellent and oil repellent. They then coated the molded pulp using the dipping coating method, so that the coated paper had excellent oil resistance and excellent water resistance and could be used for food and non-food contact applications. Tarnowiecka-Kuca et al. improved the barrier properties of paper for grease, water, oil or gas (oxygen and water vapor) by covering selected papers with biodegradable coating carriers based on polyvinyl and cellulose nanocrystals alcohol. The covering of the paper improved the barrier properties of grease, water and oil, and the best results were obtained by applying a J12 coating to Gerstar 70 g/m2 using flexo-printing technology. Salimian and Onwukwe used the plasma sputtering process to create a coating that was transparent to the visible spectrum while effectively reflecting infrared radiation above 1500 nm. By using thermal imaging and computer vision AI models, the coating label could accurately distinguish plastics associated with food packaging at the initial stage of plastic recycling. The system achieved 100% accuracy in separating food-grade plastics from other types of plastics.

Since fresh food, such as fruits, vegetables and fish, are highly perishable, they need to be preserved after harvest to extend their shelf life. The addition of antioxidants and antibacterial active substances to the edible coating can prevent the loss of water, microbial deterioration, gas exchange, respiration and so on during the storage and transportation of fresh food. Cvanić et al. summarize the latest research progress regarding the use of nano-emulsion coatings for fresh food, providing valuable information on preparation and application methods, and the application of polymers and bioactive substances. In addition, the effects of nanoemulsion coating applications on fresh fruits and vegetables during storage (shelf life, color, weight loss, etc.) are described in detail. Šuput et al. prepared a biopolymer coating containing Mentha piperita essential oil based on pumpkin cake and applied it to the preservation of Afus Ali variety of grapes; they found that the biopolymer coating could inhibit the growth of microorganisms and improve the antioxidant capacity of grapes. Liu et al. studied the effects of ozone treatment at different concentrations, treatment times and temperatures against Penicillium citrinum and Rhizopus stolonifer in spoiled fresh peeled garlic. The results showed that when the ozone concentration was 6 ppm, the time was 15 min and the temperature was 20 °C, the total number of yeast and mold colonies, the incidence of decay, the diameter of the lesion and the depth of decay were significantly reduced during storage, and that the antibacterial effect was significantly enhanced. Therefore, it extended the shelf life of fresh peeled garlic. Yang et al. used active coatings comprising flaxseed gum and sodium alginate containing carvacrol to maintain the quality of turbot (Scophthalmus maximus) during cold storage, and found that the active coating reduced the value of odor-related compounds, inhibited the growth of spoilage microorganisms, maintained the structure of proteins, and extended the shelf life of halibut in cold storage.

Jun Mei, Jing Xie, and Biao Zhang
Guest Editors

 *coatings*

Review

Application of Plant Waxes in Edible Coatings

Sabka Pashova

Commodity Science Department, University of Economics, 9002 Varna, Bulgaria; spashova@ue-varna.bg;
Tel.: +359-889-859-564

Abstract: The aim of the paper is to present edible coatings based on lipids and their application in the food industry. Therefore, this paper discusses the following: different types of plant waxes; the need for plant waxes; the advantages and disadvantages of edible coatings based on plant waxes; edible coatings based on lipids applied in the food industry; application of the most popular Carnauba wax; Candelilla Wax in the composition of edible coatings. Plant waxes are presented with their specific characteristics. Moreover, the cuticle waxes obtained from waste peels of fresh fruits and vegetables are presented; their properties and application in the composition of edible coatings are based on plant waxes. In this regard, an effective and applicable method for the industrial extraction/separation of plant wax from the cuticle and waste peels of fresh fruits and vegetables before their processing (production of wine, high-alcohol beverages, fruit-sugar preserves, vegetable preserves, juices, etc.) is proposed. Properties and possible applications of the isolated cuticle plant waxes are presented.

Keywords: plant waxes; cuticle; edible coatings

Citation: Pashova, S. Application of Plant Waxes in Edible Coatings. *Coatings* **2023**, *13*, 911. https://doi.org/10.3390/coatings13050911

Academic Editors: Jun Mei and Stefano Farris

Received: 19 December 2022
Revised: 23 April 2023
Accepted: 9 May 2023
Published: 12 May 2023

1. Introduction

Natural waxes are of plant or animal origin. They are resistant to moisture, oxidation, and microbiological decomposition. Plant waxes of commercial and industrial importance are obtained from various plant species and are used to produce cosmetic products, ink, varnishes, luster, candles, pastels, etc. Some of them are used in the composition of some edible coatings applied to various foodstuffs. By using them, the properties and quality of food products are preserved during storage.

Fresh fruits and vegetables have a natural shell in the form of skin, covered by a cuticle (a natural shell of higher plants that covers and protects the flowers, leaves, fruits, and vegetables from adverse external influences) (Figure 1) and a shell in the case of nuts [1–4]. These natural barriers regulate the exchange of oxygen, carbon dioxide, and water and reduce the loss of volatile flavor and aroma substances. Nowadays, coating emulsions are used for foods, made up of different components, including wax, shellac, beeswax, morpholine, and candelilla, usually combined with carnauba wax [5–7].

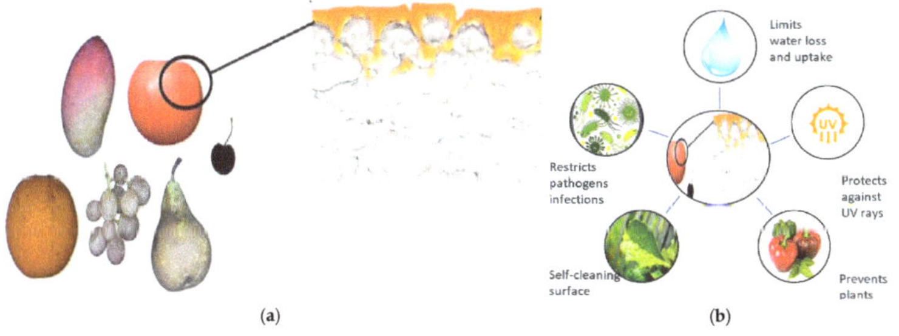

Figure 1. Plant cuticle (**a**) and its functions (**b**). Source: Fernández-Muñoz et al., 2022; Martin and Rose, 2014 [4,5].

In modern conditions, uncoated foodstuffs produced via food technology are not protected by a natural coating. Therefore, various edible coatings are applied to them, which protect the product from the adverse effects of the environment, thus preserving its properties and quality. Usually, the quality of foods decreases as a result of the loss of water and aromatic substances and as a result of the influence of oxygen on the individual components of the composition of the product. Therefore, attention should be paid to the properties of the used edible coatings for food products, regarding their permeability to oxygen, water vapor, and aromatic substances [8].

According to the EU Directive [9,10] and USA regulations [11], edible coatings are classified as food products, food components, food additives, and substances in direct contact with food or food packaging material. They are part of the foods and, therefore, should meet the requirements for the components contained in them. The components in the content of coatings must be non-toxic, and all processes related to their preparation and application to foods must comply with hygienic requirements [12,13].

Other very important requirements for edible coatings are to be safe and cover/meet the requirements of the legislation of a certain country and international recommendations, including those of the Food and Drug Administration (FDA), Codex Alimentarius Commission [14], and the European Union [15].

The US Food and Drug Administration (FDA) provides that all components that should be included in the composition of edible coatings must be safe, or the used food additives should be in a specifically defined amount.

In Europe, the components that can be included in the composition of edible coatings are considered food additives and are included in the list of additives for general use (ED 1995). In each of the cases, the use of the individual components of the composition of edible coatings is allowed, but only when the principle "in reasonable quantities" is observed. This directive has been supplemented with the introduction of specific criteria for the purity of food additives [16]. Edible films may contain components with a functional purpose, which should be written on the label of the product. In Europe, food additives must be reflected on the label, referred to the relevant category (antioxidants, preservatives, colors, etc.) with their full name or with their E-number.

Each country has its own requirements regarding the permitted additives (ED, 1995, USDA—U. S. Food and Drug Administration, 2006). According to the US requirements, organic acids—acetic, lactic, citric, malic, propionic, tartaric, and their salts—are accepted as safe and applicable for general use [17]. On the other hand, most of the essential oils used in the food, pharmaceutical, and cosmetic industries are also classified as safe and approved for use as food additives [18].

Another very important point in the regulatory framework is the presence of allergens. Most of the edible coatings contain components that can cause allergic reactions. Some of them are fresh milk, soy, fish, peanuts, oilseeds, and cereals, and some edible coatings contain milk protein (whey protein and casein), wheat protein (gluten), soy protein, and proteins from peanuts and tree nuts. Therefore, the allergens in the composition of edible coatings must be accurately and clearly presented on the label of the product on which the coating is applied [19].

The purpose of this study is to present plant waxes, lipid-based edible coatings applied in the food industry, modern trends of industrial separation of plant waxes from the cuticle of waste peels obtained from fresh fruits and vegetables before their processing, and their use in the composition of edible coatings. To achieve the research objective, scientific works and regulations were studied and subjected to systematic analysis.

2. Plant Waxes—Types

Waxes are esters of high molecular weight monohydroxy alcohols and high molecular weight carboxylic acids. They are chemically different from fats and oils, from hydrocarbon or paraffin waxes, and from synthetic waxes such as carbowax [20]. Natural waxes are used in the food and cosmetic industries and have many other industrial applications [21].

Plant waxes are waterproofing components found on the outer surface of plants. They are the main barrier against environmental stress. Some of them have gained GRAS (Generally Regarded as Safe) status approved by FDA, which allows their use in foods. Plant waxes such as candelilla wax, carnauba wax, rice bran wax, sunflower wax, etc. (Table 1) with commercial and industrial importance are derived from different plants and used for the production of cosmetics, ink, varnish, polish, candles, crayons, etc. They are also used to obtain edible coatings for various food commodities. It was found that they maintain the properties and the quality of foods during their term of storage [22].

Table 1. Plant Waxes.

Name	Plant Source of Wax	Application
Carnauba Wax	Palm tree	Coatings, cosmetics, automobiles, furniture wax, etc.
Candelilla Wax	Bush	Cosmetics and coatings
Japan Wax	Tree	Candles and varnishes
Ouricury Wax	Palm tree	Inks, varnishes
Rice Bran Wax	Rice husks	Coatings, chewing gums, candles, textiles, lubricants, cosmetics, etc.
Sunflower Wax	Seed and Seed Hulls	Cosmetics (lipsticks, mascaras, lip balms), emulsions, etc.
Laurel Wax	Fruit of the Myrica Pubescens (laurel) shrub	Scented candles Natural cosmetics and personal care products (lip balms, lipstick, lotion bars, hair pomade, ointments, salves, mascaras, creams, and lotions)

Table 1. *Cont.*

Name	Plant Source of Wax	Application
Cane Wax	Cane (dry wastes after its treatment for sugar manufacturing)	Main substitute for Carnauba wax
Cuticle Wax	Cuticular wax (obtained from the peel wastes of fresh fruit and vegetables)	Coatings, packaging, etc.
	Berry wax (obtained from the peel wastes of berry fruits)	Food packaging, cosmetic industries, and coatings

Source: own research.

The main source of Carnauba wax (CW), sometimes called the "Queen of Waxes", is the Brazilian palm tree (Coernicia cerifera Martius), also known as carnauba wax palm. The wax is found on the surface of the palm leaves. One of the unique characteristics of carnauba wax is that it contains long-chain alcohols and esters: unesterified alcohols (12%), x-hydroxy esters (14%), and esters of hydroxylated cinnamic acid (30%). Carnauba wax is one the hardest plant waxes, with melting temperatures ranging from 82.5 to 83 °C. Therefore, it is often used as a hardener to elevate the melting temperature of wax mixtures. Carnauba wax is the most commercially important plant wax. It has extensive applications in foods, confectionery coatings, cosmetics, automobiles, furniture wax, etc. [21], and CW-based emulsions, edible coatings, oleogels, etc. [23]. Food-related use of carnauba wax is outlined in the FDA regulation 21 CFR 184.1978 [22].

Candelilla wax, or The "Great Wax Rush" (Figure 2), is mainly obtained from the leaves of the plant E. antisyphilitica Zuccarini native to northern Mexican and southwest Texas [24]. Unpurified candelilla wax contains approximately 42% hydrocarbons, 39% wax, resin, and sitosteroyl esters, 8% free wax and resin acids, 6% lactones, and 5% free wax and resin alcohol. Candelilla wax is an FDA-approved food additive used for glazing in certain foods. Microemulsions of candelilla wax are used as coatings with good gloss for fruits [25].

Figure 2. Different Candelilla wax products, which differ in their presentation and refining grade. Source: Candelilla Institute, http://www.candelilla.org/?page_id=2, accessed on 1 March 2023.

Japan Wax (木蝋 Mokurō), known as sumac wax, sumach wax, vegetable wax, China green tallow, and Japan tallow, is a pale-yellow, waxy, water-insoluble solid with a gummy feel obtained from the berries and stems of the Rhus succedanea tree. It is not a true wax and contains 5% of fatty acid, resulting in more viscosity than other waxes. It thickens formulas, as well as gives a nice, malleable texture to the blend. It has been used by sumo wrestlers and geishas for centuries to shape and maintain their hairstyle. It is frequently used nowadays in haircare products. Japan wax is used in the composition of candles, furniture polishes, floor waxes, wax matches, soaps, food packaging, pharmaceuticals, cosmetics, pastels, crayons, buffing compounds, metal lubricants, adhesives, thermoplastic resins, and as a substitute for beeswax. Because it undergoes rancidification, it is not often used in foods [26].

Ouricury wax is a brown-colored wax obtained from the leaves of a Brazilian feather palm (*Syagrus coronata*) by scraping the leaf surface. Harvesting ouricury wax is more difficult than carnauba wax, as ouricury wax does not flake off the surface of the leaves. The physical properties of ouricury wax resemble carnauba wax, so it can be used as a substitute where light color is not required, e.g., in carbon paper inks, molding lubricants, and polishes [22,27].

Rice bran wax is another high-melting vegetable wax found in rice husks (*O. sativa*). It is obtained as a by-product from the de-waxing of rice bran oil [22]. The rice bran wax is a reddish brown, dark composite separated from rice bran oil and has good nutritional properties. It is separated from rice bran, which has several health-benefiting components. Rice bran wax (RBW) has been approved by the FDA as a safe food additive (21 CFR, 172.890). It has extensive applications due to its relatively low cost and abundance in Asia. It can be used as an efficient edible coating substance due to its excellent film-forming properties. The RBW has been effectively used in several food applications, such as forming oleogels, structured lipids, and edible coating of food products [28]. Food applications include fruit and vegetable coatings, confectionery, and chewing gums; other applications include candles, textiles, lubricants, cosmetics, etc. [29].

Sunflower wax is found in the seed and seed hulls of sunflower (*H. annuus*). It is obtained through the winterization of sunflower oil. Like rice bran wax, it is a hard wax. Sunflower wax has not been approved for GRAS status by the FDA. Therefore, its applications are primarily in cosmetics, such as lipsticks, mascaras, lip balms, emulsions, etc. It functions as a consistency modifier and regulates the hardness, texture, and mold release of cosmetic formulations. It can also be used as a replacement for candelilla wax, carnauba wax, and rice bran wax [21].

The cuticle is synthesized in the epidermal cells of various plants, such as leaves, fruits, stems, and flowers, as a barrier between the plant surface and the environment. The cuticle is a hydrophobic layer composed of two lipophilic components, cuticular wax, and cutin. One of the most important functions of the cuticle is due to its hydrophobic component: the waxes [30]. Cuticular wax (obtained from the leaf/peel wastes of fresh fruit and vegetables, etc.) is a continuous layer outside or within the cuticle of plants. The presence of waxes on the leaf/fruit surface affects the degree of colorization, and the effect varies according to the plant species [31]. Berry wax is isolated from the cuticle of different varieties of berry fruits. It is obtained from the peel berry wastes from the berry juice industry and is a potential source of natural waxes. Waxes find application in food packaging, cosmetic industries, coatings, etc. [32].

Only the plant waxes approved by the FDA are used in the composition of wax coatings applied in the food industry. There is a high interest and a lot of research presenting wax coating with incorporated bioactive compounds.

3. Necessity, Advantages, and Disadvantages of Edible Coatings Based on Plant Waxes

Coatings based on plant waxes used in food technology are subtle, invisible to the naked eye, and safe for consumers. Applied to food products, they slow down the loss of moisture, gases (oxygen and carbon dioxide), and volatile components of the composition, and they improve the appearance and mechanical properties of the product.

Coatings based on plant waxes are also considered as a layer of edible materials used and located between the components of the product, providing a barrier to overall exchange. In other words, edible coatings represent an integrated part of the food product that does not change its sensory properties [33,34].

Edible coatings are consumed together with the product on which they are applied. Therefore, the components included in their composition, such as antimicrobials, antioxidants, and food additives, should not affect the properties of food.

The addition of some anti-browning components in edible coatings can cause an unpleasant nuance in the aroma, especially when their concentration is high [35]. There are limited studies that present the sensor characteristics of fresh-cut fruits with applied coatings in the composition, in which food additives are included. The taste of these components is considered an extremely important point, as most of the components have a naturally bitter, astringent, or another aftertaste that will change consumer preferences. It was found that the presence of vitamin E reduces the surface gloss of coated strawberries, and this affects the appearance of the product and consumer preferences [36].

The need to use edible coatings based on plant waxes in food technology is directly related to the fact that they have some advantages over synthetic ones (Table 2). A major advantage of edible over synthetic coatings is that they can be consumed together with the food on which they are applied. Their use contributes to reducing environmental pollution. The coatings are prepared entirely from plant waxes, so their degradation is faster than polymer materials. The coatings can emphasize the organoleptic properties of the product on which they are applied since they contain various components (flavoring substances, coloring agents, and sweeteners). Coatings can be used for individual packaging of small quantities of a certain product, especially for goods that are not individually packaged, such as pears, nuts, strawberries, etc. It is possible to apply them inside heterogeneous products between individual layers. Thus, they protect the product from becoming wet and prevent the passage of moisture into it (pizza, pie, candies, etc.). Coatings can also be considered a source of antimicrobial, antioxidant, and anti-browning (cut, filleted fruit) components. They are most often applied to the surface of the product, thus preventing the penetration of unwanted components through the surface to the interior of the product. It is also possible to use edible coatings with multilayer packaging materials together with non-edible ones. In this case, the edible coating is applied as an inner layer directly on the product. There is negligible waste and environmental pollution in the production of edible coatings, and their permeability and mechanical properties are better than those of synthetic films [37–41].

The disadvantages of coatings are connected with their composition and properties, with the thickness of the layer applied on the surface of foods, the cost of the components included in the composition of coatings and the absence of requirements for the composition and application of edible coatings [42–44].

Thick coatings prohibit oxygen exchange and cause off-flavor development. Sometimes due to the surface of foods and chosen method of applying the edible coatings a thick layer of the coating can form, which reduces the properties and efficiency of the edible coatings. Some coatings are hygroscopic in nature, which helps to increase the growth of certain microorganisms, which do not allow coatings to fulfill their main purpose. For this reason, it is necessary to choose the right components (including antimicrobial agents if necessary) in the composition of edible coatings. The high cost of the components included in the composition of edible coatings increases the cost and reduces their economic efficiency. In most cases, edible coatings are more expensive than synthetic packaging, and this is one of the reasons connected with their application in the food industry. Another obstacle is the lack of sufficient information on the use of machinery and equipment for the production of edible coatings with different compositions. Finally, there are no separate requirements for the composition and use of edible coatings [43,45].

Table 2. Advantages and disadvantages of edible coatings.

Advantages	Disadvantages
Safe, not containing toxic, allergic substances; well digestible.	Thick coating can prohibit oxygen exchange and causes off-flavor development.
Improve food appearance and retention of acids, color, flavor, and sugar.	Causes anaerobic respiration of fresh fruits and vegetables due to the normal ripening during storage.
Maintain the quality of foods during storage as follows: - Slow down the loss of moisture and gases (oxygen and carbon dioxide) in foods; - Slow down the loss of volatile components from foods; - Improve the appearance and mechanical properties of food.	Some edible coatings are hygroscopic in nature, which helps to increase microbial growth.
Reduce weight loss of foods.	High cost of the components and reduced economic efficiency.
Reduce the use of plastic packaging and waste.	Difficulties in applying some edible coatings on the surface of foods.
Consumed together with foods.	No separate regulatory requirements for the use of edible coatings.

Source: own research.

4. Edible Coatings Based on Lipids Applied in the Food Industry

The increased demands of consumers nowadays raise the question of active packaging—a type of packaging that changes the conditions around the product itself, in order to preserve its quality and freshness, improve its sensory properties, and thus also guarantee food safety and storage duration.

Through the use of packaging for food products, the aim is to preserve their quality and safety from the moment of their production until their use by consumers. A very important function of packaging is the protection of goods from physical, chemical, or biological damage. The most widely distributed and used packaging material that fully meets these criteria is polyethylene and similar materials, which have been used in the food industry for more than 50 years. These materials are not only harmless, cheap, and applicable, but also suitable for packaging various goods and have good flexibility.

Nowadays, the main part of the materials used as raw materials for packaging are by-products of the processing of petroleum products and are practically non-degradable. This is the main reason why food packaging, like all other disposable packaging used to store foods for a short period of time, creates a serious global problem.

The world production of packaging materials is more than 180 million tons per year, and it is increasing every year [46]. A major limitation of using plastic packaging materials is the fact that their degradation period is long. Another point is the direct dependence of the production of polyethylene packaging on the extraction of petroleum products, and the reserves on a global scale are significantly decreasing with each subsequent year, along with increasing oil prices [47].

The properties of lipid edible coatings have been extensively investigated in recent years. By means of lipid films, unwanted loss of water vapor from foodstuffs is regulated, the quality is improved, and their shelf life is extended. By replacing synthetic packaging materials with edible lipid coatings, waste is reduced, and the recyclability of synthetic packaging materials is increased.

Various lipid components, such as animal fats and vegetable oils, are used in the production of edible coatings. Edible lipids are neutral lipids, waxes, and resins, which usually provide a coating material for fresh products, provide an effective moisture barrier property, and improve surface appearance. Resin coating is fairly effective for water loss [48]. Resins are a group of substances secreted by special plant cells in response to injury or infection in many trees and shrubs [49]. Lipid coatings have excellent moisture-protecting properties, and as a coating component, they add shine to confectionery and sugar products. Waxes are used in the composition of coatings in order to reduce respiration and moisture loss [50].

Kester and his colleagues offer an edible coating in a composition in which lipids and cellulose are combined. The edible coating is rated as a good barrier and protects the commodity from water vapor loss [51]. Cellulose-based coatings with included wax provide the best balance between the loss of aromatic substances and the prevention of natural storage losses of the goods due to water loss. These coatings are applied to fresh fruit immediately after harvest in order to preserve (reduce) the loss of aroma substances [52].

The structure of the matrix of some edible coatings includes wheat gluten and lipids, which are the main components that retain the loss of water vapor. It has been established that the combination of wheat gluten protein with diacetyl tartrate ester of monoglycerides reduces the penetration of water vapor, increases the strength of the coating, and ensures its transparency [53].

Lipid–cellulose edible coatings are designed to retain the moisture in two-component food products stored in a frozen state. It is applied between the two components of the bread and tomato sauce product. It was found that the coating successfully arrests moisture migration from the tomato sauce to the bread during storage [54].

It was established that the inclusion of plasticizers in the composition of lipid edible coatings prevents the appearance of pores and cracks after their application. For example, the addition of sunflower oil to starch-based coatings reduces the water vapor permeability of it [55].

Edible wax coatings are widely used in the food industry in order to extend the shelf-life of foods by regulating gas exchange and moisture migration. Waxes (carnauba wax, paraffin wax, etc.) are commercially used as a protective coating for fresh fruits and vegetables. They reduce moisture loss and surface abrasion during fruit handling. Generally, wax coatings are resistant to moisture loss as compared to lipid and non-lipid coating. Wax coating is effective on citrus, apple, mature green tomato, cucumber, and other vegetables, such as asparagus, beans, carrots, eggplant, okra, sweet potatoes, turnips, etc., where a shiny surface is desired. Beneficial properties of lipid-based coatings include good compatibility with other coating agents and high water vapor and gas barrier properties as compared to polysaccharide and protein-based coatings [56]. The lipid-based coating gives a greasy surface and undesirable organoleptic properties such as a waxy taste and lipid rancidity [57,58].

Carnauba wax is considered safe for human consumption by several certified organizations. European Union has authorized carnauba wax (E 903) as a food additive (coating agent). In 2012, the Panel of Additives and Nutrient Sources (ANS) on the safety of carnauba wax for human consumption jointly assessed by the Joint FAO/WHO Expert Committee on Food Additives (JECFA), and Scientific Committee on Food (SCF), and they determined 7 mg/kg body weight/day [6]. Its hydrophobic nature and meltability make it an attractive alternative to petroleum-based synthetic wax for coatings of fruits and vegetables, and in recent years, it has been explored as an additive in biopolymer-based composite packaging films. Incorporation in coatings improves their hydrophobicity, barrier against light, etc. [59,60]. Carnauba wax emulsion coating showed good commercial attributes such as glossiness, early drying nature, and reduced physiological weight loss. Its extended shelf life and enhanced antioxidant activity of packaged eggplants during storage up to 12 days [61]. Some examples connected with the application of Carnauba wax in the composition of edible coating are presented in Table 3.

Table 3. Application of Carnauba wax (CW) in composition of edible coatings.

Commodities	Coatings/Emulsions	Advantages	Source
Fresh tomatoes	Carnauba wax nanoemulsion.	Improve the gloss of tomatoes.	Miranda et al. [62].
Strawberries	Bio-nanocomposite coatings developed using arrowroot starch (AA), cellulose nanocrystals (CNC), carnauba wax nanoemulsion (CWN), and Cymbopogon martinii and Mentha spicata essential oils (CEO and MEO, respectively).	Preserve the post-harvest quality of fresh strawberries.	Gonçalves de Oliveira Filho et al. [63]
Eggplant	Poly ethylene glycol and Sodium alginate in CW emulsion.	Improve shelf life of eggplant during storage in both unpackaged and packaged.	Singh et al. [61]
Fresh-cut apples	Edible coatings/films formulated with cassava starch, glycerol, carnauba wax, and stearic acid.	Improve barrier properties of coatings applied in fresh-cut apples (respiration rate and water vapor resistance).	Chiumarelli et al. [64]
Indian jujube (*Zizyphus mauritiana Lamk.*) fruit	Carnauba wax (CW) and CW containing glycerol monolaurate (CW-GML) coating.	Shelf-life extension and quality maintenance of jujube fruit stored at 20 °C.	Chen et al. [65]
'Valencia' orange fruit	Carnauba wax-nanoclay emulsion.	Retarding respiration rates and weight loss, preserving sensory and nutritional quality of fruits.	Motamedi et al. [66]
Sweet potato (*Ipomoea batatas* (L.) *Lam.*) roots	Carnauba wax-based nanoemulsion without or with glycerol monolaurate.	Preserve food quality and extend shelf life of sweet potato roots.	Huqing et al. [67]
Fresh and minimally processed apples	Edible coatings made up of starch, carrageenan, soy proteins, corn zein, whey proteins, and waxes.	Improve functional properties of apples besides sensory and nutritional attributes.	Aayush et al. [68]
Fuji' apples	Carnauba-shellac wax-based nanoemulsion containing lemongrass oil.	Improve the quality of 'Fuji' apples during storage.	Wan-Shin et al. [69]

An aqueous emulsion containing 15% *candelilla waxes* is used for coating lemons. It has been established that lemons with a wax coating have fewer natural losses during storage. Changes in pH, acidity, and vitamin C content during the storage of fruits coated with candelilla wax were similar to those coated with commercial wax [70]. Some examples showing the application of Candelilla wax in composition of edible coatings are presented in Table 4.

Table 4. Application of Candelilla wax in composition of edible coatings.

Commodities	Coatings/Emulsions	Advantages	Source
Lemons	An aqueous emulsion containing 15% candelilla waxes.	Reduce natural losses during storage.	Paredes-Lopez et al. [70].
Strawberry	Edible films of candelilla wax alone or in combination with a *Bacillus subtilis*.	Prolong the shelf life of strawberries.	Oregel-Zamudio et al. [71].
Golden delicious apples	Edible coating formulated with candelilla wax and fermented extract of tarbush.	Improve the quality and shelf life of apples in marketing conditions.	De León-Zapata et al. [72].
Persian limes	Composite edible coatings were formulated with candelilla wax alone, and candelilla wax blended with beeswax, white mineral oil, and oleic acid (2:1 ratios).	Improve water vapor permeability, provide the lowest physiological weight loss, best dark shade green color retention, and unaltered physicochemical parameters.	Bosquez-Molina et al. [73].

Table 4. *Cont.*

Commodities	Coatings/Emulsions	Advantages	Source
Avocados	Addition of ellagic acid (at three different concentrations) into a candelilla wax matrix.	Improve the quality and shelf life of avocados.	Saucedo-Pompa et al. [74].
Apples at the industrial level	Phyto molecules of tarbush incorporated into the candelilla wax-based nanocoating.	Increase the shelf life of apples in marketing conditions and refrigeration at the industrial level.	De León-Zapata et al. [75].
Pears	Coating composed of carboxymethylcellulose (CMC), candelilla wax, and potassium sorbate (KS).	Prevent fungal infections in pears stored.	Kowalczyk et al. [76].

Paraffin wax or mineral oils have been found to be used and applied as an edible surface coating on turnips, cucumbers, sweet potatoes, eggplant, and on most citrus fruits [77].

Edible coatings based on lipids are mainly used to reduce natural losses during storage and to provide good protection of goods from the active action of oxygen from the air. Lipid applications include coatings for poultry, poultry products, shrimp, meat patties or patties, minced meat products, etc. Waxes and plant oils are added to protein- and polysaccharide-based edible coatings. They give flexibility and elasticity, improve coating properties and prevent products from sticking to each other during culinary processing.

Lipid-based edible coatings are prepared from waxes and oils. The benefits of food products coated with lipid-based coatings have been proven. Lipids not only provide hydrophobicity, cohesiveness, and flexibility but also act as an excellent barrier for the release and penetration of moisture into the respective commodity. This is due to the structure of the lipids. They prevent the passage of water vapor and preserve the quality of goods during their storage.

Edible coatings based on lipids and especially those based on plant waxes have the potential to meet consumer requirements. Their application guarantees harmlessness and preservation of the properties and quality of the product. Edible coatings partially replace some of the traditional synthetic packaging materials. Applied to food products, they protect them from unwanted processes, guarantee their quality, and extend the shelf life of the product. Lipid-based edible coatings are biodegradable, non-harmful, edible, and non-polluting to the environment.

4.1. Modern Trends, Possibility of Industrial Separation of Plant Waxes from the Cuticle of Waste Peels of Fresh Fruits and Vegetables, and Their Use in the Composition of Edible Coatings

Modern trends described and applied in food technology focused on the use of recovered, recycled, natural materials in the composition of edible coatings. In this way, the properties of foods are preserved and improved, and the efficiency of their use affects the cost price and the price of the respective goods. A major advantage of edible coatings based on plant waxes over traditional synthetic films is that they are consumed together with the food on which they are applied. Their use reduces and limits the use of natural oil reserves, which decrease every year, and contributes to preserving the ecological balance and the environment from pollution.

In this regard, the possibility of industrial extraction (separation) of plant waxes from the cuticle (peel wastes of fresh fruits and vegetables), peeled, and separated before their processing into different products (vegetable juices, fruit juices, vegetable concentrates, fruit-sugar products, wines, etc.) is proposed. The separated plant waxes should be used to obtain edible pectin-based or wax-based coatings. Lipid-based edible coatings are mainly used to extend the shelf life of foodstuffs and to preserve their quality during storage.

4.1.1. Method for Industrial Separation of Plant Waxes from the Cuticle (Waste Peels of Fresh Fruits and Vegetables)

The method should find application in the food industry and, more precisely, in the stage before using fresh fruits and vegetables for processing them into different products (vegetable juices, fruit juices, wines, vegetable concentrates, fruit and sugar products, etc.).

The present method solves the problem of full utilization and separation of plant waxes from the cuticle of fresh fruits and vegetables, which are currently not the object of attention and are not used by fresh fruits and vegetable processors. The advantage of the proposed method is the possibility of separating the plant waxes from the cuticle and using them in the composition of edible coatings based on plant waxes and emulsion coatings based on pectin or wax.

The method of industrial separation of plant waxes from the cuticle of fresh fruits and vegetables involves the following stages (Figure 3): Cleaning and sorting of the fresh fruits and vegetables. The next stage of the proposed method is to separate the peels (using suitable machines) from the surface of fresh fruits and vegetables. Plant waxes should be extracted from the separated peels with chloroform heated to 40–50 °C for 1–3 min. For this purpose, it is necessary to place the separated peels in a container and cover them with an organic solvent (chloroform). The obtained warm solution (extract containing plant waxes) should be filtrated, followed by its concentration (separation of the organic solvent from the filtrate). The obtained plant waxes finally are dried and stored in dark conditions, at a temperature of 1–3 °C.

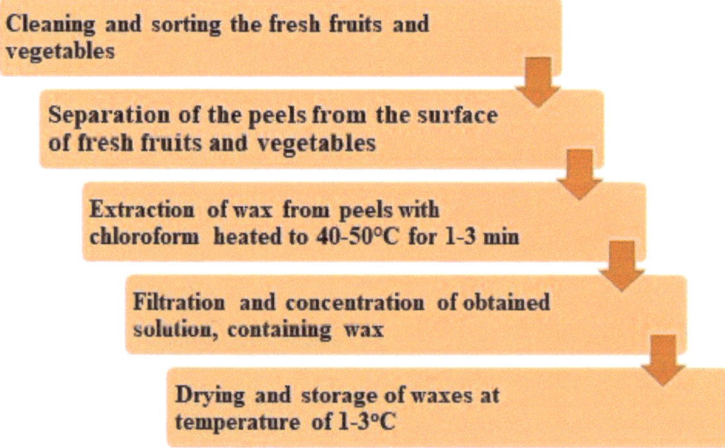

Figure 3. Stages of the industrial extraction of plant waxes from the waste peels (cuticle) of fresh fruits and vegetables.

In most plants, waxes are not present in large quantities, so they are extracted from them using organic solvents. Hexane is the most suitable solvent for the extraction of waxes, chloroform or methylene chloride may be used for the extraction of hard, high-melting waxes and for waxes containing triterpene acids [78]. According to Nawrath [79], chloroform dissolves mainly the hard wax, small amounts of the soft wax, and triterpene compounds, and cutin is insoluble in this solvent.

The proposed method makes it possible to separate the plant waxes included in the composition of the plant cuticle, as well as the surface plant waxes from the waste products (the separated peels of fresh fruits and vegetables before their processing). The amounts of plant wax obtained by the described method are suitable for use in the composition of edible coatings based on plant waxes. The extracted plant wax has a low cost, and this will also affect the price of the food products to which the edible coating will then be applied.

The obtained edible coatings based on plant waxes will reduce and limit the use of natural oil deposits and the production of slow-degrading packaging based on petroleum products. The production of such packages is currently a serious global problem, as it pollutes the environment with disposable packages thrown away by consumers every day.

The proposed method and the possibility of industrial separation of plant waxes from the peels of fresh fruits and vegetables limit the need to import expensive carnauba and candelilla waxes. The method is effective, easy to implement, and combines the economic interests of producers and the possibility of applying plant waxes to obtain edible coatings, widely used in modern conditions in the food industry to preserve the quality and shelf life of food.

4.1.2. Properties of the Isolated Plant Waxes

Studies that have been conducted determined the content of plant waxes in the cuticle of different varieties of plums and determined the influence of the amount and composition of waxes on the shelf life of plum fruits [80]. The isolated waxes were examined using a differential scanning calorimeter (DSC), infrared spectroscopy (IR spectroscopy), and X-ray structural analysis. From the conducted research, it was found that the isolated plant waxes are suitable and can be used as the main component in the composition of edible coatings, with the exception of the soft wax isolated from the cuticle of red sedge—*Prunus cerasifera* (characteristics: the melting point is 36.91 °C., i.e., it is 3.09 °C lower than the required temperature of 40 °C) [81].

Next, the rheological and optical characteristics of an aqueous solution of pectin containing plant wax and beeswax were investigated. It was established from the conducted research of plant waxes isolated from the cuticle (with the exception of a soft wax isolated from the cuticle of the blue plum variety Stanley—*Prunus domestica*) that they possess the necessary properties and can be used to obtain edible coatings. On the basis of the carried-out research, it has been proven that plant waxes isolated from the cuticle are suitable and can be used in the composition of edible pectin-based or wax-based coatings intended to extend the shelf life of food to preserve their properties and quality [82].

Modern trends applied in food technology aim to use recovered, recycled, natural plant-based waxes in the composition of edible coatings. That reduces and limits the use of natural oil deposits (petroleum products are the raw material for obtaining single-use packages that pollute the environment and degrade slowly), which decrease every day and year. This contributes to preserving the ecological balance by using more plant-based waxes for a sustainable greener economy.

5. Conclusions

As a result of the research, the following conclusions can be outlined, and directions for the application of plant waxes in the composition of edible coatings used in the food industry can be summarized.

Edible coatings are considered to be biopolymers of great variety, with properties desired by consumers, obtained from polysaccharides, lipids, and proteins, alone or in combination with other components. The advantages of edible coatings are associated with the main ingredients included in their composition, such as antimicrobial agents, aroma substances, antioxidants, coloring substances, vitamins, probiotics, etc. The disadvantages of coatings are connected with their composition and properties, the thickness of the layer of some coatings applied on the surface of foods, the cost of the components included in them, and the absence of requirements for the composition, safety, and application of edible coatings.

The necessity to use edible coatings based on plant waxes in food technology is related to their advantages over synthetic films and the possibility of preserving the properties and quality of foods on which they are applied for a longer period during their storage. Edible coatings based on lipids are widely used in food technology. Applied over food products, they protect them from unwanted processes, guarantee their quality and safety, and extend the shelf life of products. Edible lipid-based coatings are biodegradable, harmless, can be consumed with the products, and do not pollute the environment. The food industry needs edible coatings that will find application in a greater variety of foods that will extend their shelf life and preserve their quality.

Waxes isolated from plants are Carnauba Wax, Candelilla Wax, Japan Wax, Ouricury Wax, Rice Bran Wax, and Sunflower Wax. Only three of them (carnauba, candelilla, and rice bran wax), approved by the FDA, are used in the composition of edible coatings. Plant waxes isolated from the cuticle of fresh fruits and vegetables have the necessary properties and should find application in the composition of edible coatings based on plant waxes.

In this regard, an effective and applicable method for industrial separation of plant wax from peels of fresh fruits and vegetables before their processing (production of wine, high-alcohol beverages, fruit-sugar preserves, vegetable preserves, juices, etc.) is proposed. It was found that the separated waxes possess the necessary properties and should be recommended and find application for obtaining edible coatings on a pectin or wax basis. The method combines the economic interests of producers and the possibility of applying plant waxes in order to obtain edible coatings used in modern conditions in food technology to preserve the quality and shelf life of food.

In conclusion, it should be noted that consumers' demands provoke the deepening of scientific research towards new alternatives of packages not based on petroleum products, offering new ones that are recycled and have nutritional value, and one of them is ingredients (plant waxes) obtained from agricultural products. Plant waxes, isolated from the cuticle of fresh fruits and vegetables are an example of suitable, sustainable constituents obtained from agricultural products that should be used in the composition of edible coatings.

Future challenges and perspectives in the area of edible coatings on the basis of plant waxes can be presented in the following directions:

- The application of edible coatings in the food industry should be considered and presented to consumers as an innovation that supports the creation of an ecological and waste-free production by using natural plant wax; a possible way for improving the cleanliness of the environment;
- The edible coatings on the base of plant waxes should become a reality, moving from scientific research to industrial application for a greater variety of foods and having continuous growth in the future.

Funding: This research received no external funding.

Institutional Review Board Statement: Not applicable.

Informed Consent Statement: Not applicable.

Data Availability Statement: Not applicable.

Conflicts of Interest: The authors declare no conflict of interest.

References

1. Bai, J.; Plotto, A.; Baldwin, E. *Edible Coatings and Films to Improve Food Quality*; Hagenmaier, R., Bai, J., Eds.; CRC Press: Boca Raton, FL, USA, 2011; pp. 185–242.
2. Maringgal, B.; Hashim, N.; Syafinaz, I.; Amin Tawakkal, M.; Tengku Muda Mohamed, M. Recent advance in edible coating and its effect on fresh/fresh-cut fruits quality. *Trends Food Sci. Technol.* **2020**, *96*, 253–267. [CrossRef]
3. Nor, S.; Ding, P. Trends and advances in edible biopolymer coating for tropical fruit: A review. *Food Res. Int.* **2020**, *134*, 109208. [CrossRef] [PubMed]
4. Fernández-Muñoz, R.; Heredia, A.; Domínguez, E. The role of cuticle in fruit shelf-life. *Curr. Opin. Biotechnol.* **2022**, *78*, 102802. [CrossRef] [PubMed]
5. Martin, L.; Rose, J. There's more than one way to skin a fruit: Formation and functions of fruit cuticles. *J. Exp. Bot.* **2014**, *65*, 4639–4651. [CrossRef]
6. Andréa Silva de Freitas, C.; Henrique Machado de Sousa, P.; Josino Soares, D.; Ytalo Gomes da Silva, J.; Rathinaraj Benjamin, S.; Izabel Florindo Guedes, M. Carnauba wax uses in food—A review. *Food Chem.* **2019**, *291*, 38–48. [CrossRef]

7.	Puttalingamma, V. Edible coatings of carnauba wax—A novel method for preservation and extending longevity of fruits and vegetables—A review. *Int. J. Food Saf.* **2014**, *16*, 1–5.
8.	Pashova, S. *Application of Vegetable Waxes in Food Technology*; IK SAFO: Lovech, Bulgaria, 2011; pp. 147–148.
9.	ED European Parliament and Council Directive No98/72/EC 1998. Available online: https://eur-lex.europa.eu/legal-content/EN/TXT/?uri=CELEX%3A31998L0072 (accessed on 25 September 2022).
10.	ED European Parliament and Council Directive No95/2/EC 1995. Available online: https://eur-lex.europa.eu/legal-content/EN/ALL/?uri=celex%3A31995L0002 (accessed on 25 September 2022).
11.	FDA. Food Additives Permitted for Direct Addition to Food for Human Consumption 21 CFR 172, Subpart C. Coatings, Films and Related Substances. 2006. Available online: https://www.law.cornell.edu/cfr/text/21/part-172 (accessed on 10 December 2022).
12.	Guilbert, S.; Gontard, N. Edible and biodegradable food packaging. In *Foods and Packaging Materials—Chemical Interactions*; Ackermann, P., Jagerstad, M., Ohlsson, T., Eds.; The Royal Society of Chemistry: London, UK, 1995; pp. 159–168.
13.	Nussinovitch, A. *Water Soluble Polymer Applications in Foods*; Blackwell Science: Oxford, UK, 2003.
14.	General Standard for Food Additives CODEX STAN 192-1995. 1995. Revision 1997, 1999, 2001, 2003, 2004, 2005, 2006, 2007, 2008, 2009, 2010, 2011, 2012, 2013, 2014, 2015, 2016, 2018. Available online: http://www.fao.org/gsfaonline/docs/CXS_192e.pdf (accessed on 12 January 2023).
15.	EFSA. Annual Report 2012. Committed to Ensuring that Europe's Food Is Safe. 2012. Available online: https://www.efsa.europa.eu/sites/default/files/corporate_publications/files/ar12en.pdf (accessed on 15 February 2023).
16.	ED European Parliament and Council Directive No 2008/84/EC 2008. Available online: http://eur-lex.europa.eu/LexUriServ/LexUriServ.do?uri=OJ:L:2008:253:0001:0175:EN:PDF (accessed on 30 October 2022).
17.	Doores, S. Organic acids. In *Antimicrobials in Food*; Davidson, P.M., Branen, A., Eds.; Marcel Dekker Inc.: New York, NY, USA, 1993; pp. 95–136.
18.	Kabara, J. Phenols and Chelators. In *Food Preservatives*; Russell, N.J., Gould, G., Eds.; Blackie: London, UK, 1991; pp. 200–214.
19.	Franssen, L.; Krochta, J. Edible Coatings Containing Natural Antimicrobials for Processed Foods. In *Natural Antimicrobials for Minimal Processing of Foods*; Roller, S., Ed.; CRC Press: Boca Raton, FL, USA, 2003; pp. 250–262.
20.	Bodmier, R.; Hermann, J. Encyclopedia of Pharmaceutical Technology. In *Waxes*; Swarbrick, J., Boylan, J.C., Eds.; Marcel Deckker Inc.: New York, NY, USA, 1997; Volume 16, pp. 335–361.
21.	Tinto, W.; Elufioye, T.; Roach, J. *Waxes*; Pharmacognosy; Badal, S., Delgoda, R., Eds.; Academic Press: Cambridge, MA, USA, 2017; Chapter 22; pp. 443–455. [CrossRef]
22.	Lan, Y. Waxes. In *Encyclopedia of Food Chemistry*; Melton, L., Shahidi, F., Varelis, P., Eds.; Academic Press: Cambridge, MA, USA, 2019; pp. 312–316. [CrossRef]
23.	Néstor, E.; Ledesma, A.; Bautista-Hernández, I.; Rojas, R.; Aguilar-Zárate, P.; del Pilar Medina-Herrera, N.; Castro-López, C.; Cristian Guadalupe Martínez-Ávila, G. Candelilla wax: Prospective suitable applications within the food field. *LWT Food Sci. Technol.* **2022**, *159*, 113170. [CrossRef]
24.	Arato, M.; Speelman, S.; Van Huylenbroeck, G. The contribution of non-timber forest products towards sustainable rural development: The case of Candelilla wax from the Chihuahuan Desert in Mexico. *Nat. Resour. Forum* **2014**, *38*, 141–153. [CrossRef]
25.	Hagenmaier, R.; Baker, R. Edible Coatings from Candelilla Wax Microemulsions. *J. Food Sci.* **1996**, *61*, 562–565. [CrossRef]
26.	Krendlinger, E.; Wolfmeier, W. *Natural and Synthetic Waxes: Origin, Production, Technology and Applications*; Wiley-VCH: Hoboken, NJ, USA, 2022. [CrossRef]
27.	Pashova, S. Plant Waxes—Nature, Types and Application; Forum Ware International. In *Excellence in Business, Commodity Science and Tourism*; Special Issue; Bucharest Academy of Economic Studies: Bucharest, Romania; International Society of Commodity Science and Technology (IGWT): Wien, Austria, 2011; Volume 1, pp. 166–170.
28.	Modupalli, N.; Natarajan, V. Fractionation and extraction of functional compounds from rice bran wax. *Pharma Innov. J.* **2022**, *11*, 1682–1686.
29.	Sabale, V.; Sabale, P.; Lakhotiya, C. Comparative evaluation of rice bran wax as an ointment base with standard base. *Indian J. Pharm. Sci.* **2009**, *71*, 77. [CrossRef] [PubMed]
30.	Bhanot, V.; Fadanavis, S.; Panwar, J. Revisiting the architecture, biosynthesis and functional aspects of the plant cuticle: There is more scope. *Environ. Exp. Bot.* **2021**, *183*, 104364. [CrossRef]
31.	George, N. Chapter five—How Pathogens Attack Plants. In *Plant Pathology*, 5th ed.; Agrios, G.N., Ed.; Academic Press: Cambridge, MA, USA, 2005; pp. 175–205. [CrossRef]
32.	Trivedi, P.; Karppinen, K.; Klavins, L.; Kviesis, J.; Sundqvist, P.; Nguyen, N.; Heinonen, E.; Klavins, M.; Jaakola, L.; Väänänen, J.; et al. Compositional and Morphological Analyses of Wax in Northern Wild Berry Species. *Food Chem.* **2019**, *295*, 441–448. [CrossRef]
33.	Baldwin, E. Edible coatings for fresh fruits and vegetables: Past, present and future. In *Edible Coatings and Films to Improve Food Quality*; Krochta, J.M., Baldwin, E., Nisperos-Carriedo, M., Eds.; Technic: Basel, Switzerland, 1994; pp. 25–64.
34.	Park, J.; Testin, R.; Vergano, P.; Park, H.; Weller, C. Application of Laminated Edible Films to Potato Chip Packaging. *J. Food Sci.* **2006**, *61*, 766–768, 777. [CrossRef]
35.	Rojas-Grau, M.; Sobrino-Lopez, A.; Tapia, M.; Martin-Bellosso, O. Browning Inhibition in Fresh-cut 'Fuji' Apple Slices by Natural Antibrowning Agents. *J. Food Sci.* **2006**, *71*, S59–S65. [CrossRef]

36. Han, C.; Lederer, C.; McDaniel, M.; Zhao, Y. Sensory Evaluation of Fresh Strawberry (*Fragarai ananassa*) Coated with Chitosan-based Edible Coatings. *J. Food Sci.* **2005**, *70*, S172–S178. [CrossRef]
37. Guilbert, S.; Biquet, B. Edible films and coating. In *Food Packaging Tech.*; Bureau Multon, J.L., Ed.; VCH Publisher: New York, NY, USA, 1996.
38. Nesperos-Carriedo, M.; Baldwin, E.; Shaw, P. Development of an Edible Coatings for Extending Postharvest Life of Selected Fruits and Vegetables. *Proc. Annu. Meet. Fla. State Hortic. Soc.* **1992**, *104*, 122–125.
39. Park, H. Development of advanced edible coating of fruits. *Trends Food Sci. Technol.* **1999**, *10*, 250–260. [CrossRef]
40. El-Ghaouth, J.; Ponnampalam, R.; Boulet, M. Chitosan coating effect on stability of fresh Strawberries. *J. Food Sci.* **1991**, *57*, 1618–1620. [CrossRef]
41. Park, J.; Testin, R.; Rank, H.; Vergano, P.; Wlter, C. Fatty acid concentration effect on textile strength, elongation and water vapor permeability of laminated edible films. *J. Food Sci.* **1994**, *59*, 916–919. [CrossRef]
42. Okcu, Z.; Yavuz, Y.; Kerse, S. Edible Film and Coating Applications in Fruits and Vegetables. *Alınteri J. Agric. Sci.* **2018**, *33*, 221–226. [CrossRef]
43. Raghav, P.; Agarwal, N.; Saini, M. Edible coating of fruits and vegetables: A review. *Int. J. Sci. Res. Mod. Educ.* **2016**, *I*, 188–204.
44. Radev, R. Edible Films and Coatings for Food Products—Advantages and Disadvantages. Union of Scientists: Varna, Bulgaria, 2021; Volume 10, pp. 43–51.
45. Radev, R. Positive and Negative Aspects of Edible Films and Coatings for Food. In Proceedings of the Modern Commodity Expertise: Theoretical Developments: Practical Experience, Problems and Prospects, International Scientific and Practical Conference, Lviv, Ukraine, 27–28 May 2021; Volume 7, pp. 31–34.
46. Tice, P. Packaging Materials. 4. Polyethylene for Food Packaging Applications. International Life Sciences Institute Report. 2003. Available online: http://orig.ilsi.org/file/PM4_Polyethylene.pdf (accessed on 12 November 2022).
47. Weber, C.; Haugaard, V.; Festersen, R.; Bertelsen, G. Production and Application of Bio based Packaging Materials for the Food Industry. *Food Addit. Contam.* **2002**, *19*, 172–177. [CrossRef] [PubMed]
48. Morillon, V.; Debeaufort, F.; Blond, G.; Capelle, M.; Voilley, A. Factors affecting the moisture permeability of lipid based edible films: A review. *Crit. Rev. Food Sci. Nutr.* **2002**, *42*, 67–89. [CrossRef] [PubMed]
49. Debeaufort, F.; Quezada, G.; Voilley, A. Edible films and coating: Tomorrow's packaging: A review. *Crit. Rev. Food Sci. Nutr.* **1998**, *38*, 299–313. [CrossRef]
50. Avena-Bustillos, R.; Krochta, J.; Saltveit, M.; Rojas-Villegas, R.; Sauceda-Pbrez, J. Optimization of Edible Coating Formulations on Zucchini to Produce Water Loss. *J. Food Eng.* **1994**, *21*, 197–214. [CrossRef]
51. Kester, J.; Fennema, O. An Edible Film of Lipids and Cellulose Ethers: Barrier Properties to Moisture Vapor Transmission and Structural Evaluation. *J. Food Sci.* **1989**, *54*, 1383–1389. [CrossRef]
52. Nisperos-Carriedo, M.; Shaw, P.; Baldwin, E. Changes in Volatile Flavor Components of Pineapple Orange Juice as Influenced by the Application of Lipid and Composite Films. *J. Agric. Food Chem.* **1990**, *38*, 1382–1387. [CrossRef]
53. Gontard, N.; Duchez, C.; Cuq, J.; Guilbert, S. Edible Composite Films of Wheat Gluten and Lipids: Water Vapor Permeability and Other Physical Properties. *Int. J. Food Sci. Technol.* **1994**, *29*, 39–50. [CrossRef]
54. Kester, J.; Fennema, O. An Edible Film of Lipids and Cellulose Ethers: Performance in a Model Frozen-Food System. *J. Food Sci.* **1989**, *54*, 1390–1392. [CrossRef]
55. Garcia, M.; Martino, M.; Zaritzky, N. Lipid Addition to Improve Barrier Properties of Edible Starch-based Films and Coatings. *J. Food Sci.* **2000**, *65*, 941–944. [CrossRef]
56. Greener, I.; Fennema, O. Lipid-based edible films and coatings. *Lipid Technol.* **1992**, *4*, 34–38.
57. Robertson, G. *Food Packaging, Principle and Practices*, 2nd ed.; CRC Press: Boca Raton, FL, USA, 2009.
58. Vaishali, H.; Samsher, V.; Chaudhary, S.; Mithun, K. Importance of edible coating on fruits and vegetables: A review. *J. Pharmacogn. Phytochem.* **2019**, *8*, 4104–4110.
59. Devi, L.; Kalita, S.; Mukherjee, A.; Kumar, S. Carnauba wax-based composite films and coatings: Recent advancement in prolonging postharvest shelf-life of fruits and vegetables. *Trends Food Sci. Technol.* **2022**, *129*, 296–305. [CrossRef]
60. Saji, V. Wax-based artificial superhydrophobic surfaces and coatings. *Colloids Surf. A Physicochem. Eng. Asp.* **2020**, *602*, 125–132. [CrossRef]
61. Singh, S.; Khemariya, P.; Rai, A.; Rai, A.C.; Koley, T.; Singh, B. Carnauba wax-based edible coating enhances shelf-life and retain quality of eggplant (*Solanum melongena*) fruits. *LWT Food Sci. Technol.* **2016**, *74*, 420–426. [CrossRef]
62. Miranda, M.; Mori, M.; Ribeiro, M.; Spricigo, P.; Pilon, L.; Mitsuyuki, M.; Correa, D.; Ferreira, M. Carnauba wax Nano emulsion applied as an edible coating on fresh tomato for postharvest quality evaluation. *Heliyon* **2022**, *8*, e09803. [CrossRef]
63. De Oliveira Filho, J.; Albiero, B.; Calisto, Í.; Bertolo, M.; Oldoni, F.; Egea, M.; Junior, S.; de Azeredo, H.; Ferreira, M. Bio nanocomposite edible coatings based on arrowroot starch/cellulose nanocrystals/carnauba wax nano emulsion containing essential oils to preserve quality and improve shelf life of strawberry. *Int. J. Biol. Macromol.* **2022**, *219*, 812–823. [CrossRef]
64. Chiumarelli, M.; Hubinger, M. Stability, solubility, mechanical and barrier properties of cassava starch—Carnauba wax edible coatings to preserve fresh-cut apples. *Food Hydrocoll.* **2012**, *28*, 59–67. [CrossRef]
65. Huiyun, C.; Zhengxuan, S.; Huqing, Y. Effect of carnauba wax-based coating containing glycerol monolaurate on the quality maintenance and shelf-life of Indian jujube (*Zizyphus mauritiana* Lamk.) fruit during storage. *Sci. Hortic.* **2019**, *244*, 157–164. [CrossRef]

66. Motamedi, E.; Nasiri, J.; Malidarreh, T.; Kalantari, S.; Naghavi, M.; Safari, M. Performance of carnauba wax-nanoclay emulsion coatings on postharvest quality of 'Valencia' orange fruit. *Sci. Hortic.* **2018**, *240*, 170–178. [CrossRef]
67. Huqing, Y.; Xia, L.; Guoquan, L. Effect of Carnauba Wax–Based Coating Containing Glycerol Monolaurate on Decay and Quality of Sweet Potato Roots during Storage. *J. Food Prot.* **2018**, *81*, 1643–1650. [CrossRef]
68. Aayush, K.; McClements, D.; Sharma, S.; Sharma, R.; Singh, G.; Sharma, K.; Oberoi, K. Innovations in the development and application of edible coatings for fresh and minimally processed Apple. *Food Control* **2022**, *141*, 109188. [CrossRef]
69. Wan-Shin, J.; Hye-Yeon, S.; Nak-Bum, S.; Ji-Hyun, L.; Sea, M.; Kyung, S. Quality and microbial safety of 'Fuji' apples coated with carnauba-shellac wax containing lemongrass oil. *LWT Food Sci. Technol.* **2014**, *55*, 490–497. [CrossRef]
70. Paredes-Lopez, O.; Camargo-Rubio, E.; Gallardo-Navarro, Y. Use of Coatings of Candelilla Wax for the Preservation of Limes. *J. Sci. Food Agric.* **1991**, *25*, 1207–1210. [CrossRef]
71. Oregel-Zamudio, E.; Angoa-Pérez, V.; Oyoque-Salcedo, G.; Noé Aguilar-González, C.; Mena-Violante, H. Effect of candelilla wax edible coatings combined with biocontrol bacteria on strawberry quality during the shelf-life. *Sci. Hortic.* **2017**, *214*, 273–279. [CrossRef]
72. De León-Zapata, M.; Sáenz-Galindo, A.; Rojas-Molina, R.; Rodríguez-Herrera, R.; Jasso-Cantú, D.; Aguilar, C. Edible candelilla wax coating with fermented extract of tarbush improves the shelf life and quality of apples. *Food Packag. Shelf Life* **2015**, *3*, 70–75. [CrossRef]
73. Bosquez-Molina, E.; Guerrero-Legarreta, I.; Vernon-Carter, E. Moisture barrier properties and morphology of mesquite gum–candelilla wax based edible emulsion coatings. *Food Res. Int.* **2003**, *36*, 885–893. [CrossRef]
74. Saucedo-Pompa, S.; Rojas-Molina, R.; Aguilera-Carbó, A.; Saenz-Galindo, A.; de La Garza, H.; Jasso-Cantú, D.; Aguilar, C. Edible film based on candelilla wax to improve the shelf life and quality of avocado. *Food Res. Int.* **2009**, *42*, 511–515. [CrossRef]
75. De León-Zapata, M.; Ventura-Sobrevilla, J.; Salinas-Jasso, T.; Flores-Gallegos, A.; Rodríguez-Herrera, R.; Pastrana-Castro, L.; Rua-Rodríguez, M.; Aguilar, C. Changes of the shelf life of candelilla wax/tarbush bioactive based-nanocoated apples at industrial level conditions. *Sci. Hortic.* **2018**, *231*, 43–48. [CrossRef]
76. Kowalczyk, D.; Kordowska-Wiater, M.; Zięba, E.; Baraniak, B. Effect of carboxymethylcellulose/candelilla wax coating containing potassium sorbate on microbiological and physicochemical attributes of pears. *Sci. Hortic.* **2017**, *218*, 326–333. [CrossRef]
77. Lawrence, J.; Iyengar, J. Determination of Paraffin Wax and Mineral Oil on Fresh Fruits and Vegetables by High Temperature Gas Chromatography. *J. Food Saf.* **1983**, *5*, 119–129. [CrossRef]
78. Jouret, C.; Puech, J. Composition de La Cire Cuticulaire de la Prune D'ente. *Ann. Agric. Sci.* **1972**, *21*, 25–33.
79. Nawrath, C. *The Biopolymers Cutin and Suberin*; Arabidopsis Book; American Society of Plant Biologists: Rockville, MD, USA, 2002; Volume 1, p. e0021. [CrossRef]
80. Pashova, S. Research on the Cuticle and Shelf Life of Plum Fruits. Ph.D. Thesis, Varna, Bulgaria, 2007; p. 219.
81. Nikolova, K.; Panchev, I.; Kovacheva, D.; Pashova, S. Thermophysical and Optical Characteristics of Bee and Plant Waxes. *J. Optoelectron. Adv. Mater.* **2009**, *11*, 1210–1213.
82. Panchev, I.; Nikolova, K.; Pashova, S. Physical Characteristics of Wax-containing Pectin Aqueous Solutions. *J. Optoelectron. Adv. Mater.* **2009**, *11*, 1214–1217.

Review

Innovations in the Packaging of Meat and Meat Products—A Review

Marian Gil and Mariusz Rudy *

Department of Agricultural Processing and Commodity Science, Institute of Food and Nutrition Technology, College of Natural Sciences, University of Rzeszow, Zelwerowicza 4, 35-601 Rzeszow, Poland
* Correspondence: mrudy@ur.edu.pl; Tel.: +48-0-17-785-52-60

Abstract: This study aims to systematize the knowledge about innovative solutions to understand the composition of packaging materials and bioactive substances used in the packaging processes of meat and meat products, given the contemporary trends and consumer expectations. In edible packaging, the application of natural and renewable biopolymers is gaining popularity as, unlike petroleum-based plastic packaging materials, they do not cause environmental problems. Packaging using active compounds further extends the shelf life of food products compared with traditional packaging by reducing the adverse effects during storage, such as oxidation, microbial growth, and moisture loss. On the other hand, the inclusion of natural bioactive substances in packaging provides an opportunity to increase the shelf life of food products and/or decrease the use of preservatives. This direction offers a wide field for research due to the multitude of substances, their impact, and the properties of the packaged product.

Keywords: active packaging; electrospinning; bioactive substances

Citation: Gil, M.; Rudy, M. Innovations in the Packaging of Meat and Meat Products—A Review. *Coatings* **2023**, *13*, 333. https://doi.org/10.3390/coatings13020333

Academic Editors: Jun Mei and Jing Xie

Received: 24 December 2022
Revised: 23 January 2023
Accepted: 29 January 2023
Published: 1 February 2023

1. Introduction

Petroleum-based plastics are one of the most commonly used packaging materials due to their stiffness, flexibility, desirable barrier properties, inexpensiveness, and ease of processing [1]. As conventional packaging materials for meat or meat products, synthetic materials in the form of foil are used, often in combination with, e.g., cardboard outer packaging. The most commonly used synthetic plastics for meat packaging include the following: polyethylene (PE), polypropylene (PP), polyvinyl chloride (PVC), polyester (PET), polyamide (PA), polyvinylidene chloride (PVDC), and ethylenvinyl alcohol [2]. However, the mass use of such materials has resulted in serious environmental problems, such as the depletion of natural resources, garbage pollution, and global warming, as they are both nonrenewable and nondegradable [3–5].Therefore, the efforts of scientists and industry have been directed towards sustainable strategies by developing innovations in the field of packaging materials and packaging methods [6]. An expected property of new packaging materials is that they are reusable, recyclable, or biodegradable once they have served their purpose [7,8]. Therefore, the food industry is looking for an environmentally friendly replacement of non-biodegradable plastics with biodegradable plastics [9].

Active packaging (AP) is a novel packaging method that utilizes various active compounds such as antioxidants, antimicrobials, moisture absorbers, gas absorbers, and ultra-violet absorbers. These active components interact with the packaged food product or the surrounding environment to extend its shelf life by maintaining food quality, safety, and integrity. Compared with traditional packaging, packaging using active compounds further extends the shelf life of food products by reducing the harmful effects during storage, such as oxidation, microbial growth, and moisture loss [6,10].

Recently, there has been huge progress in the construction of AP systems using various methods such as dip coating [11], layer-by-layer assembly [12], electrospinning [13], solvent casting [14], extrusion [15], and homogeneous emulsification [16,17]. AP technologies can be based on either synthetic or natural materials, and some of them contain active ingredients such as antioxidants, antimicrobials, vitamins, flavors, and dyes [18]. In edible packaging, using natural and renewable biopolymers is gaining popularity as they, unlike petroleum-based plastic packaging materials, do not cause environmental problems. In edible films, substances that comply with food regulations should be present, and these films need to be economical, easy to apply, and environmentally friendly [19]. Edible films are classified based on their structural material, namely hydrocolloids (polysaccharides and proteins), lipids, and composites [20]. The limitation for edible packaging is the risk of their contamination and thus becoming inedible. Regardless, if not eaten, edible packaging is inherently biodegradable [21]. New packaging materials ensure higher functionality of the packaging, extending the shelf life and ensuring higher quality and safety of packaged meat [7]. Currently, many novel AP materials are gaining huge interest in the food industry. AP can inhibit the growth of microorganisms on the surface of the food product, enhance its nutritional and sensory properties, increase the shelf life of some food products, and decrease the environmental impact of packaging [22]. As a novel method, innovative packaging not only extends the quality and shelf life of the food product but also monitors its quality during transport and storage. AP and intelligent packaging have been adopted of late to ensure the traceability, safety, and quality of food products [23,24]. The main task of intelligent packaging is to capture and provide information about changes in the quality of packaged goods during transport and storage. They provide information about the conditions of the packaged product, without affecting its quality [25].

In contrast to traditional food packaging, functionalized packaging systems that are developed to load various bioactive compounds in matrix materials can lead to wide-ranging biological effects such as antibacterial and antioxidant effects and thus protect the food product from harmful environmental factors [26–28]. Active packages are developed by embedding a plant-based bioactive material in a polymer. Essential oils are in the spotlight as active ingredients due to their antimicrobial and antioxidant properties [29,30].

This study aimed to systematize the knowledge about innovative solutions to understand the composition of packaging materials and bioactive substances used in the packaging of meat and meat products, given the contemporary trends and consumer expectations. This will help demonstrate the positive effects of using innovative methods in the packaging of meat and its products.

2. Natural Polymers in Food Packaging

The environmental problems caused by conventional polymers have necessitated the search for alternative packaging materials. Biodegradable films based on biopolymers have become such an alternative [31]. In 2020, bio-based plastics used in food packaging amounted to 0.99 million tons, accounting for 47% of the total production of bio-based plastics [1]. The raw materials for the production of biopolymers are relatively plentiful, and the production of biopolymers consumes agricultural waste, which, together with the environmental benefits, makes the production of biopolymers profitable [32].

Biopolymers are popular in food packaging because they are edible and safer for humans. For applications in food packaging, the most frequently studied nanocomposite biomaterials are proteins, carbohydrates, and their derivatives [33–36]. To achieve an environmentally friendly alternative and promote sustainability goals, cellulose- and starch-based nanocomposites can be incorporated into packaging systems [37]. Examples of natural antioxidants for lipid food include, among others, edible films and coatings with an active coating based on cellulose derivatives, chitosan, alginate, galactomannans, or gelatin [38].

Cellulose is the most abundant biopolymer in the world, making it an ideal raw material for use in sustainable packaging materials. Cellulose ethers, such as methylcellulose, hydroxypropylcellulose, hydroxypropylmethylcellulose, and carboxymethylcellulose, are suitable for the production of packaging films [7,39]. Cellulose is obtained from natural sources such as wood, cotton and food waste, agricultural waste, cereal bran, and fruit skins [40]. Its availability from many different sources and being biodegradable, environmentally friendly, and inexpensive have made cellulose an often-preferred material in packaging [41]. Besides being edible and biodegradable, its sensory and organoleptic properties are beneficial; therefore, cellulose can be used in the encapsulation of bioactive substances to enhance the nutritional properties of food products [40].

Starch is one of the most crucial biodegradable polymers because of its abundance, low cost, biodegradability, and renewability [42,43]. Starch-based films have been used in food packaging and preservation technologies as they show excellent film-forming ability and unique gelatinization properties, along with their odorless, tasteless, and colorless nature [5,44]. So far, starch-based films have been extensively used in the packaging of different types of food products (such as meat, fruit, oil, and cheese) as they have good organoleptic and gas barrier properties [45,46]. However, starch-based materials are brittle and hydrophilic, which limits their processing and use. Starch is mixed with various synthetic and natural polymers to improve its properties; this increases the strength of the processing properties of the materials [47].

Another biopolymer is chitosan, which is derived from chitin. Chitosan films have shown good antibacterial and antioxidant performance for food packaging. The amino and hydroxyl groups in the structure of chitosan affect its antimicrobial activity against gram-positive and gram-negative bacteria [48]. Chitosan-based films have a high gas barrier. Their brittleness eliminates the use of plasticizers such as polyols (glycerin, sorbitol, and polyethylene glycol) or fatty acids (stearic and palmitic) [47].

Being a water-soluble natural polymer, gelatin is a protein of biological origin, which shows high biodegradability, biocompatibility, water absorption, nonimmunogenicity, and commercial availability. Thanks to these properties, various forms of gelatin (e.g., foils, scaffolds, capsules, filters) are used in cosmetics, pharmacy, medicine, food, and water filtration [49]. However, due to its hydrophilicity, its structure needs to be stabilized because, in the absence of biopolymer stabilization, gelatin-based materials tend to dissolve and lose their structure [50].

Among the methods available for gelatin structure stabilization, protein crosslinking is one of the most commonly used approaches to achieving hydrolytic stability of samples based on gelatin [51]. Crosslinkers such as glutaraldehyde and genipin are extensively used in this regard. However, there exist potential toxicity issues, along with the need for intensive detoxification strategies considering residual unreacted glutaraldehyde groups. In addition, the high cost of genipins is one of the primary disadvantages while using these crosslinkers [51,52]. The gelatine-based packaging material has a good oxygen barrier compared to other biopolymers and has the ability to be welded, which is important in the production of packaging. The production of gelatin foils is relatively simple; it does not require special conditions for drying and forming the foil [53]. To prevent the risk of toxicity and achieve cost-effectiveness, heat treatment of gelatin along with sugar particles has recently been introduced as an alternative chemical crosslinking method [54,55]. The resulting condensation reaction between proteins and sugar is called the Maillard reaction (MR) [50].

Gelatin-based materials show different properties (e.g., solubility, swelling, antioxidant activity, preservation of morphology after immersion) based on the degree of MR, which depends on parameters such as type of sugar, reaction time, temperature, and pH of the solution. As crosslinkers, pentoses (e.g., ribose) are more reactive than hexoses (e.g., glucose) and disaccharides (e.g., lactose) [56], whereas an increase in the percentage of sugar (up to a certain point), temperature, or pH of the solution induces a further extended response [50,57,58].

To enhance the bioavailability and stability of thiamine in raw and cooked red meat and salmon samples, the thiamine nanofiber nanocoating process has been successfully applied. Specifically, for salmon samples, this process is found to be more effective regarding bioavailability. In addition, it ensures a continuous increase in the thiamine content in red meat and fish samples under cold storage conditions for 3 days. Whereas a maximum bioavailability of 87% was reported for nanocoated red meat samples, for salmon samples, a 94% bioavailability was achieved. Therefore, given these results, in future, this nanotechnology application may play a leading role in the food industry [59].

The most successful edible protein film available on the market is a sausage casing made of collagen. The films reduce leakage and prevent discoloration and fat oxidation of thawed and chilled beef steaks. Collagen-based films are used for processed meats to increase juiciness, reduce drip. For many years, the Japanese meat industry has commercially used films and coatings based on polysaccharides. During processing, the coatings dissolve and integrate with the meat, which has a positive effect on the texture and reduces weight loss, ensuring higher yield [60].

3. Electrospinning

Nanofibers are obtained using electrospinning techniques that use electrostatic forces to form fibers and non-electrospinning techniques that use mechanical force. These include phase separation, drawing, template synthesis, self-assembly, etc. [61].

Electrospinning is a versatile, cost-effective, and convenient method to produce nano-/microfibers with a high surface-area-to-volume ratio, controlled dimensions, high load capacity, low weight, and wide-ranging flexibility [62]. Furthermore, with decades of evolution, electrospinning nanofibers can now be designed with various structures and morphologies to perform specific functions, such as uniaxial [63], hollow [64], core–shell [65], and porous structures [66].

Electrospinning is an easy and versatile nanotechnique for producing nonwoven nanofiber films. Its advantages are as follows: a high surface-area-to-volume ratio, increased porosity, small interfibrous pore size, and high gas permeability. It is widely used in natural and synthetic polymers [5,67]. Thus, electrospinning has gained interest in, among others, textiles, agriculture, water treatment, air filtration, energy storage, cosmetics, electronics and sensors, pharmaceuticals, biomedical products, and packaging [49,50,68].

Among innovative approaches to packaging, electrospinning has gained huge interest in the biomedical and the food industries, especially in meat packaging [23,69–72]. The rapid development of electrospinning has resulted in numerous applications in various fields, including biomedicine [73], food packaging [74], sensors [75], protective materials [76], textiles [77], energy [78], oil–water separation, and others [17,79]. Several applications of electrospinning have been found in food science, e.g., protecting bioactive ingredients from external factors by encapsulating them [80–82] and extending the shelf life of a food product by improving its bioavailability and controlled release of biomolecules [83–85].

Compared with traditional casting films, electrospun nanofibers show numerous unique characteristics such as high surface-area-to-volume ratio, nanoporous structure, high porosity, and high absorption capacity [62,86], which make them more sensitive to the surrounding changes in acidity/alkalinity and make it possible to control the release of the contained bioactive compounds. Thus, of late, electrospun nanofibers have gained much attention in developing food packaging films [14,87,88].

In addition, being a nonthermal process, electrospinning helps maintain the structure stability, particularly when using additives with low thermal stability at high temperatures. The process of electrospinning can be briefly divided into three steps as follows: (1) formation of a conical shape ("tailor's cone") by a charged drop of a polymer solution; (2) formation of a jet at the end of the cone if the electric field strength is sufficient to overcome the viscoelastic force of the solution; and (3) deposition of a solid jet on the collector surface and production of many fibers, with rapid volatilization of the solvent [5,89]. To obtain fibers using solution electrospinning, various materials—including synthetic and natural polymers and their combinations—can be utilized. Among them, synthetic polymers such as polystyrene and polyvinyl chloride, biocompatible and biodegradable synthetic polymers like polylactic acid and polylactic-co-glycolic acid, conductive polymers such as polyaniline and polypyrrole, and natural polymers such as chitosan, alginate, collagen, and gelatin can be directly electrospun into nanofibers [50,90–94].

The electrospinning technique has been used to develop high-performance packaging materials in the food industry, due to its unique advantages: it can produce (1) micro-/nanofibers to encapsulate unstable bioactive molecules and load with nanoparticles; (2) edible packaging nanofibers from biopolymers, which show excellent biosafety; and (3) nanofibers for the controlled release of bioactive compounds under a specific stimulus [17].

Natural polymers, especially polysaccharides and proteins, are frequently used to produce nanofibers due to their biocompatibility, nontoxicity, food-grade properties, and biodegradability [95]. In addition, their diversity of functional groups enables a wide range of active ingredients to be bound or trapped using molecular interactions [96]. Functional electrospun mats can be used to develop nanocomposite material from a diverse range of performance-enhanced plastics for packaging applications. In addition, they can be used to reinforce the physical properties of both plastics and bioplastics as transparent gas barrier layers or even as new technologies for designing bioactive packaging with antimicrobial protection and delivering nutraceuticals to food products [97]. Numerous electrospinning stimuli-responsive materials have recently been synthesized, which can achieve the controlled release of active substances, thus producing a long-term biological effect [98,99].

Nanofiber mats are promising candidates in AP [100]. In the AP industry, nanofibers are highly useful tools to protect and deliver bioactive compounds to their destination at the desired time [101,102]. Electrospun nanofibers can improve the barrier and antimicrobial properties of materials in food packaging depending upon their functional properties. These nanofibers can also be utilized as nanosensors to detect and monitor the conditions of the food product during transport and storage [103]. These biological polymers can be based on proteins, lipids, or polysaccharides [104–106]. This advanced technology is originally derived from the enrichment of antioxidants in packaging designs [107–110].

Electrospun fibers show a good capacity to charge active substances, and their huge surface area leads to a rapid response to internal and/or external factors by releasing/activating the trapped compounds in a timely manner [90,95,111].

Thus, as a new technology, electrospinning can improve the overall quality and extend the shelf life of fresh or packaged meat products [95], including (1) protecting products from microbial contamination [3,71,112], (2) preventing lipid and protein oxidation [113,114], (3) developing sensory properties [70,84], and (4) improving the functional and nutritional characteristics of meat products [22]. Electrospinning enables the incorporation of antimicrobial compounds into the matrixes/or packaging mats and allows for a functional effect on the surface of meat or products—where the microbiological activity is located—instead of mixing them directly with food [115].

Starch-based films with nanofibers show an extremely high surface activity, which makes them potential candidates for active food packaging due to their nanosize [17]. In addition, the morphology and structure of electrospun starch fibers can be easily altered to protect numerous active substances and enhance the mechanical and barrier properties [116]. Several factors like fiber orientation, additional ingredients [117,118], and final processing [119] can influence their properties required for food packaging [5].

Results show that zein-based coatings are more suitable in the packaging of food products with a high water content [120]. Yildiz et al. [121] developed an electrospun chitosan/polyethylene/curcumin nanofiber to monitor the freshness of chicken meat. Duan et al. [14] showed that curcumin-loaded nanofibers provide the ability to monitor chicken spoilage in the real world.

The challenge is to overcome the unreliability of bio-based plastics. There is a need to develop a multilayer mixture using additives [1]. In conclusion, electrospinning seems to be a promising technique with potential applications in the fields of functional food products and AP [102]. The advantage of electrospinning is its simplicity, the possibility of using it in a wide range of materials, and its low cost [61].

4. Antioxidant and Antimicrobial Compounds

Many meat products are considered highly perishable because of their high nutrient content. Temperature is the major factor in the activation of the growth of microorganisms and chemical reactions; thus, the cooling temperature has a significant impact on their properties. However, variations in the temperature during storage and transport can impair the quality of the products, e.g., by increasing microbial growth and chemical reactions such as increasing peroxides and thiobarbituric acid (TBA) values [122,123].

To increase the commercial value and safety of beef, cold storage methods and cold chain logistics have been developed and widely used. These methods are used in preserving raw beef, especially in freezing and chilling [124,125]. Freezing below $-18\ °C$ significantly extends the shelf life of meat products but degrades the quality of the meat in the freezing–thawing process. In comparison, storage at 4 °C can preserve the sensory quality of meat and lower the energy consumption; however, it cannot inhibit the growth of microbes completely, in particular some psychrophiles, so the shelf life of the products is limited [28,124].

The meat industry is interested in achieving packaging durability goals and producing modern solutions based on bio-based, biodegradable, compostable, recyclable, or reusable materials [126]. Increasing demand for meat has urged significant advances in meat packaging, guaranteeing healthy and safe products. Meanwhile, the safety and quality of meat are dependent on the packaging materials and technologies applied [112,127].

Innovations in food packaging nanomaterials are primarily attributable to their following distinct characteristics: excellent optical, barrier, and thermal properties, antimicrobial activity, and advanced sensing properties affecting their chemical, physical, and biological potential unlike their bulk counterparts [37,128].

Nanomaterials consisting of TiO_2 [129,130], SiO_2 [131,132], AgNPs [133], graphene [134], and nanocellulose [135] possess remarkable characteristics such as high catalytic activity and conductivity, which make them quintessential candidates for biosensory abilities [37].

Exemplary electrochemical immunosensors that are appropriate for the detection of *Salmonella* in meat samples have recently been found in the literature [136]. For example, graphene is a fully reliable biosensing nanomaterial that can be easily integrated with smart packaging systems. Graphene-based nanofibers and electrodes are applied in the development of a flexible detector for ethanol [137], histamine [138], and ammonia [37,139]. Among these films, pigment-based natural colorimetric films have gained considerable attention due to their nontoxicity, biocompatibility, nature of pH sensing, and others [140,141]. These pH-sensitive colorimetric films can show visible color changes while reacting with non-neutral volatile gases generated from high-protein degraded food products, which can provide visual information about the quality and microbial contamination of the food product [14,86,142,143].

To delay lipid oxidation and reduce chemical additives causing health disorders, functional packaging using natural antioxidants is applied to extend the shelf life of meat products [112,144,145].

Antioxidant and antimicrobial compounds used in food packaging are of different origins: natural, such as essential oils, nisin, curcumin, α-tocopherol and vitamins, phenolic-rich plant and pomace extracts, allyl isothiocyanate, and chitosan [146,147]; synthetic antioxidants, such as butylhydroxytoluene and its analogs, butylhydroxyanisole, and t-butylhydroxyquinone [23]; or antimicrobial, such as organic acids (acetic, sorbic and ascorbic, benzoic and propane), nitrites, and nitrates [148,149].

Thymol, which is the primary component of thyme oil (classified as Generally Recognized As Safe by United States Food and Drug Administration), is a promising alternative to chemical preservatives with good antimicrobial and antioxidant properties [150]. Although thymol's potential as a food preservative has been widely discussed, its use in film/coating formulations is highly limited due to its high volatility and hydrophobicity [151]. Given these issues, particular attention has been paid to the encapsulation of plant-derived bioactive compounds in biopolymer nanocarriers [26,28]. Lin et al. [152] used gelatin nanofibers that contain thyme essential oil/ε-polylysine β-cyclodextrin nanoparticles to control the growth of *Campylobacter jejuni* on the surface of poultry with no effects on the sensory and textural properties and color. The packaged chicken samples showed lower aerobic bacteria counts, total volatile basic nitrogen, trimethylamine and TBA content, and pH values [123].

Cinnamaldehyde (3-phenyl-2-propenal), a component of natural cinnamon oil with a common flavor, is one of the important antioxidant and antimicrobial agents. It can be used to improve the quality of food products and extend their shelf life. Its sensitivity to heat, light, humidity, oxygen, and liquid form at room temperature necessitates its encapsulation. Zein nanofiber mass containing 1000 ppm loaded with cinnamaldehyde showed good bactericidal activity against *Staphylococcus aureus* PTCC 1337 (Persian Type Culture Collection (PTCC)) and *Escherichia coli* O157:H7 with no significant adverse effects on texture or color in nitrite-reduced sausages [123]. The number of *E. coli* and *S. aureus* (colony-forming unit/g samples) decreased in all sausages during storage due to the presence of zein nanofibers with cinnamaldehyde as an antibacterial agent and nitrates [123]. Many studies [84,153–155] reported that cinnamaldehyde, zein nanofibers with cinnamaldehyde, and nitrites show long-term growth inhibition of *S. aureus* and *E. coli*. After 10 days of storage, samples with packages containing phase change materials used for temperature buffering did not contain *E. coli* and *S. aureus* bacteria [123].

Using unstable substances in AP, positive results are observed in nanoencapsulation techniques, including nanoparticles, nanoemulsions, and nanocapsules. This prevents the degradation of, for example, saffron bioactive compounds under adverse conditions until they are delivered for physiological purposes [156]. In this context, electrospinning and electrospraying have recently gained increased interest in encapsulating bioactive ingredients and food packaging. These methods are simple, versatile, nonthermal, and thus highly suitable for the encapsulation of heat-sensitive compounds [157–159]. Studies [159] have indicated that electroyarn containing 30% zein and 10% saffron extract show great potential in extending the shelf life of seafood products and delaying their spoilage during cold storage.

An overview of sample compositions of novel packaging materials, as well as the bioactive substances used and the spectrum of their effects on the quality of packaged food products, is presented in Table 1.

Table 1. Examples and effects of using bioactive substances in packaging materials.

Substance	Matrix	Positive Effects Obtained	Product	Source
		Antimicrobial effect		
Tea tree oil	Nanofiber membrane	Inhibition of 99.99% *Salmonella* after 4 days of operation without affecting the sensory quality	Chicken meat	[160]
Cinnamon essential oil (as core)	Encapsulated in Eudragit L100 (as a shell) by coaxial electrospinning technology	Controlled release, good antibacterial efficacy against *E. coli* and *S. aureus*	Pork loin	[161]
Pomegranate peel extract (PE)	Electrospun chitosan/polyethylene oxide (CS/PEO) active nanofibers/active CS/PEO/PE nanofibers	Effective inhibition of *E. coli* O157:H7 on samples at 4 and 25 °C for 7 and 10 days, respectively, compared to control packaging	Beef	[112]
Thyme (EO)	Silk fibroin nanofibers	*Salmonella typhimurium* reduction from 6.64 to 2.24 log CFU/g	Chicken meat	[70]

Table 1. *Cont.*

Substance	Matrix	Positive Effects Obtained	Product	Source
Oregano (EO)	Sodium alginate foil	A reduction in *Listeria* population of approximately 1.5 log at 8 °C and 12 °C at the end of storage and almost 2.5 log at 4 °C	Ham	[162]
Chitosan	Electrospun fibers based on chitosan and poly(ethylene oxide) CS/PEO	The ability to maintain safety and extend the shelf life by a week	Fresh red meat	[163]
Gallic acid + chitosan or carvacrol + chitosan	Starch foil	Complete inhibition of the growth of *Listeria monocytogenes* for 4 weeks of storage, starch films filled with chitosan or chitosan and carvacrol delayed the growth of the microbiota by 1–2 weeks	Ham	[164]
Electrospun gelatin-glycerine-ε-polylysine nanofibers	Gelatine	Growth inhibition of *L. monocytogenes*	Beef	[165]
Lemon (LEO)	Thermally stable and porous vermiculite (VML), LEO/VML complex, coupled with konjac glucomannan-grafted-poly(acrylic acid)/polyvinyl alcohol composite	Long-term LEO control release effectively inhibiting *E. coli* growth during storage, thus extending the shelf life of chilled pork by 3 days	Pork	[166]
Methyl ferulate	Zein	Effectively inhibition of microorganism growth in fish meat and slowing down of the production and accumulation of alkaline substances, thus controlling the increase in pH and maintaining freshness	Fish	[167]
Thyme EO/ε-polylysine β-cyclodextrin nanoparticles	Gelatin nanofibers	Controls the growth of *C. jejuni* on the surface of poultry without affecting the sensory evaluation	Poultry meat	[152]
Eugenol	Gelatin nanofibers	Strong antibacterial activity/growth retardation of total mesophilic aerobic and total psychrophilic bacteria	Meat products	[102]
Covered with poly-caprolactone/chitosan nonwoven fabric (film 1) covered with polycaprolac-tone/chitosan nonwoven fabric reinforced with *Colombian propolis* extract (film 2)	Linear low-density polyethylene film	Improving color stability and microbiological stability of pork samples	Pork	[168]
Antioxidant effect				
Rosemary extract	Low-density polyethylene	Significant inhibition of lipid oxidation	Pork patties	[169]

Table 1. *Cont.*

Substance	Matrix	Positive Effects Obtained	Product	Source
Chitosan	Gelatin foil	Delaying the oxidation of fats and the formation of methemoglobin	Beef	[170]
Cinnamon (85%) + rosemary essential oil (15%)	Whey protein	Significant inhibition of lipid oxidation	Salami	[171]
Green tea extract	Polyamide	Very good antioxidant capacity and extending the shelf life from 6 to 23 days	Minced meat	[172]
Antioxidant + antimicrobial action				
Beetroot peel extract	Gelatin–sodium alginate coating	Minimum inhibitory concentration of 2.5 mg/mL against Gram-positive bacteria (*S. aureus* and *E. coli*) and Gram-negative bacteria (*Salmonella enterica* and *L. monocytogenes*); delaying chemical oxidation and improving sensory characteristics	Beef meat	[173]
Lactobacillus plantarum postbiotics	Bacterial nanocellulose	Reduction (~5 log cycles) in the number of *L. monocytogenes* in minced meat. *L. plantarum* postbiotics showed moderate antioxidant activity in meat	Minced meat	[174]
Anethum graveolens (EO)	Plantago major seed mucosa	Action against *E. coli, S. aureus,* and fungi extending the shelf life of meat from 6 to 18 days; and inhibition of the growth of bacteria and slowing down of oxidative changes	Beef meat	[175]
Clove and argan oils	Poly(lactic acid) films coated with chitosan oil	Low oxygen permeability, high radical scavenging activity, and strong growth inhibition of *L. monocytogenes, S. typhimurium,* and *E. coli*	Beef meat	[176]
Aqueous green tea extract	Chitosan coating	Improvement in physicochemical properties (pH, color, and lipid oxidation) and microbiological properties of samples during storage; the inclusion of 0.1% and 0.5% green tea water extract in the 1% chitosan coating effectively retards the formation of malondialdehyde and microbial growth, while having a beneficial effect on the pH and intensity of red pork color	Pork cutlet with bone	[177]

Table 1. *Cont.*

Substance	Matrix	Positive Effects Obtained	Product	Source
ZnO nanoparticles with propolis	Composite film based on pullulan/chitosan (PLN/CTS)	Strong antibacterial activity against *E. coli* and *L. monocytogenes*: in meat samples wrapped in PLN/CTS/ZnO/PPS foil before packaging, the value of the total aerobic bacteria count (TABC) remained at the level of 6.7 Log CFU/g after 8 days of storage, controls showed a rapid increase (TABC) of ~6 Log CFU/g after 6 days and finally ~9 Log CFU/g within 8 days; excellent antioxidant activity: after 15 days of storage, while the peroxide values (PV) of packaged meat in the control group increased sharply to 22 meq/kg, meat wrapped in PLN/CTS/ZnO/PPS film showed a much lower peroxide count of ~10 meq/kg, showing approximately 55% reduced lipid oxidation	Pork loin	[178]
Catechin and lysozyme	Gelatin foil	Extending the shelf life and reducing the total number of bacteria, yeasts, and molds. Effective inhibition of lipid oxidation and microbial growth	Minced pork	[179]
Origanum virens (EO)	Whey protein concentrate (WPC)	Inhibition of total microbial load, higher acidity, and protection against discoloration; the EO-WPC film had a positive effect on the retardation of chain reactions of fat oxidation in alheiras	Traditional Portuguese sausages (paínhos and alheiras)	[180]
Terminalia arjuna extract	Maltodextrin and calcium alginate	Lipid oxidation was inhibited, and the number of yeasts and molds was reduced	Chevon sausages	[181]
Ethanol propolis extract	Chitosan film enriched with cellulose nanoparticle	*Pseudomonas* spp., *LAB* (lactic acid bacteria), and *Enterobacteriaceae* slow down the growth of microorganisms and the oxidation of lipids and proteins	Ground beef	[182]
Resveratrol	Gelatin/zein mats	Good antibacterial activity against *E. coli* and *S. aureus*, antioxidant activity to inhibit discoloration, and extended shelf life	Pork	[183]
Curcumin (CUR)	Packaging nanofibers based on gelatin/chitosan (GA/CS)	Inclusion of CUR significantly improved the antioxidant and antimicrobial activity of GA/CS/CUR nanofibers	Meat and seafood	[184]

Table 1. *Cont.*

Substance	Matrix	Positive Effects Obtained	Product	Source
Cloves (CL) and cinnamon (CI)	Corn starch (CS)	Inclusion of CL and CI EO in CS film at 3% significantly reduced the microbial population and thiobarbituric acid reactive substances (TBARS) values in raw meat during refrigerated storage	Beef	[185]
Spice EO (*Laurus nobilis*, LEO; and *Rosmarinus officinalis*, REO)	Polyvinyl alcohol electroyarn	Active packaging coatings containing LEO and REO extended the shelf life by reducing the process of lipid oxidation and reducing the number of *Listeria* during cold storage	Chicken breast fillets	[69]

5. Summary

The introduction of new technologies in food packaging has made the packaging market dynamic. This involves many changes in, among others, the verification of the usability of new materials in industrial conditions, especially in terms of their impact on the quality and safety of packaged food products. A promising direction is using natural polymers for this purpose, which offers a possibility to solve the problems as a result of the generation of huge amounts of waste by the food industry. However, the inclusion of natural bioactive substances in packaging provides an opportunity to extend the shelf life of food products and/or reduce the use of food preservatives. This provides a wide field for research due to the multitude of substances and the spectrum of their impact, together with the properties of the packaged product.

Author Contributions: Conceptualization, M.G. and M.R.; introduction, M.G. and M.R.; methodology, M.G. and M.R.; resources, M.G. and M.R.; writing—original draft preparation, M.G. and M.R.; writing—review and editing, M.G. and M.R. All authors have read and agreed to the published version of the manuscript.

Funding: This research received no external funding.

Institutional Review Board Statement: Not applicable.

Informed Consent Statement: Not applicable.

Data Availability Statement: Not applicable.

Conflicts of Interest: The authors declare no conflict of interest.

References

1. Sid, S.; Mor, R.S.; Kishore, A.; Sharanagat, V.S. Bio-Sourced Polymers as Alternatives to Conventional Food Packaging Materials: A Review. *Trends Food Sci. Technol.* **2021**, *115*, 87–104. [CrossRef]
2. Cenci-Goga, B.T.; Iulietto, M.F.; Sechi, P.; Borgogni, E.; Karama, M.; Grispoldi, L. New Trends in Meat Packaging. *Microbiol. Res.* **2020**, *11*, 56–67. [CrossRef]
3. Nilsen-Nygaard, J.; Fernández, E.N.; Radusin, T.; Rotabakk, B.T.; Sarfraz, J.; Sharmin, N.; Sivertsvik, M.; Sone, I.; Pettersen, M.K. Current Status of Biobased and Biodegradable Food Packaging Materials: Impact on Food Quality and Effect of Innovative Processing Technologies. *Compr. Rev. Food Sci. Food Saf.* **2021**, *20*, 1333–1380. [CrossRef] [PubMed]
4. Zubair, M.; Ullah, A. Recent Advances in Protein Derived Bionanocomposites for Food Packaging Applications. *Crit. Rev. Food Sci. Nutr.* **2020**, *60*, 406–434. [CrossRef] [PubMed]
5. Zhu, W.; Zhang, D.; Liu, X.; Ma, T.; He, J.; Dong, Q.; Din, Z.; Zhou, J.; Chen, L.; Hu, Z.; et al. Improving the Hydrophobicity and Mechanical Properties of Starch Nanofibrous Films by Electrospinning and Cross-Linking for Food Packaging Applications. *LWT* **2022**, *169*, 114005. [CrossRef]
6. Li, X.; Zhang, R.; Hassan, M.M.; Cheng, Z.; Mills, J.; Hou, C.; Realini, C.E.; Chen, L.; Day, L.; Zheng, X.; et al. Active Packaging for the Extended Shelf-Life of Meat: Perspectives from Consumption Habits, Market Requirements and Packaging Practices in China and New Zealand. *Foods* **2022**, *11*, 2903. [CrossRef]

7.   Schumann, B.; Schmid, M. Packaging Concepts for Fresh and Processed Meat—Recent Progresses. *Innov. Food Sci. Emerg. Technol.* **2018**, *47*, 88–100. [CrossRef]
8.   Ingrao, C.; Giudice, A.L.; Bacenetti, J.; Khaneghah, A.M.; Sant'Ana, A.S.; Rana, R.; Siracusa, V. Foamy Polystyrene Trays for Fresh-Meat Packaging: Life-Cycle Inventory Data Collection and Environmental Impact Assessment. *Food Res. Int.* **2015**, *76*, 418–426. [CrossRef]
9.   Agarwal, A.; Shaida, B.; Rastogi, M.; Singh, N.B. Food Packaging Materials with Special Reference to Biopolymers-Properties and Applications. *Chem. Afr.* **2022**, 1–28. [CrossRef]
10.   Realini, C.E.; Marcos, B. Active and Intelligent Packaging Systems for a Modern Society. *Meat Sci.* **2014**, *98*, 404–419. [CrossRef]
11.   Jung, S.; Cui, Y.; Barnes, M.; Satam, C.; Zhang, S.; Chowdhury, R.A.; Adumbumkulath, A.; Sahin, O.; Miller, C.; Sajadi, S.M.; et al. Multifunctional Bio-Nanocomposite Coatings for Perishable Fruits. *Adv. Mater.* **2020**, *32*, 1908291. [CrossRef]
12.   Hu, B.; Chen, L.; Lan, S.; Ren, P.; Wu, S.; Liu, X.; Shi, X.; Li, H.; Du, Y.; Ding, F. Layer-by-Layer Assembly of Polysaccharide Films with Self-Healing and Antifogging Properties for Food Packaging Applications. *ACS Appl. Nano Mater.* **2018**, *1*, 3733–3740. [CrossRef]
13.   Liu, M.; Wang, F.; Liang, M.; Si, Y.; Yu, J.; Ding, B. In Situ Green Synthesis of Rechargeable Antibacterial N-Halamine Grafted Poly(Vinyl Alcohol) Nanofibrous Membranes for Food Packaging Applications. *Compos. Commun.* **2020**, *17*, 147–153. [CrossRef]
14.   Duan, N.; Li, Q.; Meng, X.; Wang, Z.; Wu, S. Preparation and Characterization of K-Carrageenan/Konjac Glucomannan/TiO$_2$ Nanocomposite Film with Efficient Anti-Fungal Activity and Its Application in Strawberry Preservation. *Food Chem.* **2021**, *364*, 130441. [CrossRef]
15.   Huang, G.; Li, Y.; Qin, Z.; Liang, Q.; Xu, C.; Lin, B. Hybridization of Carboxymethyl Chitosan with MOFs to Construct Recyclable, Long-Acting and Intelligent Antibacterial Agent Carrier. *Carbohydr. Polym.* **2020**, *233*, 115848. [CrossRef]
16.   Li, S.; Sun, J.; Yan, J.; Zhang, S.; Shi, C.; McClements, D.J.; Liu, X.; Liu, F. Development of Antibacterial Nanoemulsions Incorporating Thyme Oil: Layer-by-Layer Self-Assembly of Whey Protein Isolate and Chitosan Hydrochloride. *Food Chem.* **2021**, *339*, 128016. [CrossRef]
17.   Min, T.; Zhou, L.; Sun, X.; Du, H.; Zhu, Z.; Wen, Y. Electrospun Functional Polymeric Nanofibers for Active Food Packaging: A Review. *Food Chem.* **2022**, *391*, 133239. [CrossRef]
18.   Liu, Y.; Wang, S.; Zhang, R.; Lan, W.; Qin, W. Development of Poly(Lactic Acid)/Chitosan Fibers Loaded with Essential Oil for Antimicrobial Applications. *Nanomaterials* **2017**, *7*, 194. [CrossRef]
19.   Yildirim-Yalcin, M.; Tornuk, F.; Toker, O.S. Recent Advances in the Improvement of Carboxymethyl Cellulose-Based Edible Films. *Trends Food Sci. Technol.* **2022**, *129*, 179–193. [CrossRef]
20.   Dhall, R.K. Advances in Edible Coatings for Fresh Fruits and Vegetables: A Review. *Crit. Rev. Food Sci. Nutr.* **2013**, *53*, 435–450. [CrossRef]
21.   Holman, B.W.B.; Kerry, J.P.; Hopkins, D.L. Meat Packaging Solutions to Current Industry Challenges: A Review. *Meat Sci.* **2018**, *144*, 159–168. [CrossRef] [PubMed]
22.   Zhang, C.; Li, Y.; Wang, P.; Zhang, H. Electrospinning of Nanofibers: Potentials and Perspectives for Active Food Packaging. *Compr. Rev. Food Sci. Food Saf.* **2020**, *19*, 479–502. [CrossRef] [PubMed]
23.   Gagaoua, M.; Pinto, V.Z.; Göksen, G.; Alessandroni, L.; Lamri, M.; Dib, A.L.; Boukid, F. Electrospinning as a Promising Process to Preserve the Quality and Safety of Meat and Meat Products. *Coatings* **2022**, *12*, 644. [CrossRef]
24.   Bhargava, N.; Sharanagat, V.S.; Mor, R.S.; Kumar, K. Active and Intelligent Biodegradable Packaging Films Using Food and Food Waste-Derived Bioactive Compounds: A Review. *Trends Food Sci. Technol.* **2020**, *105*, 385–401. [CrossRef]
25.   Osmólska, E.; Stoma, M.; Starek-Wójcicka, A. Application of Biosensors, Sensors, and Tags in Intelligent Packaging Used for Food Products—A Review. *Sensors* **2022**, *22*, 9956. [CrossRef]
26.   Rehman, A.; Jafari, S.M.; Aadil, R.M.; Assadpour, E.; Randhawa, M.A.; Mahmood, S. Development of Active Food Packaging via Incorporation of Biopolymeric Nanocarriers Containing Essential Oils. *Trends Food Sci. Technol.* **2020**, *101*, 106–121. [CrossRef]
27.   Smaoui, S.; Hlima, H.B.; Tavares, L.; Ennouri, K.; Braiek, O.B.; Mellouli, L.; Abdelkafi, S.; Khaneghah, A.M. Application of Essential Oils in Meat Packaging: A Systemic Review of Recent Literature. *Food Control* **2022**, *132*, 108566. [CrossRef]
28.   Dai, J.; Hu, W.; Yang, H.; Li, C.; Cui, H.; Li, X.; Lin, L. Controlled Release and Antibacterial Properties of PEO/Casein Nanofibers Loaded with Thymol/β-Cyclodextrin Inclusion Complexes in Beef Preservation. *Food Chem.* **2022**, *382*, 132369. [CrossRef]
29.   Raut, J.S.; Karuppayil, S.M. A Status Review on the Medicinal Properties of Essential Oils. *Ind. Crops Prod.* **2014**, *62*, 250–264. [CrossRef]
30.   Doğan, C.; Doğan, N.; Gungor, M.; Eticha, A.K.; Akgul, Y. Novel Active Food Packaging Based on Centrifugally Spun Nanofibers Containing Lavender Essential Oil: Rapid Fabrication, Characterization, and Application to Preserve of Minced Lamb Meat. *Food Packag. Shelf Life* **2022**, *34*, 100942. [CrossRef]
31.   Said, N.S.; Sarbon, N.M. Response Surface Methodology (RSM) of Chicken Skin Gelatin Based Composite Films with Rice Starch and Curcumin Incorporation. *Polym. Test.* **2020**, *81*, 106161. [CrossRef]
32.   Popović, S.Z.; Lazić, V.L.; Hromiš, N.M.; Šuput, D.Z.; Bulut, S.N. Chapter 8—Biopolymer Packaging Materials for Food Shelf-Life Prolongation. In *Biopolymers for Food Design*; Grumezescu, A.M., Holban, A.M., Eds.; Handbook of Food Bioengineering; Academic Press: Cambridge, MA, USA, 2018; pp. 223–277, ISBN 978-0-12-811449-0.

33. Hosseini, S.M.H.; Ghiasi, F.; Jahromi, M. 12—Nanocapsule Formation by Complexation of Biopolymers. In *Nanoencapsulation Technologies for the Food and Nutraceutical Industries*; Jafari, S.M., Ed.; Academic Press: Cambridge, MA, USA, 2017; pp. 447–492, ISBN 978-0-12-809436-5.

34. Shankar, S.; Wang, L.-F.; Rhim, J.-W. Preparation and Properties of Carbohydrate-Based Composite Films Incorporated with CuO Nanoparticles. *Carbohydr. Polym.* **2017**, *169*, 264–271. [CrossRef]

35. Shankar, S.; Tanomrod, N.; Rawdkuen, S.; Rhim, J.-W. Preparation of Pectin/Silver Nanoparticles Composite Films with UV-Light Barrier and Properties. *Int. J. Biol. Macromol.* **2016**, *92*, 842–849. [CrossRef]

36. Babaremu, K.; Oladijo, O.P.; Akinlabi, E. Biopolymers: A Suitable REPLACEMENT for Plastics in Product Packaging. *Adv. Ind. Eng. Polym. Res.* **2023**, *in press*. [CrossRef]

37. Ahmad, A.; Qurashi, A.; Sheehan, D. Nano Packaging—Progress and Future Perspectives for Food Safety, and Sustainability. *Food Packag. Shelf Life* **2023**, *35*, 100997. [CrossRef]

38. Ruan, C.; Zhang, Y.; Wang, J.; Sun, Y.; Gao, X.; Xiong, G.; Liang, J. Preparation and Antioxidant Activity of Sodium Alginate and Carboxymethyl Cellulose Edible Films with Epigallocatechin Gallate. *Int. J. Biol. Macromol.* **2019**, *134*, 1038–1044. [CrossRef]

39. Robertson, G.L. *Food Packaging: Principles and Practice*; CRC Press: Boca Raton, FL, USA, 2016; ISBN 9780429105401. [CrossRef]

40. Liu, Y.; Ahmed, S.; Sameen, D.E.; Wang, Y.; Lu, R.; Dai, J.; Li, S.; Qin, W. A Review of Cellulose and Its Derivatives in Biopolymer-Based for Food Packaging Application. *Trends Food Sci. Technol.* **2021**, *112*, 532–546. [CrossRef]

41. Zhong, Y.; Godwin, P.; Jin, Y.; Xiao, H. Biodegradable Polymers and Green-Based Antimicrobial Packaging Materials: A Mini-Review. *Adv. Ind. Eng. Polym. Res.* **2020**, *3*, 27–35. [CrossRef]

42. Onyeaka, H.; Obileke, K.; Makaka, G.; Nwokolo, N. Current Research and Applications of Starch-Based Biodegradable Films for Food Packaging. *Polymers* **2022**, *14*, 1126. [CrossRef]

43. Su, C.; Li, D.; Wang, L.; Wang, Y. Biodegradation Behavior and Digestive Properties of Starch-Based Film for Food Packaging—A Review. *Crit. Rev. Food Sci. Nutr.* **2022**, 1–23. [CrossRef]

44. Dang, K.M.; Yoksan, R. Morphological Characteristics and Barrier Properties of Thermoplastic Starch/Chitosan Blown Film. *Carbohydr. Polym.* **2016**, *150*, 40–47. [CrossRef] [PubMed]

45. Pelissari, F.M.; Ferreira, D.C.; Louzada, L.B.; dos Santos, F.; Corrêa, A.C.; Moreira, F.K.V.; Mattoso, L.H. Chapter 10—Starch-Based Edible Films and Coatings: An Eco-Friendly Alternative for Food Packaging. In *Starches for Food Application*; Clerici, M.T.P.S., Schmiele, M., Eds.; Academic Press: Cambridge, MA, USA, 2019; pp. 359–420, ISBN 978-0-12-809440-2.

46. Thakur, R.; Pristijono, P.; Scarlett, C.J.; Bowyer, M.; Singh, S.P.; Vuong, Q.V. Starch-Based Films: Major Factors Affecting Their Properties. *Int. J. Biol. Macromol.* **2019**, *132*, 1079–1089. [CrossRef] [PubMed]

47. Wiszumirska, K.; Czarnecka-Komorowska, D.; Kozak, W.; Biegańska, M.; Wojciechowska, P.; Jarzębski, M.; Pawlak-Lemańska, K. Characterization of Biodegradable Food Contact Materials under Gamma-Radiation Treatment. *Materials* **2023**, *16*, 859. [CrossRef] [PubMed]

48. Priyadarshi, R.; Rhim, J.-W. Chitosan-Based Biodegradable Functional Films for Food Packaging Applications. *Innov. Food Sci. Emerg. Technol.* **2020**, *62*, 102346. [CrossRef]

49. Okutan, N.; Terzi, P.; Altay, F. Affecting Parameters on Electrospinning Process and Characterization of Electrospun Gelatin Nanofibers. *Food Hydrocoll.* **2014**, *39*, 19–26. [CrossRef]

50. Etxabide, A.; Akbarinejad, A.; Chan, E.W.C.; Guerrero, P.; de la Caba, K.; Travas-Sejdic, J.; Kilmartin, P.A. Effect of Gelatin Concentration, Ribose and Glycerol Additions on the Electrospinning Process and Physicochemical Properties of Gelatin Nanofibers. *Eur. Polym. J.* **2022**, *180*, 111597. [CrossRef]

51. Krishnakumar, G.S.; Sampath, S.; Muthusamy, S.; John, M.A. Importance of Crosslinking Strategies in Designing Smart Biomaterials for Bone Tissue Engineering: A Systematic Review. *Mater. Sci. Eng. C* **2019**, *96*, 941–954. [CrossRef]

52. Kwak, H.W.; Park, J.; Yun, H.; Jeon, K.; Kang, D.-W. Effect of Crosslinkable Sugar Molecules on the Physico-Chemical and Antioxidant Properties of Fish Gelatin Nanofibers. *Food Hydrocoll.* **2021**, *111*, 106259. [CrossRef]

53. Hanani, Z.A.N.; Roos, Y.H.; Kerry, J.P. Use and Application of Gelatin as Potential Biodegradable Packaging Materials for Food Products. *Int. J. Biol. Macromol.* **2014**, *71*, 94–102. [CrossRef]

54. Kchaou, H.; Benbettaïeb, N.; Jridi, M.; Abdelhedi, O.; Karbowiak, T.; Brachais, C.-H.; Léonard, M.-L.; Debeaufort, F.; Nasri, M. Enhancement of Structural, Functional and Antioxidant Properties of Fish Gelatin Films Using Maillard Reactions. *Food Hydrocoll.* **2018**, *83*, 326–339. [CrossRef]

55. Etxabide, A.; Vairo, C.; Santos-Vizcaino, E.; Guerrero, P.; Pedraz, J.L.; Igartua, M.; de la Caba, K.; Hernandez, R.M. Ultra Thin Hydro-Films Based on Lactose-Crosslinked Fish Gelatin for Wound Healing Applications. *Int. J. Pharm.* **2017**, *530*, 455–467. [CrossRef]

56. Etxabide, A.; Kilmartin, P.A.; Maté, J.I.; Prabakar, S.; Brimble, M.; Naffa, R. Analysis of Advanced Glycation End Products in Ribose-, Glucose- and Lactose-Crosslinked Gelatin to Correlate the Physical Changes Induced by Maillard Reaction in Films. *Food Hydrocoll.* **2021**, *117*, 106736. [CrossRef]

57. Etxabide, A.; Urdanpilleta, M.; Gómez-Arriaran, I.; de la Caba, K.; Guerrero, P. Effect of PH and Lactose on Cross-Linking Extension and Structure of Fish Gelatin Films. *React. Funct. Polym.* **2017**, *117*, 140–146. [CrossRef]

58. Stevenson, M.; Long, J.; Seyfoddin, A.; Guerrero, P.; de la Caba, K.; Etxabide, A. Characterization of Ribose-Induced Crosslinking Extension in Gelatin Films. *Food Hydrocoll.* **2020**, *99*, 105324. [CrossRef]

59. Yaman, M.; Sar, M.; Ceylan, Z. A Nanofiber Application for Thiamine Stability and Enhancement of Bioaccessibility of Raw, Cooked Salmon and Red Meat Samples Stored at 4 °C. *Food Chem.* **2022**, *373*, 131447. [CrossRef]
60. Shaikh, S.; Yaqoob, M.; Aggarwal, P. An Overview of Biodegradable Packaging in Food Industry. *Curr. Res. Food Sci.* **2021**, *4*, 503–520. [CrossRef]
61. Alghoraibi, I.; Alomari, S. Different Methods for Nanofiber Design and Fabrication. In *Handbook of Nanofibers*; Barhoum, A., Bechelany, M., Makhlouf, A., Eds.; Springer International Publishing: Cham, Switzerland, 2018; pp. 1–46, ISBN 978-3-319-42789-8.
62. Xue, J.; Wu, T.; Dai, Y.; Xia, Y. Electrospinning and Electrospun Nanofibers: Methods, Materials, and Applications. *Chem. Rev.* **2019**, *119*, 5298–5415. [CrossRef]
63. Liu, P.; Wu, S.; Zhang, Y.; Zhang, H.; Qin, X. A Fast Response Ammonia Sensor Based on Coaxial PPy–PAN Nanofiber Yarn. *Nanomaterials* **2016**, *6*, 121. [CrossRef]
64. Das, S.K.; Afzal, M.A.F.; Srivastava, S.; Patil, S.; Sharma, A. Enhanced Electrical Conductivity of Suspended Carbon Nanofibers: Effect of Hollow Structure and Improved Graphitization. *Carbon* **2016**, *108*, 135–145. [CrossRef]
65. He, P.; Zhong, Q.; Ge, Y.; Guo, Z.; Tian, J.; Zhou, Y.; Ding, S.; Li, H.; Zhou, C. Dual Drug Loaded Coaxial Electrospun PLGA/PVP Fiber for Guided Tissue Regeneration under Control of Infection. *Mater. Sci. Eng. C* **2018**, *90*, 549–556. [CrossRef]
66. Katsogiannis, K.A.G.; Vladisavljević, G.T.; Georgiadou, S. Porous Electrospun Polycaprolactone (PCL) Fibres by Phase Separation. *Eur. Polym. J.* **2015**, *69*, 284–295. [CrossRef]
67. Thenmozhi, S.; Dharmaraj, N.; Kadirvelu, K.; Kim, H.Y. Electrospun Nanofibers: New Generation Materials for Advanced Applications. *Mater. Sci. Eng. B* **2017**, *217*, 36–48. [CrossRef]
68. Kny, E.; Ghosal, K.; Thomas, S. (Eds.) *Electrospinning: From Basic Research to Commercialization*; Soft Matter Series; The Royal Society of Chemistry: London, UK, 2018; ISBN 978-1-78801-100-6.
69. Göksen, G.; Fabra, M.J.; Pérez-Cataluña, A.; Ekiz, H.I.; Sanchez, G.; López-Rubio, A. Biodegradable Active Food Packaging Structures Based on Hybrid Cross-Linked Electrospun Polyvinyl Alcohol Fibers Containing Essential Oils and Their Application in the Preservation of Chicken Breast Fillets. *Food Packag. Shelf Life* **2021**, *27*, 100613. [CrossRef]
70. Lin, L.; Liao, X.; Cui, H. Cold Plasma Treated Thyme Essential Oil/Silk Fibroin Nanofibers against Salmonella Typhimurium in Poultry Meat. *Food Packag. Shelf Life* **2019**, *21*, 100337. [CrossRef]
71. Surendhiran, D.; Cui, H.; Lin, L. Encapsulation of Phlorotannin in Alginate/PEO Blended Nanofibers to Preserve Chicken Meat from Salmonella Contaminations. *Food Packag. Shelf Life* **2019**, *21*, 100346. [CrossRef]
72. Forghani, S.; Almasi, H.; Moradi, M. Electrospun Nanofibers as Food Freshness and Time-Temperature Indicators: A New Approach in Food Intelligent Packaging. *Innov. Food Sci. Emerg. Technol.* **2021**, *73*, 102804. [CrossRef]
73. Zhang, C.; Li, Y.; Wang, P.; Zhang, A.; Feng, F.; Zhang, H. Electrospinning of Bilayer Emulsions: The Role of Gum Arabic as a Coating Layer in the Gelatin-Stabilized Emulsions. *Food Hydrocoll.* **2019**, *94*, 38–47. [CrossRef]
74. Ge, L.; Zhao, Y.; Mo, T.; Li, J.; Li, P. Immobilization of Glucose Oxidase in Electrospun Nanofibrous Membranes for Food Preservation. *Food Control* **2012**, *26*, 188–193. [CrossRef]
75. Ding, Y.; Wang, Y.; Su, L.; Bellagamba, M.; Zhang, H.; Lei, Y. Electrospun Co3O4 Nanofibers for Sensitive and Selective Glucose Detection. *Biosens. Bioelectron.* **2010**, *26*, 542–548. [CrossRef]
76. Xiong, J.; Shao, W.; Wang, L.; Cui, C.; Jin, Y.; Yu, H.; Han, P.; Gao, Y.; Liu, F.; Ni, Q.; et al. PAN/FPU Composite Nanofiber Membrane with Superhydrophobic and Superoleophobic Surface as a Filter Element for High-Efficiency Protective Masks. *Macromol. Mater. Eng.* **2021**, *306*, 2100371. [CrossRef]
77. Karagoz, S.; Kiremitler, N.B.; Sarp, G.; Pekdemir, S.; Salem, S.; Goksu, A.G.; Onses, M.S.; Sozdutmaz, I.; Sahmetlioglu, E.; Ozkara, E.S.; et al. Antibacterial, Antiviral, and Self-Cleaning Mats with Sensing Capabilities Based on Electrospun Nanofibers Decorated with ZnO Nanorods and Ag Nanoparticles for Protective Clothing Applications. *ACS Appl. Mater. Interfaces* **2021**, *13*, 5678–5690. [CrossRef]
78. Yan, Y.; Liu, X.; Yan, J.; Guan, C.; Wang, J. Electrospun Nanofibers for New Generation Flexible Energy Storage. *ENERGY Environ. Mater.* **2021**, *4*, 502–521. [CrossRef]
79. Lee, M.W.; An, S.; Latthe, S.S.; Lee, C.; Hong, S.; Yoon, S.S. Electrospun Polystyrene Nanofiber Membrane with Superhydrophobicity and Superoleophilicity for Selective Separation of Water and Low Viscous Oil. *ACS Appl. Mater. Interfaces* **2013**, *5*, 10597–10604. [CrossRef]
80. Al-Moghazy, M.; Mahmoud, M.; Nada, A.A. Fabrication of Cellulose-Based Adhesive Composite as an Active Packaging Material to Extend the Shelf Life of Cheese. *Int. J. Biol. Macromol.* **2020**, *160*, 264–275. [CrossRef]
81. Bruni, G.P.; de Oliveira, J.P.; Gómez-Mascaraque, L.G.; Fabra, M.J.; Martins, V.G.; da Rosa Zavareze, E.; López-Rubio, A. Electrospun β-Carotene–Loaded SPI:PVA Fiber Mats Produced by Emulsion-Electrospinning as Bioactive Coatings for Food Packaging. *Food Packag. Shelf Life* **2020**, *23*, 100426. [CrossRef]
82. Zhang, R.; Lan, W.; Ji, T.; Sameen, D.E.; Ahmed, S.; Qin, W.; Liu, Y. Development of Polylactic Acid/ZnO Composite Membranes Prepared by Ultrasonication and Electrospinning for Food Packaging. *LWT* **2021**, *135*, 110072. [CrossRef]
83. Coelho, S.C.; Estevinho, B.N.; Rocha, F. Encapsulation in Food Industry with Emerging Electrohydrodynamic Techniques: Electrospinning and Electrospraying—A Review. *Food Chem.* **2021**, *339*, 127850. [CrossRef]
84. Karim, M.; Fathi, M.; Soleimanian-Zad, S. Nanoencapsulation of Cinnamic Aldehyde Using Zein Nanofibers by Novel Needle-Less Electrospinning: Production, Characterization and Their Application to Reduce Nitrite in Sausages. *J. Food Eng.* **2021**, *288*, 110140. [CrossRef]

85. Luo, X.; Lim, L.-T. Curcumin-Loaded Electrospun Nonwoven as a Colorimetric Indicator for Volatile Amines. *LWT* **2020**, *128*, 109493. [CrossRef]
86. Guo, M.; Wang, H.; Wang, Q.; Chen, M.; Li, L.; Li, X.; Jiang, S. Intelligent Double-Layer Fiber Mats with High Colorimetric Response Sensitivity for Food Freshness Monitoring and Preservation. *Food Hydrocoll.* **2020**, *101*, 105468. [CrossRef]
87. Deng, L.; Zhang, X.; Li, Y.; Que, F.; Kang, X.; Liu, Y.; Feng, F.; Zhang, H. Characterization of Gelatin/Zein Nanofibers by Hybrid Electrospinning. *Food Hydrocoll.* **2018**, *75*, 72–80. [CrossRef]
88. Shekarforoush, E.; Ajalloueian, F.; Zeng, G.; Mendes, A.C.; Chronakis, I.S. Electrospun Xanthan Gum-Chitosan Nanofibers as Delivery Carrier of Hydrophobic Bioactives. *Mater. Lett.* **2018**, *228*, 322–326. [CrossRef]
89. Bhushani, J.A.; Anandharamakrishnan, C. Electrospinning and Electrospraying Techniques: Potential Food Based Applications. *Trends Food Sci. Technol.* **2014**, *38*, 21–33. [CrossRef]
90. Kerr-Phillips, T.; Travas-Sejdic, J. Conducting Polymers: Electrospun Materials. In *Encyclopedia of Polymer Applications*; CRC Press: Boca Raton, FL, USA; Taylor and Francis: New York, NY, USA, 2019; pp. 602–623.
91. Chan, E.W.C.; Bennet, D.; Baek, P.; Barker, D.; Kim, S.; Travas-Sejdic, J. Electrospun Polythiophene Phenylenes for Tissue Engineering. *Biomacromolecules* **2018**, *19*, 1456–1468. [CrossRef] [PubMed]
92. Beikzadeh, S.; Akbarinejad, A.; Swift, S.; Perera, J.; Kilmartin, P.A.; Travas-Sejdic, J. Cellulose Acetate Electrospun Nanofibers Encapsulating Lemon Myrtle Essential Oil as Active Agent with Potent and Sustainable Antimicrobial Activity. *React. Funct. Polym.* **2020**, *157*, 104769. [CrossRef]
93. Contreras, A.; Raxworthy, M.J.; Wood, S.; Tronci, G. Hydrolytic Degradability, Cell Tolerance and On-Demand Antibacterial Effect of Electrospun Photodynamically Active Fibres. *Pharmaceutics* **2020**, *12*, 711. [CrossRef]
94. Hernandez, J.L.; Doan, M.-A.; Stoddard, R.; VanBenschoten, H.M.; Chien, S.-T.; Suydam, I.T.; Woodrow, K.A. Scalable Electrospinning Methods to Produce High Basis Weight and Uniform Drug Eluting Fibrous Biomaterials. *Front. Biomater. Sci.* **2022**, *1*, 1–13. [CrossRef]
95. Lamri, M.; Bhattacharya, T.; Boukid, F.; Chentir, I.; Dib, A.L.; Das, D.; Djenane, D.; Gagaoua, M. Nanotechnology as a Processing and Packaging Tool to Improve Meat Quality and Safety. *Foods* **2021**, *10*, 2633. [CrossRef]
96. Hemmati, F.; Bahrami, A.; Esfanjani, A.F.; Hosseini, H.; McClements, D.J.; Williams, L. Electrospun Antimicrobial Materials: Advanced Packaging Materials for Food Applications. *Trends Food Sci. Technol.* **2021**, *111*, 520–533. [CrossRef]
97. Shepa, I.; Mudra, E.; Dusza, J. Electrospinning through the Prism of Time. *Mater. Today Chem.* **2021**, *21*, 100543. [CrossRef]
98. Kamsani, N.H.; Haris, M.S.; Pandey, M.; Taher, M.; Rullah, K. Biomedical Application of Responsive 'Smart' Electrospun Nanofibers in Drug Delivery System: A Minireview. *Arab. J. Chem.* **2021**, *14*, 103199. [CrossRef]
99. Colino, C.I.; Lanao, J.M.; Gutierrez-Millan, C. Recent Advances in Functionalized Nanomaterials for the Diagnosis and Treatment of Bacterial Infections. *Mater. Sci. Eng. C* **2021**, *121*, 111843. [CrossRef]
100. Arkoun, M.; Daigle, F.; Holley, R.A.; Heuzey, M.C.; Ajji, A. Chitosan-Based Nanofibers as Bioactive Meat Packaging Materials. *Packag. Technol. Sci.* **2018**, *31*, 185–195. [CrossRef]
101. Leidy, R.; Ximena, Q.-C.M. Use of Electrospinning Technique to Produce Nanofibres for Food Industries: A Perspective from Regulations to Characterisations. *Trends Food Sci. Technol.* **2019**, *85*, 92–106. [CrossRef]
102. Yilmaz, M.T.; Hassanein, W.S.; Alkabaa, A.S.; Ceylan, Z. Electrospun Eugenol-Loaded Gelatin Nanofibers as Bioactive Packaging Materials to Preserve Quality Characteristics of Beef. *Food Packag. Shelf Life* **2022**, *34*, 100968. [CrossRef]
103. Moreira, J.B.; de Morais, M.G.; de Morais, E.G.; da Silva Vaz, B.; Costa, J.A.V. Chapter 14—Electrospun Polymeric Nanofibers in Food Packaging. In *Impact of Nanoscience in the Food Industry*; Grumezescu, A.M., Holban, A.M., Eds.; Handbook of Food Bioengineering; Academic Press: Cambridge, MA, USA, 2018; pp. 387–417, ISBN 978-0-12-811441-4.
104. Haghighi, H.; Licciardello, F.; Fava, P.; Siesler, H.W.; Pulvirenti, A. Recent Advances on Chitosan-Based Films for Sustainable Food Packaging Applications. *Food Packag. Shelf Life* **2020**, *26*, 100551. [CrossRef]
105. Sharif, N.; Golmakani, M.-T.; Hajjari, M.M.; Aghaee, E.; Ghasemi, J.B. Antibacterial Cuminaldehyde/Hydroxypropyl-β-Cyclodextrin Inclusion Complex Electrospun Fibers Mat: Fabrication and Characterization. *Food Packag. Shelf Life* **2021**, *29*, 100738. [CrossRef]
106. Yu, Z.; Dhital, R.; Wang, W.; Sun, L.; Zeng, W.; Mustapha, A.; Lin, M. Development of Multifunctional Nanocomposites Containing Cellulose Nanofibrils and Soy Proteins as Food Packaging Materials. *Food Packag. Shelf Life* **2019**, *21*, 100366. [CrossRef]
107. Delosière, M.; Durand, D.; Bourguet, C.; Terlouw, E.M.C. Lipid Oxidation, Pre-Slaughter Animal Stress and Meat Packaging: Can Dietary Supplementation of Vitamin E and Plant Extracts Come to the Rescue? *Food Chem.* **2020**, *309*, 125668. [CrossRef]
108. Khumkomgool, A.; Saneluksana, T.; Harnkarnsujarit, N. Active Meat Packaging from Thermoplastic Cassava Starch Containing Sappan and Cinnamon Herbal Extracts via LLDPE Blown-Film Extrusion. *Food Packag. Shelf Life* **2020**, *26*, 100557. [CrossRef]
109. Panrong, T.; Karbowiak, T.; Harnkarnsujarit, N. Thermoplastic Starch and Green Tea Blends with LLDPE Films for Active Packaging of Meat and Oil-Based Products. *Food Packag. Shelf Life* **2019**, *21*, 100331. [CrossRef]
110. Smaoui, S.; Hlima, H.B.; Tavares, L.; Braïek, O.B.; Ennouri, K.; Abdelkafi, S.; Mellouli, L.; Khaneghah, A.M. Application of Eco-Friendly Active Films and Coatings Based on Natural Antioxidant in Meat Products: A Review. *Prog. Org. Coat.* **2022**, *166*, 106780. [CrossRef]
111. Wen, P.; Wen, Y.; Zong, M.-H.; Linhardt, R.J.; Wu, H. Encapsulation of Bioactive Compound in Electrospun Fibers and Its Potential Application. *J. Agric. Food Chem.* **2017**, *65*, 9161–9179. [CrossRef] [PubMed]

112. Surendhiran, D.; Li, C.; Cui, H.; Lin, L. Fabrication of High Stability Active Nanofibers Encapsulated with Pomegranate Peel Extract Using Chitosan/PEO for Meat Preservation. *Food Packag. Shelf Life* **2020**, *23*, 100439. [CrossRef]
113. Domínguez, R.; Pateiro, M.; Gagaoua, M.; Barba, F.J.; Zhang, W.; Lorenzo, J.M. A Comprehensive Review on Lipid Oxidation in Meat and Meat Products. *Antioxidants* **2019**, *8*, 429. [CrossRef]
114. Domínguez, R.; Barba, F.J.; Gómez, B.; Putnik, P.; Kovačević, D.B.; Pateiro, M.; Santos, E.M.; Lorenzo, J.M. Active Packaging Films with Natural Antioxidants to Be Used in Meat Industry: A Review. *Food Res. Int.* **2018**, *113*, 93–101. [CrossRef]
115. Amna, T.; Yang, J.; Ryu, K.-S.; Hwang, I.H. Electrospun Antimicrobial Hybrid Mats: Innovative Packaging Material for Meat and Meat-Products. *J. Food Sci. Technol.* **2015**, *52*, 4600–4606. [CrossRef]
116. Mihindukulasuriya, S.D.F.; Lim, L.-T. Nanotechnology Development in Food Packaging: A Review. *Trends Food Sci. Technol.* **2014**, *40*, 149–167. [CrossRef]
117. Wang, H.; Kong, L.; Ziegler, G.R. Aligned Wet-Electrospun Starch Fiber Mats. *Food Hydrocoll.* **2019**, *90*, 113–117. [CrossRef]
118. Wang, H.; Kong, L.; Ziegler, G.R. Fabrication of Starch—Nanocellulose Composite Fibers by Electrospinning. *Food Hydrocoll.* **2019**, *90*, 90–98. [CrossRef]
119. Cai, J.; Zhang, D.; Zhou, R.; Zhu, R.; Fei, P.; Zhu, Z.-Z.; Cheng, S.-Y.; Ding, W.-P. Hydrophobic Interface Starch Nanofibrous Film for Food Packaging: From Bioinspired Design to Self-Cleaning Action. *J. Agric. Food Chem.* **2021**, *69*, 5067–5075. [CrossRef]
120. Alehosseini, A.; Gómez-Mascaraque, L.G.; Martínez-Sanz, M.; López-Rubio, A. Electrospun Curcumin-Loaded Protein Nanofiber Mats as Active/Bioactive Coatings for Food Packaging Applications. *Food Hydrocoll.* **2019**, *87*, 758–771. [CrossRef]
121. Yildiz, E.; Sumnu, G.; Kahyaoglu, L.N. Monitoring Freshness of Chicken Breast by Using Natural Halochromic Curcumin Loaded Chitosan/PEO Nanofibers as an Intelligent Package. *Int. J. Biol. Macromol.* **2021**, *170*, 437–446. [CrossRef]
122. Gokoglu, N. Novel Natural Food Preservatives and Applications in Seafood Preservation: A Review. *J. Sci. Food Agric.* **2019**, *99*, 2068–2077. [CrossRef]
123. Karim, M.; Fathi, M.; Soleimanian-Zad, S.; Spigno, G. Development of Sausage Packaging with Zein Nanofibers Containing Tetradecane Produced via Needle-Less Electrospinning Method. *Food Packag. Shelf Life* **2022**, *33*, 100911. [CrossRef]
124. Cheng, S.; Wang, X.; Yang, H.; Lin, R.; Wang, H.; Tan, M. Characterization of Moisture Migration of Beef during Refrigeration Storage by Low-Field NMR and Its Relationship to Beef Quality. *J. Sci. Food Agric.* **2020**, *100*, 1940–1948. [CrossRef]
125. Frank, D.; Zhang, Y.; Li, Y.; Luo, X.; Chen, X.; Kaur, M.; Mellor, G.; Stark, J.; Hughes, J. Shelf Life Extension of Vacuum Packaged Chilled Beef in the Chinese Supply Chain. A Feasibility Study. *Meat Sci.* **2019**, *153*, 135–143. [CrossRef]
126. Madanayake, N.H.; Hossain, A.; Adassooriya, N.M. Nanobiotechnology for Agricultural Sustainability, and Food and Environmental Safety. *Qual. Assur. Saf. Crops Foods* **2021**, *13*, 20–36. [CrossRef]
127. Wang, C.; Chang, T.; Dong, S.; Zhang, D.; Ma, C.; Chen, S.; Li, H. Biopolymer Films Based on Chitosan/Potato Protein/Linseed Oil/ZnO NPs to Maintain the Storage Quality of Raw Meat. *Food Chem.* **2020**, *332*, 127375. [CrossRef]
128. Chen, Y.; Fan, Z.; Zhang, Z.; Niu, W.; Li, C.; Yang, N.; Chen, B.; Zhang, H. Two-Dimensional Metal Nanomaterials: Synthesis, Properties, and Applications. *Chem. Rev.* **2018**, *118*, 6409–6455. [CrossRef]
129. Qiu, M.; Sun, P.; Liu, Y.; Huang, Q.; Zhao, C.; Li, Z.; Mai, W. Visualized UV Photodetectors Based on Prussian Blue/TiO_2 for Smart Irradiation Monitoring Application. *Adv. Mater. Technol.* **2018**, *3*, 1700288. [CrossRef]
130. Xu, X.; Chen, J.; Cai, S.; Long, Z.; Zhang, Y.; Su, L.; He, S.; Tang, C.; Liu, P.; Peng, H.; et al. A Real-Time Wearable UV-Radiation Monitor Based on a High-Performance p-CuZnS/n-TiO_2 Photodetector. *Adv. Mater.* **2018**, *30*, 1803165. [CrossRef] [PubMed]
131. Gu, Z.; Fu, A.; Ye, L.; Kuerban, K.; Wang, Y.; Cao, Z. Ultrasensitive Chemiluminescence Biosensor for Nuclease and Bacterial Determination Based on Hemin-Encapsulated Mesoporous Silica Nanoparticles. *ACS Sens.* **2019**, *4*, 2922–2929. [CrossRef] [PubMed]
132. Hu, J.; Shen, Z.; Tan, L.; Yuan, J.; Gan, N. Electrochemical Aptasensor for Simultaneous Detection of Foodborne Pathogens Based on a Double Stirring Bars-Assisted Signal Amplification Strategy. *Sens. Actuators B Chem.* **2021**, *345*, 130337. [CrossRef]
133. Zhai, X.; Li, Z.; Shi, J.; Huang, X.; Sun, Z.; Zhang, D.; Zou, X.; Sun, Y.; Zhang, J.; Holmes, M.; et al. A Colorimetric Hydrogen Sulfide Sensor Based on Gellan Gum-Silver Nanoparticles Bionanocomposite for Monitoring of Meat Spoilage in Intelligent Packaging. *Food Chem.* **2019**, *290*, 135–143. [CrossRef]
134. Krishnan, S.K.; Singh, E.; Singh, P.; Meyyappan, M.; Nalwa, H.S. A Review on Graphene-Based Nanocomposites for Electrochemical and Fluorescent Biosensors. *RSC Adv.* **2019**, *9*, 8778–8881. [CrossRef]
135. Golmohammadi, H.; Morales-Narváez, E.; Naghdi, T.; Merkoçi, A. Nanocellulose in Sensing and Biosensing. *Chem. Mater.* **2017**, *29*, 5426–5446. [CrossRef]
136. Soares, R.R.A.; Hjort, R.G.; Pola, C.C.; Parate, K.; Reis, E.L.; Soares, N.F.F.; McLamore, E.S.; Claussen, J.C.; Gomes, C.L. Laser-Induced Graphene Electrochemical Immunosensors for Rapid and Label-Free Monitoring of Salmonella Enterica in Chicken Broth. *ACS Sens.* **2020**, *5*, 1900–1911. [CrossRef]
137. Wu, K.-L.; Jiang, B.-B.; Cai, Y.-M.; Wei, X.-W.; Li, X.-Z.; Cheong, W.-C. Efficient Electrocatalyst for Glucose and Ethanol Based on Cu/Ni/N-Doped Graphene Hybrids. *ChemElectroChem* **2017**, *4*, 1419–1428. [CrossRef]
138. Zhou, S.; Zhang, L.; Xie, L.; Zeng, J.; Qiu, B.; Yan, M.; Liang, Q.; Liu, T.; Liang, K.; Chen, P.; et al. Interfacial Super-Assembly of Nanofluidic Heterochannels from Layered Graphene and Alumina Oxide Arrays for Label-Free Histamine-Specific Detection. *Anal. Chem.* **2021**, *93*, 2982–2987. [CrossRef]
139. Minitha, C.R.; Anithaa, V.S.; Subramaniam, V.; Rajendra Kumar, R.T. Impact of Oxygen Functional Groups on Reduced Graphene Oxide-Based Sensors for Ammonia and Toluene Detection at Room Temperature. *ACS Omega* **2018**, *3*, 4105–4112. [CrossRef]

140. Wu, C.; Li, Y.; Sun, J.; Lu, Y.; Tong, C.; Wang, L.; Yan, Z.; Pang, J. Novel Konjac Glucomannan Films with Oxidized Chitin Nanocrystals Immobilized Red Cabbage Anthocyanins for Intelligent Food Packaging. *Food Hydrocoll.* **2020**, *98*, 105245. [CrossRef]

141. Mohammadalinejhad, S.; Almasi, H.; Moradi, M. Immobilization of Echium Amoenum Anthocyanins into Bacterial Cellulose Film: A Novel Colorimetric PH Indicator for Freshness/Spoilage Monitoring of Shrimp. *Food Control* **2020**, *113*, 107169. [CrossRef]

142. Wu, C.; Sun, J.; Chen, M.; Ge, Y.; Ma, J.; Hu, Y.; Pang, J.; Yan, Z. Effect of Oxidized Chitin Nanocrystals and Curcumin into Chitosan Films for Seafood Freshness Monitoring. *Food Hydrocoll.* **2019**, *95*, 308–317. [CrossRef]

143. Alizadeh-Sani, M.; Tavassoli, M.; McClements, D.J.; Hamishehkar, H. Multifunctional Halochromic Packaging Materials: Saffron Petal Anthocyanin Loaded-Chitosan Nanofiber/Methyl Cellulose Matrices. *Food Hydrocoll.* **2021**, *111*, 106237. [CrossRef]

144. Ham, J.; Lim, W.; Whang, K.-Y.; Song, G. Butylated Hydroxytoluene Induces Dysregulation of Calcium Homeostasis and Endoplasmic Reticulum Stress Resulting in Mouse Leydig Cell Death. *Environ. Pollut.* **2020**, *256*, 113421. [CrossRef]

145. Yang, C.; Lim, W.; Bazer, F.W.; Song, G. Propyl Gallate Induces Cell Death and Inhibits Invasion of Human Trophoblasts by Blocking the AKT and Mitogen-Activated Protein Kinase Pathways. *Food Chem. Toxicol.* **2017**, *109*, 497–504. [CrossRef]

146. Domingues, J.M.; Teixeira, M.O.; Teixeira, M.A.; Freitas, D.; da Silva, S.F.; Tohidi, S.D.; Fernandes, R.D.V.; Padrão, J.; Zille, A.; Silva, C.; et al. Inhibition of Escherichia Virus MS2, Surrogate of SARS-CoV-2, via Essential Oils-Loaded Electrospun Fibrous Mats: Increasing the Multifunctionality of Antivirus Protection Masks. *Pharmaceutics* **2022**, *14*, 303. [CrossRef]

147. Aminzare, M.; Hashemi, M.; Ansarian, E.; Bimakr, M.; Hassanzad, A.; Hassan, A. Using Natural Antioxidants in Meat and Meat Products as Preservatives: A Review. *Adv. Anim. Vet. Sci.* **2019**, *7*, 417–426. [CrossRef]

148. Kara, H.H.; Xiao, F.; Sarker, M.; Jin, T.Z.; Sousa, A.M.M.; Liu, C.-K.; Tomasula, P.M.; Liu, L. Antibacterial Poly (Lactic Acid) (PLA) Films Grafted with Electrospun PLA/Allyl Isothiocyanate Fibers for Food Packaging. *J. Appl. Polym. Sci.* **2016**, *133*, 1–8. [CrossRef]

149. Zaitoon, A.; Lim, L.-T.; Scott-Dupree, C. Activated Release of Ethyl Formate Vapor from Its Precursor Encapsulated in Ethyl Cellulose/Poly (Ethylene Oxide) Electrospun Nonwovens Intended for Active Packaging of Fresh Produce. *Food Hydrocoll.* **2021**, *112*, 106313. [CrossRef]

150. Escobar, A.; Pérez, M.; Romanelli, G.; Blustein, G. Thymol Bioactivity: A Review Focusing on Practical Applications. *Arab. J. Chem.* **2020**, *13*, 9243–9269. [CrossRef]

151. Da Silva, B.D.; Bernardes, P.C.; Pinheiro, P.F.; Fantuzzi, E.; Roberto, C.D. Chemical Composition, Extraction Sources and Action Mechanisms of Essential Oils: Natural Preservative and Limitations of Use in Meat Products. *Meat Sci.* **2021**, *176*, 108463. [CrossRef] [PubMed]

152. Lin, L.; Zhu, Y.; Cui, H. Electrospun Thyme Essential Oil/Gelatin Nanofibers for Active Packaging against Campylobacter Jejuni in Chicken. *LWT* **2018**, *97*, 711–718. [CrossRef]

153. Xing, F.; Hua, H.; Selvaraj, J.N.; Zhao, Y.; Zhou, L.; Liu, X.; Liu, Y. Growth Inhibition and Morphological Alterations of Fusarium Verticillioides by Cinnamon Oil and Cinnamaldehyde. *Food Control* **2014**, *46*, 343–350. [CrossRef]

154. Chen, H.; Hu, X.; Chen, E.; Wu, S.; McClements, D.J.; Liu, S.; Li, B.; Li, Y. Preparation, Characterization, and Properties of Chitosan Films with Cinnamaldehyde Nanoemulsions. *Food Hydrocoll.* **2016**, *61*, 662–671. [CrossRef]

155. Stanojević, D.; Comic, L.; Stefanović, O.D.; Solujić-Sukdolak, S. Antimicrobial Effects of Sodium Benzoate, Sodium Nitrite Andpotassium Sorbate and Their Synergistic Action in Vitro. *Bulg. J. Agric. Sci.* **2009**, *15*, 308–312.

156. Mirhadi, E.; Nassirli, H.; Malaekeh-Nikouei, B. An Updated Review on Therapeutic Effects of Nanoparticle-Based Formulations of Saffron Components (Safranal, Crocin, and Crocetin). *J. Pharm. Investig.* **2020**, *50*, 47–58. [CrossRef]

157. Ansarifar, E.; Moradinezhad, F. Encapsulation of Thyme Essential Oil Using Electrospun Zein Fiber for Strawberry Preservation. *Chem. Biol. Technol. Agric.* **2022**, *9*, 2. [CrossRef]

158. Cetinkaya, T.; Mendes, A.C.; Jacobsen, C.; Ceylan, Z.; Chronakis, I.S.; Bean, S.R.; García-Moreno, P.J. Development of Kafirin-Based Nanocapsules by Electrospraying for Encapsulation of Fish Oil. *LWT* **2021**, *136*, 110297. [CrossRef]

159. Najafi, Z.; Cetinkaya, T.; Bildik, F.; Altay, F.; Yeşilçubuk, N.Ş. Nanoencapsulation of Saffron (*Crocus sativus* L.) Extract in Zein Nanofibers and Their Application for the Preservation of Sea Bass Fillets. *LWT* **2022**, *163*, 113588. [CrossRef]

160. Cui, H.; Bai, M.; Li, C.; Liu, R.; Lin, L. Fabrication of Chitosan Nanofibers Containing Tea Tree Oil Liposomes against Salmonella Spp. in Chicken. *LWT* **2018**, *96*, 671–678. [CrossRef]

161. Zhang, J.; Zhang, J.; Huang, X.; Shi, J.; Muhammad, A.; Zhai, X.; Xiao, J.; Li, Z.; Povey, M.; Zou, X. Study on Cinnamon Essential Oil Release Performance Based on PH-Triggered Dynamic Mechanism of Active Packaging for Meat Preservation. *Food Chem.* **2023**, *400*, 134030. [CrossRef]

162. Pavli, F.; Argyri, A.A.; Skandamis, P.; Nychas, G.-J.; Tassou, C.; Chorianopoulos, N. Antimicrobial Activity of Oregano Essential Oil Incorporated in Sodium Alginate Edible Films: Control of Listeria Monocytogenes and Spoilage in Ham Slices Treated with High Pressure Processing. *Materials* **2019**, *12*, 3726. [CrossRef]

163. Arkoun, M.; Daigle, F.; Heuzey, M.-C.; Ajji, A. Mechanism of Action of Electrospun Chitosan-Based Nanofibers against Meat Spoilage and Pathogenic Bacteria. *Molecules* **2017**, *22*, 585. [CrossRef]

164. Zhao, Y.; Teixeira, J.S.; Saldaña, M.D.A.; Gänzle, M.G. Antimicrobial Activity of Bioactive Starch Packaging Films against Listeria Monocytogenes and Reconstituted Meat Microbiota on Ham. *Int. J. Food Microbiol.* **2019**, *305*, 108253. [CrossRef]

165. Lin, L.; Gu, Y.; Cui, H. Novel Electrospun Gelatin-Glycerin-ε-Poly-Lysine Nanofibers for Controlling Listeria Monocytogenes on Beef. *Food Packag. Shelf Life* **2018**, *18*, 21–30. [CrossRef]

166. Li, X.; Xiao, N.; Xiao, G.; Bai, W.; Zhang, X.; Zhao, W. Lemon Essential Oil/Vermiculite Encapsulated in Electrospun Konjac Glucomannan-Grafted-Poly (Acrylic Acid)/Polyvinyl Alcohol Bacteriostatic Pad: Sustained Control Release and Its Application in Food Preservation. *Food Chem.* **2021**, *348*, 129021. [CrossRef]
167. Li, T.; Shen, Y.; Chen, H.; Xu, Y.; Wang, D.; Cui, F.; Han, Y.; Li, J. Antibacterial Properties of Coaxial Spinning Membrane of Methyl Ferulate/Zein and Its Preservation Effect on Sea Bass. *Foods* **2021**, *10*, 2385. [CrossRef]
168. Vargas Romero, E.; Lim, L.-T.; Suárez Mahecha, H.; Bohrer, B.M. The Effect of Electrospun Polycaprolactone Nonwovens Containing Chitosan and Propolis Extracts on Fresh Pork Packaged in Linear Low-Density Polyethylene Films. *Foods* **2021**, *10*, 1110. [CrossRef]
169. Bolumar, T.; LaPeña, D.; Skibsted, L.H.; Orlien, V. Rosemary and Oxygen Scavenger in Active Packaging for Prevention of High-Pressure Induced Lipid Oxidation in Pork Patties. *Food Packag. Shelf Life* **2016**, *7*, 26–33. [CrossRef]
170. Cardoso, G.P.; Dutra, M.P.; Fontes, P.R.; Ramos, A.D.L.S.; de Miranda Gomide, L.A.; Ramos, E.M. Selection of a Chitosan Gelatin-Based Edible Coating for Color Preservation of Beef in Retail Display. *Meat Sci.* **2016**, *114*, 85–94. [CrossRef] [PubMed]
171. Ribeiro-Santos, R.; de Melo, N.R.; Andrade, M.; Azevedo, G.; Machado, A.V.; Carvalho-Costa, D.; Sanches-Silva, A. Whey Protein Active Films Incorporated with a Blend of Essential Oils: Characterization and Effectiveness. *Packag. Technol. Sci.* **2018**, *31*, 27–40. [CrossRef]
172. Borzi, F.; Torrieri, E.; Wrona, M.; Nerín, C. Polyamide Modified with Green Tea Extract for Fresh Minced Meat Active Packaging Applications. *Food Chem.* **2019**, *300*, 125242. [CrossRef]
173. Chaari, M.; Elhadef, K.; Akermi, S.; Ben Akacha, B.; Fourati, M.; Chakchouk Mtibaa, A.; Ennouri, M.; Sarkar, T.; Shariati, M.A.; Rebezov, M.; et al. Novel Active Food Packaging Films Based on Gelatin-Sodium Alginate Containing Beetroot Peel Extract. *Antioxidants* **2022**, *11*, 2095. [CrossRef]
174. Yordshahi, A.S.; Moradi, M.; Tajik, H.; Molaei, R. Design and Preparation of Antimicrobial Meat Wrapping Nanopaper with Bacterial Cellulose and Postbiotics of Lactic Acid Bacteria. *Int. J. Food Microbiol.* **2020**, *321*, 108561. [CrossRef]
175. Behbahani, B.A.; Shahidi, F.; Yazdi, F.T.; Mortazavi, S.A.; Mohebbi, M. Use of Plantago Major Seed Mucilage as a Novel Edible Coating Incorporated with Anethum Graveolens Essential Oil on Shelf Life Extension of Beef in Refrigerated Storage. *Int. J. Biol. Macromol.* **2017**, *94*, 515–526. [CrossRef]
176. Stoleru, E.; Vasile, C.; Irimia, A.; Brebu, M. Towards a Bioactive Food Packaging: Poly (Lactic Acid) Surface Functionalized by Chitosan Coating Embedding Clove and Argan Oils. *Molecules* **2021**, *26*, 4500. [CrossRef]
177. Montaño-Sánchez, E.; Torres-Martínez, B.D.M.; Vargas-Sánchez, R.D.; Huerta-Leidenz, N.; Sánchez-Escalante, A.; Beriain, M.J.; Torrescano-Urrutia, G.R. Effects of Chitosan Coating with Green Tea Aqueous Extract on Lipid Oxidation and Microbial Growth in Pork Chops during Chilled Storage. *Foods Basel Switz.* **2020**, *9*, 766. [CrossRef]
178. Roy, S.; Priyadarshi, R.; Rhim, J.-W. Development of Multifunctional Pullulan/Chitosan-Based Composite Films Reinforced with ZnO Nanoparticles and Propolis for Meat Packaging Applications. *Foods* **2021**, *10*, 2789. [CrossRef]
179. Kaewprachu, P.; Osako, K.; Benjakul, S.; Rawdkuen, S. Quality Attributes of Minced Pork Wrapped with Catechin–Lysozyme Incorporated Gelatin Film. *Food Packag. Shelf Life* **2015**, *3*, 88–96. [CrossRef]
180. Catarino, M.D.; Alves-Silva, J.M.; Fernandes, R.P.; Gonçalves, M.J.; Salgueiro, L.R.; Henriques, M.F.; Cardoso, S.M. Development and Performance of Whey Protein Active Coatings with Origanum Virens Essential Oils in the Quality and Shelf Life Improvement of Processed Meat Products. *Food Control* **2017**, *80*, 273–280. [CrossRef]
181. Kalem, I.K.; Bhat, Z.F.; Kumar, S.; Noor, S.; Desai, A. The Effects of Bioactive Edible Film Containing Terminalia Arjuna on the Stability of Some Quality Attributes of Chevon Sausages. *Meat Sci.* **2018**, *140*, 38–43. [CrossRef]
182. Shahbazi, Y.; Shavisi, N. A Novel Active Food Packaging Film for Shelf-Life Extension of Minced Beef Meat. *J. Food Saf.* **2018**, *38*, e12569. [CrossRef]
183. Li, L.; Wang, H.; Chen, M.; Jiang, S.; Cheng, J.; Li, X.; Zhang, M.; Jiang, S. Gelatin/Zein Fiber Mats Encapsulated with Resveratrol: Kinetics, Antibacterial Activity and Application for Pork Preservation. *Food Hydrocoll.* **2020**, *101*, 105577. [CrossRef]
184. Duan, M.; Sun, J.; Huang, Y.; Jiang, H.; Hu, Y.; Pang, J.; Wu, C. Electrospun Gelatin/Chitosan Nanofibers Containing Curcumin for Multifunctional Food Packaging. *Food Sci. Hum. Wellness* **2023**, *12*, 614–621. [CrossRef]
185. Radha Krishnan, K.; Babuskin, S.; Rakhavan, K.R.; Tharavin, R.; Azhagu Saravana Babu, P.; Sivarajan, M.; Sukumar, M. Potential Application of Corn Starch Edible Films with Spice Essential Oils for the Shelf Life Extension of Red Meat. *J. Appl. Microbiol.* **2015**, *119*, 1613–1623. [CrossRef]

Article

Effect of Storage Conditions on the Physicochemical Characteristics of Bilayer Edible Films Based on Iron Yam–Pea Starch Blend and Corn Zein

Xiaoyong Song

College of Energy and Power Engineering, North China University of Water Resources and Electric Power, Zhengzhou 450011, China; songxiaoyong@ncwu.edu.cn; Tel./Fax: +86-371-6579-0043

Abstract: Edible iron yam–pea starch-based bilayer films with a water and oil proof layer of corn zein were prepared by casting method and stored under normal temperature (25 °C; relative humidity (RH): 43%, 54%, 65%), refrigeration (4 °C) and freezing (−17 °C) for 150 days. The mechanical properties, scanning electron microscope (SEM) micrographs, oxygen and water vapor permeability, color, transmittance, haze, water content were systematically evaluated after 0, 30, 60, 90, 120 and 150 days. The transmittance, haze, and water content of bilayer films changed greatly within 150 days, which indirectly indicated the changes of the internal microstructure of the film matrix. The results were further verified by SEM analysis. Water and oxygen resistance gradually become worse. At 25 °C and 54% RH, the barrier performances were relatively strong. Films had relatively good tensile strength at normal temperature and high humidity, and relatively good elongation at break at low temperature and high humidity. SEM observation showed that there was no interlayer separation during storage. The internal network structure disappeared and reappeared again. The changes of internal microstructure also verified the changes of barrier and mechanical properties of bilayer films.

Keywords: bilayer edible film; iron yam–pea starch; corn zein; physicochemical characterization; storage

Citation: Song, X. Effect of Storage Conditions on the Physicochemical Characteristics of Bilayer Edible Films Based on Iron Yam–Pea Starch Blend and Corn Zein. *Coatings* **2022**, *12*, 1524. https://doi.org/10.3390/coatings12101524

Academic Editor: Domingo Martínez-Romero

Received: 14 September 2022
Accepted: 6 October 2022
Published: 12 October 2022

Publisher's Note: MDPI stays neutral with regard to jurisdictional claims in published maps and institutional affiliations.

1. Introduction

With the increasing demand of consumers for green and healthy food, the research on degradable and edible packaging materials has become a hot spot [1–7]. Biodegradable polymers have been widely used in film production [8–13]. The enhancement of people's environmental protection concept has led to changes in the manufacturing methods of food packaging materials [14–17]. However, both physical and chemical properties of edible films based on biological macromolecules, are easily affected by the changes of the external environmental conditions. These changes may affect its practical application and consumer acceptability [18,19]. Therefore, for a long storage time, the high stability of film performance is usually required. Studies on the behavior of edible films during storage have been reported in the last years. Such as caseinate-based films [20], whey protein isolate-based films [21], cassava starch films [22], jumbo squid myofibrillar protein-based films [23], gellan gum edible films [24], prebiotic edible films [25], fish gelatin edible films [26], triticale flour films [27], glycerol plasticized–soybean protein films [28], whey protein edible films [18,29–31], chitosan films [32], squid mantle muscle films [19], Soybean protein films [28], glutenin films [33].

To the best of our knowledge, however, there is no work related to the storage-induced changes of bilayer edible films based on iron yam–pea starch blend and corn zein. Among various natural polymeric materials, starch is considered as a good edible film material because of its wide source, low price and biodegradability [34]. However, the starch based biodegradable film has the disadvantages of brittleness and low water resistance [14,35,36]. Corn zein films can compensate for these shortcomings of starch films through hydrophobic amino acids [37,38], excellent film forming ability [39], low oxygen permeability [40]. However, poor mechanical properties and high cost of zein limit its practical application [41]. In short, the inherent properties of starch and protein films restrict the use of starch and protein in pure form for intended purpose [42,43]. The preliminary work of our research team shows that the preparation of starch and protein bilayer film, making full use of the advantages of the two film-forming substrates, can be used as one of the choices of food inner packaging materials (such as instant food seasoning bags) [44]. The iron yam (*Dioscorea opposita Thunb*) is the only Chinese yam variety protected by the national origin protection list [45]. The content of iron yam starch is up to 28%, which can be used as the base material for the development of edible film [46]. Corn zein contains more hydrophobic amino acids and is a good material for preparing waterproof film [47].

In this study, a bilayer film with an outer iron yam–pea starch layer and an inner oil and water-resistant corn zein layer was developed and stored under normal temperature (25 °C; RH: 43%, 54%, 65%), refrigeration (4 °C) and freezing (-17 °C) for 150 days. The physicochemical properties and microstructures were analyzed every 30 days. The effects of temperature, humidity and time on the mechanical properties, color, transmittance, water content, water resistance, and oxygen resistance were studied. On this basis, the cross-sectional microstructures of the bilayer films were analyzed by SEM, and the effects of the microstructures of the bilayer films on their chemical properties were studied.

2. Materials and Methods

2.1. Materials

Iron yam (*Dioscorea opposita Thunb*) starch was purchased from Hangzhou Mei Yan Trade (Hangzhou, China). Pea starch was purchased from Shanghai Luyuan Starch (Shanghai, China) and corn zein (CZ, 91.1% protein content) was purchased from Dulai Biotechnology (Nanjing, China). Absolute ethyl alcohol ($\geq$99.7%), D-Sorbitol ($\geq$98.0%), polyethylene glycol-400 ($\geq$99.0%), citric acid ($\geq$99.5%,) and glycerol ($\geq$99.0%,) were purchased from Tianjing Chemical Reagent (Luoyang, China). All other chemicals were analytical grade.

2.2. Film Preparation and Storage

The bilayer films were prepared following Song and Wang [46]. Our previous study found that IYP10/CZ3 (10 mL iron yam-pea starch forming solution with 3 mL corn zein solution) bilayers have good mechanical and barrier properties [46]. Therefore, IYP10/CZ3 bilayers were selected as the research object and stored for 150 days at five different conditions: normal temperature (25 °C) and 43% RH, 54% RH, 65% RH, refrigeration (4 °C), freezing (-17 °C), respectively. For five storage conditions the film characterizations were performed on days: 0, 30, 60, 90, 120 and 150.

2.3. Color

The measurement of color was carried out by a Lab-Scan II colorimeter (Hunter Lab, Inc., Reston, VA, USA) according to the method of Zuo et al. [44]. The color was presented in terms of L^*, a^* and b^* values. The whitish index (*WI*) of film was calculated using the following equation.

$$WI = 100 - \left[(100 - L^*)^2 + a^{*2} + b^{*2} \right]^{0.5} \tag{1}$$

2.4. Transmittance and Haze

The measurements of transmittance and haze were carried out by a transmittance haze meter (WGT-S, Shanghai Yanhe Scientific Instrument Co., Ltd., Shanghai, China) according to the method of Zuo et al. [44]. These were made six times for each test.

2.5. Water Content

The measurements of water content (*WC*) were carried out according to the method of Zuo et al. [44] and calculated as the following equation:

$$WC = (M_0 - M) / M_0 \tag{2}$$

where M_0—the initial mass (g), M—the bone-dry mass (g). WC was expressed as g H_2O/g dry solids. Measurements were carried out three times.

2.6. Water Vapor Permeability

The measurements of water vapor permeability (WVP) were carried out according to the modified method of Song et al. [48]. The measurements were carried out five times.

2.7. Oxygen Permeability

The peroxide value (POV) of soybean oil was expressed by oxygen permeability (OP) [47]. We put 20 g of soybean oil into a 250 mL conical flask, sealed the bilayer film (D = 9 cm) at the mouth of the conical flask, and stored it at 60 °C for 10 days for measurement. The measurement was to be made in triplicate.

2.8. Mechanical Properties

The measurements of tensile properties were carried out by a TA-XT2i Texture Analyzer (Stable Microsystems, Godalming, UK) [48]. Each sample was measured nine times.

2.9. Scanning Electron Microscope (SEM)

A field emission scanning electron microscope (Nova NanoSEM400, FEI company, Hillsboro, OR, USA) was used to analyze the microstructure of the film, and the working voltage was 3 kV. We exposed the samples to P_2O_5 at 25 °C and 0% RH for 2 weeks and soaked the fragments (50 × 60 mm^2) in liquid nitrogen, then installed them on the metal grid, and gold plated them under a vacuum.

2.10. Statistical Analysis

The ANOVA program of SPSS software (version 16.0 (IBM, Chicago, IL, USA) was used for analysis of variance. LSD test was used to test the difference of mean value, and $p < 0.05$ was considered statistically significant.

3. Results and Discussion

3.1. Color

The color stability of the edible film is essential for practical application. It is of great significance to study the color change of the bilayer film during storage to determine the shelf life of convenient food packaging materials [49,50]. As shown in Table 1, the changes of L^*, a^*, b^* and WI values of bilayer films during storage were analyzed. It can be seen that with the prolongation of storage time, the L^* value of bilayer film did not change obviously on the whole, and a small decrease occurred after 120 days. Moreover, the change of temperature and humidity did not cause a big change of L^* value of bilayer film during the same storage period, which indicated that the brightness of bilayer film was well-preserved [4]. The a^* value decreased slightly with time but increased significantly with the increase of temperature and relative humidity. At 25 °C and 65% of relative humidity, the a^* value reached the maximum values of -1.50 (30 d), -1.66 (60 d), -1.81 (90 d), -1.69 (120 d) and -2.05 (150 d), respectively. Indicating that the green of the bilayer film was gradually weakening. In addition, with the increase of temperature and relative humidity, the b^* value of bilayer films decreased significantly, and the yellowness of bilayer films decreased gradually. The increase of a^* value and decrease of b^* value of bilayer film might be due to the loosening of the internal structure of bilayer film with the increase of storage temperature and relative humidity [18,29]. Some corn zein entered the iron yam pea starch blend film. The variation of WI value of bilayer film was consistent with that of L^*, a^* and b^* value. Therefore, the following conclusions could be drawn from the experiments: different storage environments affected the appearance color of bilayer film to a certain extent [19]. However, under this experimental condition, with the storage time prolonged, the same storage conditions did not lead to significant changes in the color of the bilayer film for 150 days' worth of storage periods.

Table 1. Change of color of bilayer film during storage.

	Storage Conditions	Storage Time				
		30 d	60 d	90 d	120 d	150 d
L^*	−18 °C	94.47 ± 0.32 [b,A]	94.31 ± 0.24 [ab,A]	94.28 ± 0.31 [a,A]	94.28 ± 0.54 [ab,A]	93.77 ± 0.10 [ab,B]
	−5 °C	94.62 ± 0.14 [b,A]	94.53 ± 0.40 [b,A]	94.64 ± 0.22 [b,B]	94.46 ± 0.56 [b,A]	93.44 ± 0.54 [a,A]
	25 °C, RH:43%	94.67 ± 0.33 [a,A]	94.47 ± 0.07 [a,A]	94.22 ± 0.31 [a,AB]	94.35 ± 0.59 [a,A]	94.17 ± 0.31 [a,B]
	25 °C, RH:54%	94.75 ± 0.30 [b,A]	94.33 ± 0.22 [b,A]	94.33 ± 0.28 [b,B]	94.37 ± 0.12 [b,A]	93.75 ± 0.41 [a,AB]
	25 °C, RH:65%	94.44 ± 0.16 [a,A]	94.34 ± 0.38 [a,A]	94.29 ± 0.24 [a,B]	94.06 ± 0.27 [a,A]	94.08 ± 0.29 [a,AB]
a^*	−18 °C	−2.90 ± 0.12 [a,A]	−3.11 ± 0.20 [a,A]	−3.15 ± 0.04 [a,A]	−3.17 ± 0.15 [a,A]	−3.19 ± 0.19 [a,A]
	−5 °C	−2.70 ± 0.10 [bc,A]	−2.76 ± 0.01 [c,B]	−2.88 ± 0.20 [c,B]	−3.02 ± 0.16 [ab,A]	−3.16 ± 0.13 [a,A]
	25 °C, RH:43%	−2.63 ± 0.08 [a,B]	−2.66 ± 0.10 [a,B]	−2.70 ± 0.17 [a,B]	−2.72 ± 0.14 [a,B]	−2.72 ± 0.16 [a,B]
	25 °C, RH:54%	−2.14 ± 0.07 [b,C]	−2.31 ± 0.13 [ab,C]	−2.30 ± 0.16 [ab,C]	−2.42 ± 0.13 [ab,C]	−2.63 ± 0.33 [a,B]
	25 °C, RH:65%	−1.50 ± 0.18 [c,D]	−1.66 ± 0.15 [bc,D]	−1.69 ± 0.09 [ab,D]	−1.82 ± 0.15 [bc,D]	−2.05 ± 0.14 [a,C]
b^*	−18 °C	17.37 ± 0.42 [b,D]	17.07 ± 0.53 [b,D]	16.87 ± 0.67 [b,D]	16.66 ± 0.26 [b,C]	15.50 ± 0.37 [a,C]
	−5 °C	16.59 ± 0.58 [b,CD]	16.57 ± 0.84 [b,D]	15.53 ± 0.40 [ab,C]	14.87 ± 0.48 [a,B]	14.65 ± 0.71 [a,C]
	25 °C, RH:43%	16.01 ± 0.52 [a,C]	14.98 ± 0.72 [a,C]	14.95 ± 0.61 [a,B]	14.91 ± 0.78 [a,B]	14.66 ± 0.87 [a,BC]
	25 °C, RH:54%	14.10 ± 0.82 [a,B]	14.06 ± 0.33 [a,B]	13.49 ± 0.79 [a,B]	13.32 ± 0.84 [a,A]	13.32 ± 0.78 [a,AB]
	25 °C, RH:65%	12.47 ± 0.55 [a,A]	12.13 ± 0.54 [a,A]	12.01 ± 0.87 [a,A]	11.96 ± 0.45 [a,A]	11.57 ± 1.01 [a,A]
WI	−18 °C	81.54 ± 0.49 [a,A]	81.69 ± 0.57 [a,A]	81.74 ± 0.73 [a,A]	82.05 ± 0.21 [a,A]	83.18 ± 0.38 [a,A]
	−5 °C	82.32 ± 0.55 [a,AB]	82.34 ± 0.72 [a,A]	83.35 ± 0.45 [ab,B]	83.43 ± 0.49 [b,B]	84.04 ± 0.68 [ab,A]
	25 °C, RH:43%	82.92 ± 0.44 [a,B]	83.71 ± 0.65 [a,B]	83.80 ± 0.50 [a,BC]	83.87 ± 0.76 [a,B]	83.99 ± 0.92 [a,AB]
	25 °C, RH:54%	84.66 ± 0.85 [a,C]	84.80 ± 0.24 [a,C]	85.04 ± 0.65 [a,C]	85.17 ± 0.71 [a,C]	85.35 ± 0.62 [a,B]
	25 °C, RH:65%	86.17 ± 0.54 [a,D]	86.50 ± 0.55 [a,D]	86.51 ± 0.76 [a,D]	86.68 ± 0.38 [a,C]	86.96 ± 1.02 [a,C]

Note: Data were mean ± standard deviation. The different lowercase letters indicate significant differences ($p < 0.05$) in the color with varying storage temperature and RH; the different capital letters indicate significant differences ($p < 0.05$) in the color on different storage time.

3.2. Transmittance and Haze

The change of the transmittance of the bilayer film with time under different storage conditions is shown in Figure 1. It can be seen from Figure 1A that the transmittance did not change significantly during the storage period, except that at 60th and 150th day, the transmittance fluctuated to varying degrees at low temperature (4 °C) and high humidity (RH 65%). The transmittance reflected the change of film color to a certain extent, and the research results showed that there was good compatibility between the bilayer film and the biomolecules in the bilayer film [21]. Therefore, the experimental results will be further analyzed and verified in the follow-up SEM study.

Haze and transmittance are two different concepts, haze mainly reflects the gloss and transparency of film matrix [4]. Appropriate haze helps to improve consumer acceptability of film packaging materials. Figure 1B shows that the crystallinity of the film was greater than 0 (3.12%–11.92%) in different storage periods due to its own crystallinity. It can also be seen that the haze of the film increased with time, indicating that the biocompatibility of the bilayer film changed in varying degrees during storage [29,48]. At the same time, plasticizers such as glycerol and polyethylene glycol-400 migrated to the surface of the film during storage, which increased the roughness of the film surface, and might also be one of the reasons for the increase of haze.

3.3. Water Content

Water is a good plasticizer for packaging materials based on biological macromolecules, especially for edible polysaccharide films. The change of water content in the film indirectly affects its barrier and mechanical properties [30]. Under different storage conditions, the change of water content of the bilayer film with time was shown in Figure 2. The water content decreased significantly from 0 to 30 days ($p < 0.05$), probably due to the migration of PEG-400 during storage, which led to the decrease of water holding capacity of the bilayer film. Compared with other storage environments, the film water content was relatively low at 40% RH and 4 °C, indicating that the storage environment had a greater impact on the water content of edible biomacromolecule-based film, which was also one of the main factors limiting its practical application. [32]. However, the water content generally increased with the extension of storage time, reaching 13.19% (−17 °C), 8.84% (4 °C), 7.41% (25 °C, RH 43%), 9.29% (25 °C, RH 54%) and 14.39% (25 °C, RH 65%) after 150 days, which might be due to the oxidation of protein film and the regeneration of starch film during storage, and the increase of cross-linking between water molecules and crystalline structure in films [40].

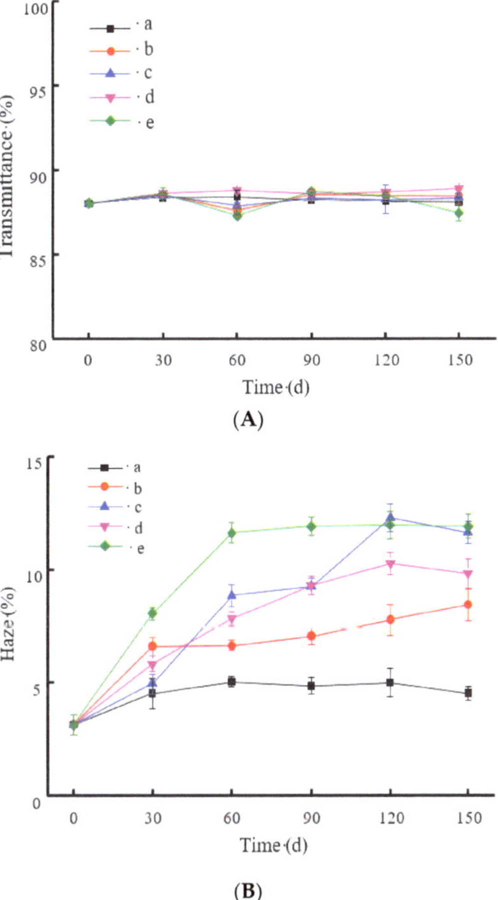

Figure 1. Change of transmittance (**A**) and haze (**B**) of double-layer film during storage. A (−17 °C), b (4 °C), c (25 °C, RH 43%), d (25 °C, RH 54%), e (25 °C, RH 65%).

3.4. Barrier Properties

The change of humidity in storage environment has a great influence on the properties of starch-based film matrix with strong hydrophilicity, especially its water resistance. In different storage environments, the variation of WVP with storage time is shown in Figure 3A. At normal temperature (25 °C) and high relative humidity (RH 65%), the water resistance of bilayer films became worse after 150 days of storage because its WVP reached 11.88×10^{-11} g/m·s·Pa. This was mainly because water played a certain role in plasticizing the bilayer film [46]. Excessive water made the internal structure of the film became loose, and water molecules were easier to penetrate through the film matrix. Under freezing (−17 °C) and refrigeration (4 °C) conditions, the WVP value reached 13.80×10^{-11} g/m·s·Pa and 12.11×10^{-11} g/m·s·Pa, respectively, after 150 days of storage, which were higher than that of WVP at normal temperature. This was mainly due to the closer bonding between polymers and the smaller pore size, which hindered the transfer of water molecules [44]. The WVP of bilayer film reached 10.08×10^{-11} g/m·s·Pa after 120 days of storage at 25 °C and 54% RH, indicating that the bilayer film has good water resistance.

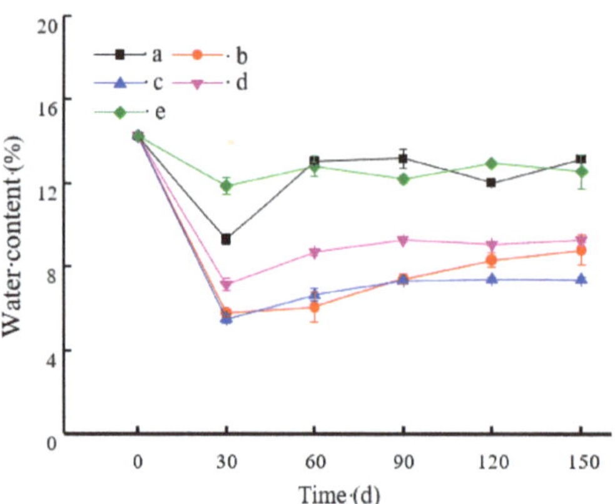

Figure 2. Water content of double-layer membrane during storage. A (−17 °C), b (4 °C), c (25 °C, RH 43%), d (25 °C, RH 54%), e (25 °C, RH 65%).

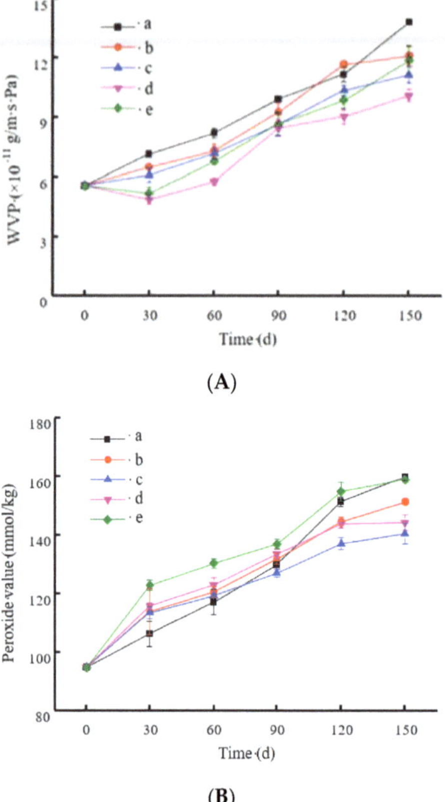

(**A**)

(**B**)

Figure 3. Water vapor permeability (WVP, **A**) and peroxide values (POV, **B**) of bilayer films during storage. A (−18 °C), b (4 °C), c (25 °C, RH 43%), d (25 °C, RH 54%), e (25 °C, RH 65%).

As shown in Figure 3B, the POV value of the contents of bilayer film packaging increased significantly after aging for 10 days ($p < 0.05$). It showed that the oxygen resistance tends to be worse in different storage environments. POV reached 140.47 mmol/kg and 158.97 mmol/kg after 150 days aging at normal temperature (25 °C) and RH was 43% and 65%, respectively. It can be clearly seen that the bilayer film had good oxygen resistance in low humidity storage environment, but poor oxygen resistance in high humidity conditions. The possible reason is that in the low humidity environment, the humidity difference between the bilayer film and external environment led to the loss of some water in the bilayer film, the internal structure of the film became compact, and the oxidation resistance was improved [33,48]. In the high humidity environment, the swelling phenomenon inside the bilayer film made its internal structure loose and the oxygen permeability increased. Oxygen permeability of film matrix usually increases with the increase of temperature [46]. Under the experimental conditions, the oxygen permeability at the early stage of storage conformed to this change rule, but at the later stage, the change rule was contrary. This might be because the process of oxygen molecules passing through films could be divided into two aspects: diffusion and dissolution. The change of water content resulted in the above phenomena.

3.5. Mechanical Properties

The shelf life of bilayer film as packaging material is directly determined by the change of mechanical properties during storage [18]. Figure 4 shows the changes of TS (Figure 4A) and E (Figure 4B) of bilayer films under different storage conditions. It can be seen that at normal temperature (25 °C) and RH of 43%, 54% and 65% for 30 days, TS increased to 10.22 Mpa, 11.04 Mpa and 11.07 Mpa, respectively, while E decreased significantly ($p < 0.05$). This might be due to the decrease of internal voidage and the compactness of internal structure of bilayer films during the early storage period. With the prolongation of storage time, TS of bilayer films decreased significantly ($p < 0.05$), E increased significantly ($p < 0.05$). After 150 days of storage, TS of bilayer films decreased to 5.56 Mpa, but the lowest E value could still reach 40.92%. This might be due to the migration of plasticizers in different degrees during storage, which improved the plasticizing effect [51]. It can also be seen from Figure 4 that under different storage conditions, the variation of TS and E of bilayer films was different in the same storage period. Under the conditions of normal temperature (25 °C) and high humidity (RH 65%), the bilayer films had relatively good tensile strength. Under the conditions of low temperature (−17 °C and 4 °C) and high humidity (RH 65%), the bilayer films had relatively good toughness. After 150 days of storage, the mechanical properties of bilayer film were still better than those of Ciannamea et al. [52].

3.6. Scanning Electron Microscope (SEM)

The changes of barrier and mechanical properties of bilayer film reflect the change of its internal microstructure to a certain extent [46]. The scanning electron micrographs of IYP-CZ bilayer films obtained on days 0, 30 and 150, for different storage environments, are shown in Figure 5. From the micrographs obtained on day 30, we can see that the internal network structure of bilayer films no longer exited and became denser under the conditions of freezing and normal temperature, and there was no obvious boundary between bilayer films. However, the cross section of the bilayer film was relatively rough in the freezing environment (−17 °C). In the cold storage environment (4 °C), there were still some network structures in the cross section of the bilayer films, and the transition between the bilayer films was natural. Figure 5 also shows that after 150 days of storage, network structure appeared again in the cross section of bilayer films, but the two phases of starch film and protein film became more easily distinguished. At the same time, the 2000-fold cross-section SEM analysis showed that some cracks appeared in the bilayer film after 150 days of storage, which also explained the reason why the barrier performance of the bilayer films became worse even though its internal structure became more compact at the later stage of storage [52]. At the same time, the changes in bilayer films structure observed by SEM also verified the changes of water content and mechanical properties [43].

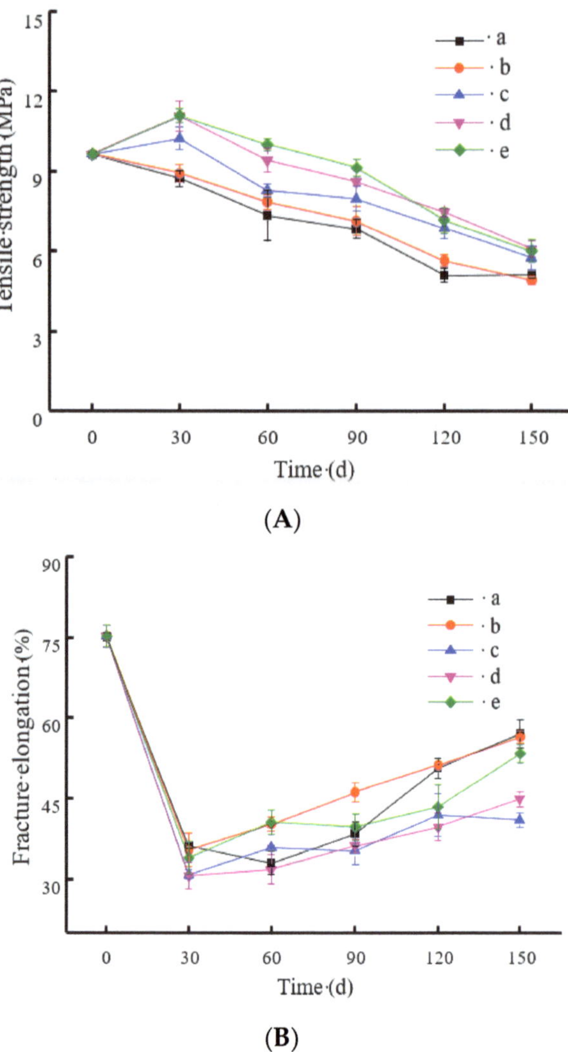

Figure 4. Tensile strength (**A**) and fracture elongation (**B**) of bilayer films during storage. A (−17 °C), b (4 °C), c (25 °C, RH 43%), d (25 °C, RH 54%), e (25 °C, RH 65%).

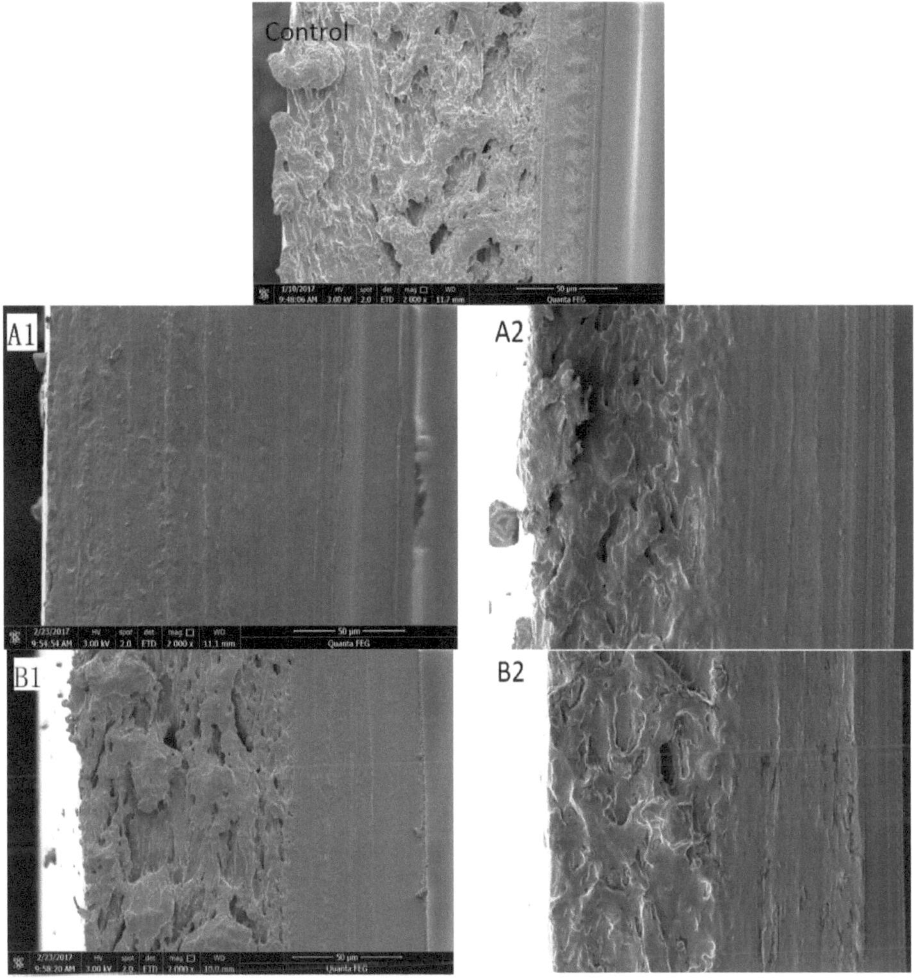

Figure 5. *Cont.*

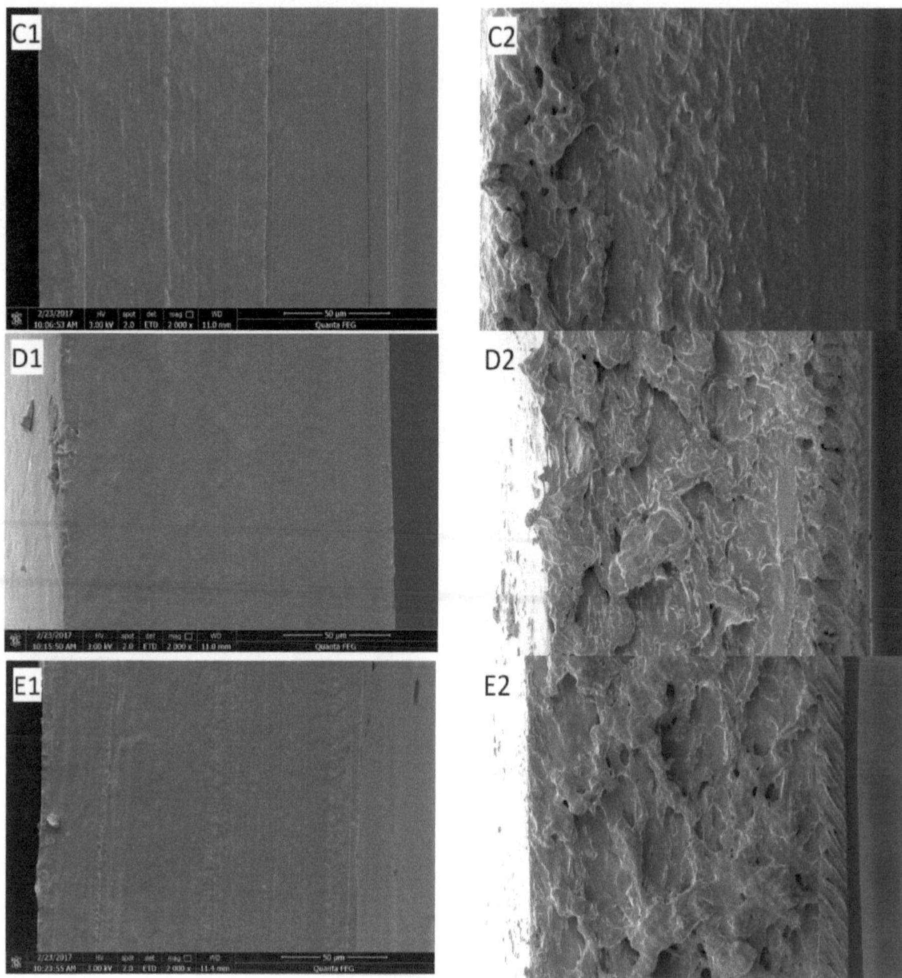

Figure 5. Microstructure changes of cross section of bilayer film during storage (SEM). Storage for 1 day: Control; Storage for 30 days: (**A1**) −17 °C, (**B1**) 4 °C, (**C1**) 25 °C, RH 43%, (**D1**) 25 °C, RH 54%, (**E1**) 25 °C, RH 65%; Storage for 150 days: (**A2**) −17 °C, (**B2**) 4 °C, (**C2**) 25 °C, RH 43%, (**D2**) 25 °C, RH 54%, (**E2**) 25 °C, RH 65%.

4. Conclusions

The disadvantages of single-layer starch film and single-layer corn zein film in practical application were overcome, and the formation mechanism of iron yam–pea starch blend and corn zein bilayer film (IYP/CZ) was explored by using the respective advantages of iron yam–pea starch and corn zein as film materials. The bilayer film IYP/CZ was stored for 150 days under the conditions of freezing, cold storage and normal temperature. Every 30 days, the physicochemical properties and microstructure were analyzed to study the effects of different temperatures, humidity and time on the appearance color, light transmittance, haze, moisture content, water resistance, oxygen resistance and mechanical properties of the bilayer film. On this basis, the microstructure of the cross section was analyzed by SEM to study the effects of the microstructure of the bilayer film on its physicochemical properties. These results revealed that we developed the iron yam–pea starch blend/corn zein bilayer films with satisfactory barrier, mechanical properties and storage properties. This work provides a new method for developing inner packaging materials for convenient food seasoning.

Funding: This work was funded by the Postdoctoral Science Foundation of China (2015M582184), the National Natural Science Foundation of China (31301586), the Youth Science and Technology Innovation Talents Support Program of North China University of Water Resources and Electric Power (70442).

Institutional Review Board Statement: Not applicable.

Informed Consent Statement: Not applicable.

Data Availability Statement: The datasets used and/or analyzed during the current study are available from the corresponding author on request.

Conflicts of Interest: The author declares no conflict of interest.

References

1. Arifin, H.R.; Djali, M.; Nurhadi, B.; Hasim, S.A.; Hilmi, A.; Puspitasari, A.V. Improved properties of corn starch-based bio-nanocomposite film with different types of plasticizers reinforced by nanocrystalline cellulose. *Int. J. Food Prop.* **2022**, *25*, 509–521. [CrossRef]
2. Chawla, R.; Sivakumar, S.; Kaur, H. Antimicrobial edible films in food packaging: Current scenario and recent nanotechnological advancements- a review. *Carbohydr. Polym. Technol. Appl.* **2021**, *2*, 100024. [CrossRef]
3. Hasan, M.; Rusman, R.; Khaldun, I.; Ardana, L.; Mudatsir, M.; Fansuri, H. Active edible sugar palm starch-chitosan films carrying extra virgin olive oil: Barrier, thermo-mechanical, antioxidant, and antimicrobial properties. *Int. J. Biol. Macromol.* **2020**, *163*, 766–775. [CrossRef]
4. Peng, Y.; Li, Y.F. Combined effects of two kinds of essential oils on physical, mechanical and structural properties of chitosan films. *Food Hydrocoll.* **2014**, *36*, 287–293. [CrossRef]
5. Pirnia, M.; Shirani, K.; Yazdi, F.T.; Moratazavi, S.A.; Mohebbi, M. Characterization of antioxidant active biopolymer bilayer film based on gelatin-frankincense incorporated with ascorbic acid and Hyssopus officinalis essential oil. *Food Chem. X* **2022**, *14*, 100300. [CrossRef] [PubMed]
6. Xin, Y.; Jin, Z.; Chen, F.; Lai, S.; Yang, H.S. Effect of chitosan coatings on the evolution of sodium carbonate-soluble pectin during sweet cherry softening under non-isothermal conditions. *Int. J. Biol. Macromol.* **2020**, *154*, 267–275. [CrossRef]
7. Yadav, A.; Kumar, N.; Upadhyay, A.; Sethi, S.; Singh, A. Edible coating as postharvest management strategy for shelf-life extension of fresh tomato (Solanum lycopersicum L.): An overview. *J. Food Sci.* **2022**, *87*, 2256–2290. [CrossRef]
8. Danyxa, P.H.; Carolina, M.J.; Alex, L.C.; Silvia, G. Edible cassava starch films carrying rosemary antioxidant extracts for potential use as active food packaging. *Food Hydrocoll.* **2017**, *63*, 488–495.
9. Olaimat, A.N.; Sawalha, A.G.A.; Al-Nabulsi, A.A.; Osaili, T.; Al-Biss, B.A.; Ayyash, M.; Holley, R.A. Chitosan–ZnO nanocomposite coating for inhibition of *Listeria* monocytogenes on the surface and within white brined cheese. *J. Food Sci.* **2022**, *87*, 3151–3162. [CrossRef]
10. Podshivalov, A.; Zakharova, M.; Glazacheva, E.; Uspenskaya, M. Gelatin/potato starch edible biocomposite films: Correlation between morphology and physical properties. *Carbohydr. Polym.* **2017**, *157*, 1162–1172. [CrossRef]
11. Singh, T.P.; Chauhan, G.; Mendiratta, S.K.; Agrawal, R.K.; Arora, S.; Verma, A.K.; Rajkumar, V. In vitro antioxidant and antimicrobial activities of clove extract and its effectiveness in bio-composite film on storage stability of goat meat balls. *J. Food Sci.* **2022**, *87*, 2083–2095. [CrossRef] [PubMed]
12. Thakur, R.; Saberi, B.; Pristijono, P.; Golding, J.; Stathopoulos, C.; Scarlett, C.; Bowyer, M.; Vuong, Q. Characterization of rice starch-ι-carrageenan biodegradable edible film. Effect of stearic acid on the film properties. *Int. J. Biol. Macromol.* **2016**, *93*, 952–960. [CrossRef]
13. Wang, W.H.; Wang, K.; Xiao, J.D.; Liu, Y.W.; Zhao, Y.; Liu, A.J. Performance of high amylose starch-composited gelatin films influenced by gelatinization and concentration. *Int. J. Biol. Macromol.* **2017**, *94*, 258–265. [CrossRef] [PubMed]
14. Das, M.; Chowdhury, T. Heat sealing property of starch based self-supporting edible films. *Food Packag. Shelf Life* **2016**, *9*, 64–68. [CrossRef]
15. Jaramillo, C.M.; Gutiérrez, T.J.; Goyanes, S.; Bernal, C.; Famá, L. Biodegradability and plasticizing effect of yerba mate extract on cassava starch edible films. *Carbohydr. Polym.* **2016**, *151*, 150–159. [CrossRef]
16. Saberi, B.; Thakur, R.; Vuong, Q.V.; Chockchaisawasdee, S.; Golding, J.B.; Scarlett, C.J.; Stathopoulos, C.E. Optimization of physical and optical properties of biodegradable edible films based on pea starch and guar gum. *Ind. Crops Prod.* **2016**, *86*, 342–352. [CrossRef]
17. Peng, Y.; Yin, L.; Li, Y.F. Combined effects of lemon essential oil and surfactants on physical and structural properties of chitosan films. *Int. J. Food Sci. Technol.* **2013**, *48*, 44–50. [CrossRef]
18. Piccirilli, G.N.; Soazo, M.; Pérez, L.M.; Delorenzi, N.J.; Verdini, R.A. Effect of storage conditions on the physicochemical characteristics of edible films based on whey protein concentrate and liquid smoke. *Food Hydrocoll.* **2019**, *87*, 221–228. [CrossRef]

19. Leerahawong, A.; Tanaka, M.; Okazaki, E.; Osako, K. Stability of the physical properties of plasticized edible films from squid (Todarodes pacificus) mantle muscle during storage. *J. Food Sci.* **2012**, *77*, E159–E165. [CrossRef]
20. Matsakidou, A.; Tsimidou, M.Z.; Kiosseoglou, V. Storage behavior of caseinate-based films incorporating maize germ oil bodies. *Food Res. Int.* **2019**, *116*, 1031–1040. [CrossRef]
21. Osés, J.; Idoya, F.P.; Mendoza, M.; Maté, J.I. Stability of the mechanical properties of edible films based on whey protein isolate during storage at different relative humidity. *Food Hydrocoll.* **2009**, *23*, 125–131. [CrossRef]
22. Famá, L.; Goyanes, S.; Gerschenson, L. Influence of storage time at normal temperature on the physicochemical properties of cassava starch films. *Carbohydr. Polym.* **2007**, *70*, 265–273. [CrossRef]
23. Nuria, B.P.; Fernando, F.M.; Pilar, M. Jumbo squid (*Dosidicus gigas*) myofibrillar protein concentrate for edible packaging films and storage stability. *LWT-Food Sci. Technol.* **2014**, *55*, 543–550.
24. Xiao, G.N.; Zhu, Y.B.; Wang, L.X.; You, Q.; Huo, P.; You, Y.R. Production and Storage of Edible Film Using Gellan Gum. *Procedia Environ. Sci.* **2011**, *8*, 756–763. [CrossRef]
25. Soukoulis, C.; Behboudi-Jobbehdar, S.; Yonekura, L.; Parmenter, C.; Fisk, I.D. Stability of Lactobacillus rhamnosus GG in prebiotic edible films. *Food Chem.* **2014**, *159*, 302–308. [CrossRef]
26. Neira, L.M.; Martucci, J.F.; Stejskal, N.; Ruseckaite, R.A. Time-dependent evolution of properties of fish gelatin edible films enriched with carvacrol during storage. *Food Hydrocoll.* **2019**, *94*, 304–310. [CrossRef]
27. Borneo, R.; Alba, N.; Aguirre, A. New films based on triticale flour: Properties and effects of storage time. *J. Cereal Sci.* **2016**, *68*, 82–87. [CrossRef]
28. Ciannamea, E.M.; Stefani, P.M.; Ruseckaite, R.A. Storage-induced changes in functional properties of glycerol plasticized–soybean protein concentrate films produced by casting. *Food Hydrocoll.* **2015**, *45*, 247–255. [CrossRef]
29. Schmid, M.; Merzbacher, S.; Müller, K. Time-dependent crosslinking of whey protein based films during storage. *Mater. Lett.* **2018**, *215*, 8–10. [CrossRef]
30. Ket-On, A.; Pongmongkol, N.; Somwangthanaroj, A.; Janjarasskul, T.; Tananuwong, K. Properties and storage stability of whey protein edible film with spice powders. *J. Food Sci. Technol.* **2016**, *53*, 2933–2942. [CrossRef]
31. Schmid, M.; Reichert, K.; Hammann, F.; Stäbler, A. Storage time-dependent alteration of molecular interaction–property relationships of whey protein isolate based films and coatings. *J. Mater. Sci.* **2015**, *50*, 4396–4404. [CrossRef]
32. Leceta, I.; Peñalba, M.; Arana, P.; Guerrero, P.; de la Caba, K. Ageing of chitosan films: Effect of storage time on structure and optical, barrier and mechanical properties. *Eur. Polym. J.* **2015**, *66*, 170–179. [CrossRef]
33. Hernández-Muñoz, P.; López-Rubio, A.; Del-Valle, V.; Almenar, E.; Gavara, R. Mechanical and water barrier properties of glutenin films influenced by storage time. *J. Agric. Food Chem.* **2004**, *52*, 79–83. [CrossRef] [PubMed]
34. Matta, E.; Tavera-Quiroz, M.J.; Bertola, N. Isomalt-Plasticized Methylcellulose-Based Films as Carriers of Ascorbic Acid. *Food Bioprocess Technol.* **2020**, *13*, 2186–2199. [CrossRef]
35. Friedrich, J.C.C.; Silva, O.A.; Faria, M.G.I.; Colauto, N.B.; Gazzin, Z.C.; Colauto, G.A.L.; Caetano, J.; Dragunski, D.C. Improved antioxidant activity of a starch and gelatin-based biodegradable coating containing Tetradenia riparia extract. *Int. J. Biol. Macromol.* **2020**, *165*, 1038–1046. [CrossRef]
36. Basiak, E.; Lenart, A.; Debeaufort, F. Effect of starch type on the physico-chemical properties of edible films. *Int. J. Biol. Macromol.* **2017**, *98*, 348–356. [CrossRef] [PubMed]
37. Wang, X.J.; Zheng, X.Q.; Liu, X.L.; Kopparapu, N.K.; Cong, W.S.; Deng, Y.P. Preparation of glycosylated zein and retarding effect on lipid oxidation of ground pork. *Food Chem.* **2017**, *227*, 335–341. [CrossRef] [PubMed]
38. Yin, Y.C.; Yin, S.W.; Yang, X.Q.; Tang, C.H.; Wen, S.H.; Chen, Z.; Xiao, B.J.; Wu, L.Y. Surface modification of sodium caseinate films by zein coatings. *Food Hydrocoll.* **2014**, *36*, 1–8. [CrossRef]
39. Ghanbarzadeh, B.; Oromiehi, A.R. Biodegradable biocomposite films based on whey protein and zein: Barrier, mechanical properties and AFM analysis. *Int. J. Biol. Macromol.* **2008**, *43*, 209–215. [CrossRef] [PubMed]
40. Fund, T.; Isa, D.A.; Banu, O. Water vapor and oxygen-barrier performance of corn–zein coated polypropylene films. *J. Food Eng.* **2010**, *96*, 342–347.
41. Dong, S.; Gao, A.; Xu, H.; Chen, Y. Effects of Dielectric Barrier Discharges (DBD) Cold PlasmaTreatment on Physicochemical and Structural Propertiesof Zein Powders. *Food Bioprocess Technol.* **2017**, *10*, 434–444. [CrossRef]
42. Nagar, M.; Sharanagat, V.S.; Kumar, Y.; Singh, L. Development and characterization of elephant foot yam starch–hydrocolloids based edible packaging film: Physical, optical, thermal and barrier properties. *J. Food Sci. Technol.* **2020**, *57*, 1331–1341. [CrossRef] [PubMed]
43. Saberi, B.; Thakur, R.; Bhuyan, D.J.; Vuong, Q.V.; Chockchaisawasdee, S.; Golding, J.B.; Scarletta, C.J.; Stathopoulosc, C.E. Development of edible blend films with good mechanical and barrier properties from pea starch and guar gum. *Starch Starke* **2017**, *69*, 1–15. [CrossRef]
44. Zuo, G.J.; Song, X.Y.; Chen, F.S.; Shen, Z.X. Physical and structural characterization of edible bilayer films made with zein and corn-wheat starch. *J. Saudi Soc. Agric. Sci.* **2017**, *18*, 324–331. [CrossRef]
45. Zhang, Y.Z.; Li, J.; Zhao, J.; Bian, W.; Li, Y.; Wang, X.J. Adsorption behavior of modified Iron stick yam skin with Polyethyleneimine as a potential biosorbent for the removal of anionic dyes in single and ternary systems at low temperature. *Bioresour. Technol.* **2016**, *222*, 285–293. [CrossRef] [PubMed]

46. Song, X.Y.; Wang, Y.Q. Development and characterization of edible bilayer films based on iron yam-pea starch blend and corn zein. *Food Sci. Technol.* **2021**, *41*, 684–694. [CrossRef]
47. Cho, S.Y.; Lee, S.Y.; Rhee, C. Edible oxygen barrier bilayer film pouches from corn zein and soy protein isolate for olive oil packaging. *LWT-Food Sci. Technol.* **2010**, *43*, 1234–1239. [CrossRef]
48. Song, X.Y.; Cheng, L.M.; Tan, L. Edible iron yam and maize starch convenient food flavoring packaging films with lemon essential oil as plasticization. *Food Sci. Technol.* **2019**, *39*, 971–979. [CrossRef]
49. Erahimi, B.; Mohammadi, R.; Rouhi, M.; Mortazavian, A.M.; Aliabadi, S.S.; Koushki, M.R. Survival of probiotic bacteria in carboxymethyl cellulose-based edible film and assessment of quality parameters. *LWT-Food Sci. Technol.* **2018**, *87*, 54–60. [CrossRef]
50. Pérez, L.M.; Piccirilli, G.N.; Delorenzi, N.J.; Verdini, R.A. Effect of different combinations of glycerol and/or trehalose on physical and structural properties of whey protein concentrate-based edible films. *Food Hydrocoll.* **2016**, *56*, 352–359. [CrossRef]
51. Korhonen, K.; Smolander, E.; Korhonen, O.; Ketolainen, J.; Laitinen, R. Effect of storage on the physical stability of thin polymethacrylate-perphenazine films. *Eur. J. Pharm. Sci.* **2017**, *104*, 293–301. [CrossRef] [PubMed]
52. Ciannamea, E.M.; Espinosa, J.P.; Stefani, P.M.; Ruseckaite, R.A. Long-term stability of compression-molded soybean protein concentrate films stored under specific conditions. *Food Chem.* **2018**, *243*, 448–452. [CrossRef] [PubMed]

 *coatings*

Article

Production and Characterization of Active Pectin Films with Olive or Guava Leaf Extract Used as Soluble Sachets for Chicken Stock Powder

Mohammed Sabbah [1,*], Asmaa Al-Asmar [2,3], Duaa Younis [1], Fuad Al-Rimawi [4], Michela Famiglietti [5] and Loredana Mariniello [5]

1   Department of Nutrition and Food Technology, An-Najah National University, P.O. Box 7, Nablus P400, Palestine; d.yunis@najah.edu
2   Department of Biology and Biotechnology, Faculty of Science, An-Najah National University, P.O. Box 7, Nablus P400, Palestine; a.alasmar@najah.edu
3   Energy, Water and Food Security Research Center, An-Najah National University, P.O. Box 7, Nablus P400, Palestine
4   Department of Chemistry, Al-Quds University, Abu Dis, Jerusalem P144, Palestine; falrimawi@staff.alquds.edu
5   Department of Chemical Sciences, University of Naples "Federico II", 80126 Naples, Italy; michela.famiglietti@unina.it (M.F.); loredana.mariniello@unina.it (L.M.)
*   Correspondence: m.sabbah@najah.edu; Tel.: +790-92345115

Abstract: The goal of this study was to improve the functionality of two pectin (PEC) edible films by incorporating olive leaf extract (OLE) or guava leaf extract (GLE). Different concentrations of OLE or GLE (0.1 and 0.2% w/v) were used, and 30% glycerol was added as a plasticizer. The obtained films were evaluated for their mechanical properties, antioxidant activity, thickness, color, opacity, permeability to gases and water vapor, moisture content, and moisture uptake. Soluble sachets were then prepared and filled with chicken stock powder. The results indicated that incorporating OLE or GLE into the PEC films significantly increased their opacity, greenness, and antioxidant activity, which increased from 8.5% in the control to 83.9% when 0.2% GLE was added. Additionally, the films had lower water vapor permeability than the control film. The moisture uptake of the films was also significantly increased when GLE was added. Furthermore, the developed sachets were tested in real-life scenarios, mirroring their intended usage in households. After being introduced to boiling water, the sachets rapidly dissolved within seconds. These results suggest that OLE or GLE, as natural additives, can be used to improve the functionality and activity of edible films.

Keywords: functional edible packaging; food preservation; pectin films; food wrapping; olive byproducts; plant leaf; circular economy

Citation: Sabbah, M.; Al-Asmar, A.; Younis, D.; Al-Rimawi, F.; Famiglietti, M.; Mariniello, L. Production and Characterization of Active Pectin Films with Olive or Guava Leaf Extract Used as Soluble Sachets for Chicken Stock Powder. *Coatings* **2023**, *13*, 1253. https://doi.org/10.3390/coatings13071253

Academic Editor: Jun Mei, Jing Xie and Biao Zhang

Received: 13 June 2023
Revised: 14 July 2023
Accepted: 14 July 2023
Published: 16 July 2023

1. Introduction

Today, bioplastics, edible films, and coatings are becoming the most effective ways to reduce food packaging's adverse impact on the environment [1,2]. More than 368 million tons of plastics was used in 2019, and a significant amount of this quantity ended up in landfills and the oceans, polluting our air, water, and soil [3,4]. Several researchers have worked to find the best combination between low-price film-based materials and their activation with different plant extracts or additives that potentially have antioxidant, antimicrobial, and anti-biofilm activities [5], to enhance the food's shelf-life or other functionality for food applications [6–9]. As a result of short shelf-life and inadequate packaging, food waste remains a significant global challenge, emphasizing the need for innovative and sustainable packaging solutions [10–16]. Edible films and coatings are prepared mainly from hydrocolloidal materials (e.g., carbohydrates and proteins, or a combination of both sources). Carbohydrate-material-based films show good mechanical properties, although they are poor in terms of permeability. Meanwhile, edible films prepared mainly from protein show good barrier properties against gases, although they have poor mechanical properties [15,17–25].

In recent years, the development of edible and coating solutions based on pectin has increased due to several reasons. First, they are classified as food-grade materials and are non-toxic [26]. They can also be obtained from renewable available resources such as citrus peel, apple pomace, and sugar beet [27]. They also have good film formation ability, even without plasticizers [28]. Additionally, they can be used in different food products as stabilizers or thickeners, and they have good water solubility, biocompatibility, and compostability [27].

Commercial pectin is derived from a variety of plant sources, including apple pomace (14%), and citrus peel (85%) [6,29]. Fruit processing industries generate significant quantities of fruit waste that, if not utilized, ends up in landfills, contributing to environmental degradation due to microbial decay and greenhouse gas emissions [30]. However, pectin, which is a complex polysaccharide, represents the highest percentage of plant mass composition (about 35% in dicotyledonous plant cells and approximately 5% in woody tissues [31], whereas in grass plants it is about 2%–10%) [32]. Recently, Chandel et al. [27] reported on pectin obtained from 26 sources, along with the extraction methods that were used to produce commercial pectin for different applications, e.g., food, pharmaceuticals [33], cosmetics, and food packaging [15,17,19,28,34].

Pectin film alone does not possess strong functionality in terms of antimicrobial or antioxidant activity, which may potentially discourage manufacturers from using it for food wrappings or coatings. Lately, many researchers have worked to improve the functionality of pectin film due to its availability and low price, by adding essential oils, functionalized nanoparticles [35], and plant extracts. Adding plant extracts or essential oils such as mint [36], lemon [37], oregano [38], orange [37], clove bud [39], cinnamon [40], or lime peel extract [41] to the edible films proved that pectin-based films can be used as food packaging to extend the shelf-life of food products.

Extracts of plant leaves represent a promising approach that can be incorporated, as they have antioxidant and antimicrobial activity. Therefore, olive (*Olea europaea* L.) leaf extracts (OLEs) have been used for centuries in folk medicine as therapeutic infusions, and they are among the main byproducts generated from the olive oil industry [42]. One of the major phenolic compounds found in olive leaves is oleuropein, a glucoside ester of elenolic acid and hydroxytyrosol. Oleuropein has been reported to have significant health benefits, including antioxidant, cholesterol-lowering, cardioprotective, anti-inflammatory, hypoglycemic, and antimicrobial properties [42]. The wide profile of polyphenols in olive leaves gives OLE great potential as a natural antioxidant. OLE has been incorporated as an antioxidant in several food matrices, such as edible oils, frying oils, table olives, meat, and meat products [43–46]. Moreover, Elsayed et al. [47] utilized an OLE coating incorporated with a zinc/selenium oxide nanocomposite to enhance the postharvest quality of green bean pods; they concluded that by using 3% OLE with zinc, the green bean pods' shelf-life increased to 28 days in cold storage, and all attributes improved compared to the other treatments or controls. In addition, guava (*Psidium guajava* L.) leaf extract (GLE) contains several bioactive compounds, especially phenolic compounds that contribute to antioxidant and anti-inflammatory properties [48,49]. However, quercetin is the most potent antioxidant found in GLE [48,50]. Guava leaves are traditionally used to treat gastrointestinal ailments (e.g., diarrhea, stomach pain, gastroenteritis, indigestion, and dysentery) and dermatological problems (e.g., skin infection, skin aging, and ulcers).

Pectin sachets were prepared from lime peel pectin integrated with coconut water and lime peel extract, and they were used to retard soybean oil's oxidation; the obtained results indicated that the pectin sachets retarded the soybean oil's oxidation for 30 days of storage [41].

Due to the lack of previous works, as well as the potential functionality of OLE and GLE that contribute to the shelf-life and preservation of food, both materials were used in this study as film functionality materials. The main objective of this project was therefore to utilize OLE and GLE at different concentrations in pectin-based films, and to evaluate their contributions to the obtained films, which were used to prepare soluble sachets that were used in this study as filling sachets for chicken stock powder.

2. Materials and Methods

2.1. Materials

Olive leaves were collected from olive trees in Ramallah/West Bank, Palestine, in October 2021. The leaves were collected, washed with tap water, and then dried for 30 days at room temperature (25–30 °C). The leaves were then ground until a fine powder (0.1 to 0.6 mm) was obtained.

The guava leaf samples used in this study were obtained from guava trees in Qalqilya/West Bank, Palestine. The leaf samples were washed with tap water and dried for 30 days in a dry place at room temperature (25–30 °C). The samples were milled and sieved after drying. Particles ranging from 0.1 to 0.6 mm in size were chosen for the extractions. Prior to the extraction, the samples were placed in plastic bags and stored in a dark, dry area. Food-grade vegetable glycerol (GLY) was purchased from Heartlandvapes LLC., Owasso, OK, USA.

2.2. Preparation of Olive and Guava Leaves Extracts

OLE and GLE were extracted following the methodology outlined by Annegowda et al. [51], with certain modifications. Then, 10 g of leaf powder from each material was combined with 100 mL of ethanol in a beaker. The mixture was subjected to sonication for a duration of 2 h, while maintaining the solvent surface at room temperature. Subsequently, the residues were dissolved in ethanol and subjected to another round of extraction. The resulting extracts were filtered using vacuum filtration and Whatman filter paper, followed by concentration using a rotary evaporator set at 50 °C. After that, the crude extract was dried in a freeze-dryer. Finally, the obtained extract was sieved through a stainless steel sieve (with a mesh size of 425 μm) provided by Octagon Digital Endecotts Limited (Lombard Road, London, UK) until a fine powder was obtained. The extraction yield of OLE was about 9.0%, while that of GLE was 11.0%.

2.3. Film Preparation with Olive and Guava Leaf Extracts

Citrus PEC stock solution (2.0 g) was added to 100 mL of distilled water and completely solubilized. A film-forming solution (FFS) containing 0.1 and 0.2 w/v of OLE or GLE was added and continuously stirred for 30 min at room temperature. Then, GLY (30% w/v relative to PEC) was added to the FFS as a plasticizer. The concentrations of both OLE and GLE were chosen in accordance with the preliminary investigation, wherein varying concentrations of 0%, 0.05%, 0.1%, 0.2%, and 0.4% (w/v) were employed. The obtained film samples were subsequently analyzed for their mechanical properties and color characteristics. After careful evaluation, it was found that the lower concentrations (0.05% w/v) did not show any significant effect on the film properties. However, the higher concentrations (0.4% w/v) had an unwanted effect on the color of the films, which became more greenish, opaque, and rigid. Therefore, the concentrations of 0.1% and 0.2% (w/v) were deemed suitable and selected for further experimentation. The final volume was adjusted to 50 mL using distilled water, and then the FFSs were poured onto 8 cm diameter polystyrene Petri dishes and dried in a drying chamber at 30 °C for 12 h. The dried films were peeled off from the Petri dishes and then placed inside a desiccator (50%–55% RH) containing a saturated solution of $Mg(NO_3)_2 \cdot 6H_2O$ for 2 h before analysis of the properties of the prepared films.

2.4. Film Characterization

The effects of OLE and GLE at two different concentrations (0.1, 0.2% w/v) on the films' mechanical properties—including tensile strength (TS), elongation at break (EB), and

Young's modulus (YM)—were measured using a universal testing instrument (Brookfield CT3 Texture Analyzer, model CT3 50K, Brookfield, Chandler, AZ, USA), as described by [8]. The dry films were used for testing the mechanical properties. The films were cut into strips (10 mm wide) and then loaded between the grips of the CT3 Texture Analyzer and tested, with an initial grip separation of about 50 mm and a speed of 0.5 mm/s. At least six strips from each film were tested, and the experiment was repeated three times at a different intervals.

The thickness of the obtained films was tested using a micrometer screw gauge (0–25 mm), where at least 6 different points from the whole films were measured.

The ability of the films to scavenge DPPH free radicals was tested as described by Famiglietti et al. [43]. Briefly, 20 mg of each film was solubilized in 500 mL of distilled water, and then 100 mL of each solution was mixed with 900 mL of DPPH (0.05 mg/mL methanolic solution). After the samples were kept in the dark for 30 min at 25 °C, the absorbance at 517 nm was measured using a UV–visible Spectrophotometer (SmartSpec 3000 Bio-Rad, Segrate, Milan, Italy), where methanol was used as a blank and a sample with methanol was added to DPPH solution as a control. The antioxidant activity of the films was calculated based on the following equation:

$$\%DPPH \text{ scavenging activity} = (A_{control} - A_{sample}/A_{control}) \times 100 \tag{1}$$

where A is the absorbance value of the control and the sample at 517 nm. Each sample was measured in triplicate.

Tonyali et al. [52] and Famiglietti et al. [43] provided insights into measuring the films' opacity. In accordance with their methods, the films were measured six times for each sample. The opacity of the films was then calculated using the following equation:

$$\text{Opacity (mm}^{-1}) = A600/x \tag{2}$$

where $A600$ is the absorbance of the sample at 600 nm, and x is the film thickness (mm).

The moisture content of the film was determined as described by Galus and Lenart [53], by measuring the mass loss of one gram after 24 h of oven drying at 105 °C and expressing it as a percentage. The weight gain of each specimen after 24 h at 50% RH was used to determine the ability of the specimen to absorb moisture. A total of 3 specimens (2 cm^{-2}) were cut from the films and weighed ($W1$), and then the films were dried in the oven at 105 °C for 24 h before being weighed again ($W2$). Following conditioning at 25 °C and 50% RH for 24 h, the film samples were weighed once more ($W3$) after being placed in a desiccator over a saturated solution of $Mg(NO_3)_2$ 6H$_2$O.

Water content and uptake were calculated according to the following formulae:

$$\text{Water content (\%)} = \frac{W1 - W2}{W1} \times 100 \tag{3}$$

$$\text{Water uptake (\%)} = \frac{W3 - W2}{W3} \times 100 \tag{4}$$

The color values of the developed films were measured with a colorimeter (chroma Meter Konica Minolta CR-400, Japan) using the CIE color scale to evaluate the parameters L* (lightness/brightness, which ranges from 0 to 100), a* (redness/greenness), and b* (yellowness/blueness) (the two chromatic components, which range from −120 to 120) [54].

The films' barrier properties against water vapor (WV), O_2, and CO_2 were analyzed by means of a MultiPerm apparatus (ExtraSolution s.r.l, Pisa, Italy). The measurements were performed in duplicate for each film (50% RH, 25 °C) according to ASTM F1249-13 [55], ASTM D3985-05 [56], and ASTM F2476-05 [57]. Before testing, the film specimens were conditioned for 24 h at 50% RH and placed in aluminum masks to reduce the film test area to 2 cm^2.

2.5. Sachet Application Experiment

The obtained films (control, pectin with 0.2% OLE, and pectin with 0.2% GLE), were folded and heat-sealed on two sides using a kitchen sealer machine, and then the sachets were filled to the top with about 18–20 g of chicken stock powder before being heat-sealed. Finally, boiling water was prepared, and the sachets was added to the boiling water and stirred until they completely dissolved.

2.6. Statistical Analysis

The obtained data were statistically analyzed using JMP software (SAS Institute, Cary, NC, USA, version 5.0). The data were subjected to analysis of variance (ANOVA), and the means were compared using the Tukey–Kramer HSD test. Differences were considered to be significant at $p < 0.05$.

3. Results and Discussion

3.1. Effect of Incorporation of OLE and GLE in Pectin-Based Films on Thickness and Opacity

Pectin-based films modified with olive leaf extract (OLE) or guava leaf extract (GLE) were obtained at different concentrations (0.1% and 0.2% w/v) in the presence of 30% glycerol (GLY) as a plasticizer. Figure 1A shows the visual observation of the obtained films compared to the control film containing only pectin (PEC). The addition of GLE or OLE increased the greenish color of the pectin films, with a higher intensity observed in the films containing GLE. Previous research has shown that GLE significantly increased the film color compared to the control [58]. This suggests that the differences in the chemical structure of natural pigments and phenolic compounds between GLE and the control may be responsible for the differences in film color [59,60].

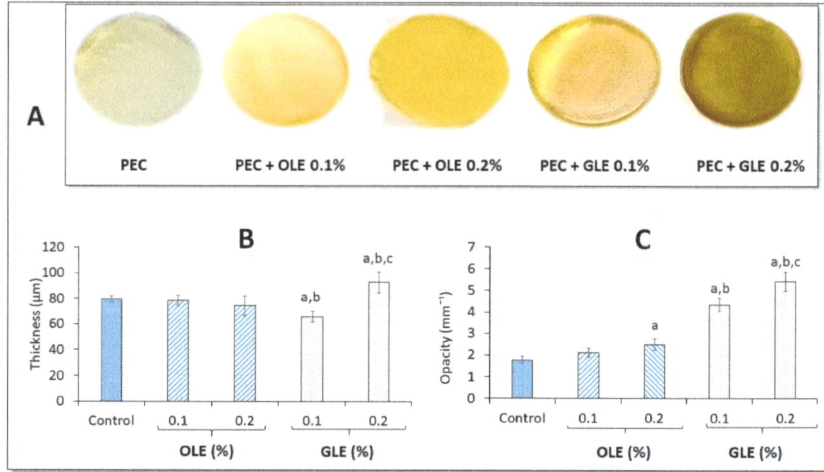

Figure 1. Effects of using OLE and GLE at different concentrations (0.1 w/v and 0.2 w/v) on pectin-based films: observation (**A**), thickness (**B**), and opacity (**C**). The values that are significantly different from the control are indicated by "a" ($p < 0.05$); the values indicated by "b" are significantly different ($p < 0.05$) from films obtained with different materials at the same concentration; the values indicated by "c" are significantly different ($p < 0.05$) from the films obtained with the same material at different concentrations.

Film thickness was also determined, and the results (Figure 1B) indicated that modifying the pectin with OLE did not change the film thickness compared to the control. However, when 0.2% GLE was used, the film thickness significantly increased compared to the control and the films obtained with OLE. This can be explained by the ability of

GLE to increase the intermolecular forces among pectin polymer chains [61]. Moreover, there is no clear explanation for why GLE has different effects on film thickness at different concentrations. In fact, it is thought that the GLE interacts with the pectin molecules differently depending on the concentration. We could hypothesize that when the GLE is present at low concentrations (0.1% w/v), it may be able to penetrate between the pectin molecules and disrupt their interactions. The films would become less thin as a result. At higher concentrations (0.2% w/v), the GLE may not be able to penetrate between the pectin molecules, instead forming a layer on the surface. The films would be thickened as a result.

Additionally, the opacity of the films significantly increased with increasing concentrations of OLE or GLE. Notably, in the films containing 0.2% OLE, the opacity was significantly higher compared to the control film. Similarly, when GLE was used, the results indicated that the film's opacity was twice that of the control film or the films containing OLE (Figure 1C).

3.2. Films' Mechanical Properties

The obtained films were also characterized for their mechanical properties, including tensile strength (TS), elongation at break (EB), and Young's modulus (YM). The results, reported in Figure 2, showed that the addition of either OLE or GLE at different concentrations did not significantly affect the TS value. However, the EB result was significantly lower than that of the control films only in the pectin films prepared with 0.1% GLE. The YM of the films significantly decreased in the films prepared with either OLE or GLE compared to the control film containing only PEC. This indicates that increasing the concentration of OLE or GLE led to a significant decrease in YM.

Luo et al. [62] investigated the incorporation of different concentrations of GLE into a sodium-alginate-based material, and they found that adding GLE to sodium alginate significantly increased the TS and significantly decreased the EB compared to the control film obtained using only sodium alginate. This obtained result is consistent with the results obtained recently by Chou et al. [58], where they incorporated GLE into fish skin gelatin. Moreover, according to some studies, excessive proportions of the extracts may result in uneven dispersion in the mixture [62,63]. Plant oils reduce the intermolecular forces between polymer chains, decreasing the films' strength and increasing their flexibility [61].

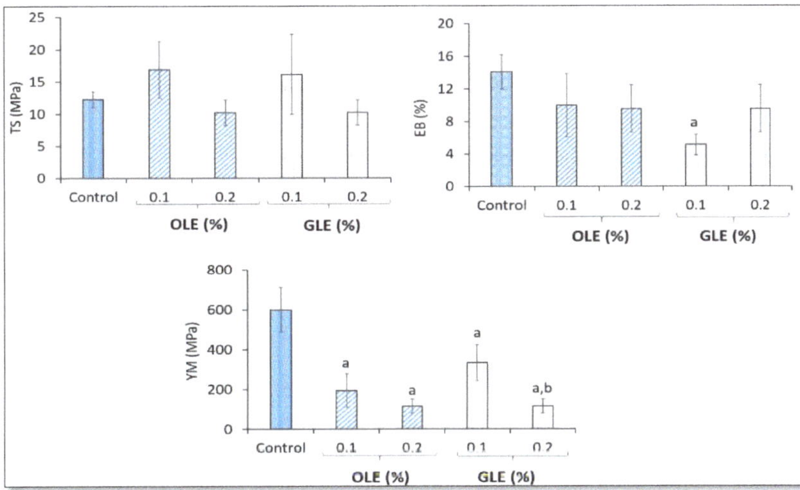

Figure 2. Effect of using OLE and GLE at different concentrations (0.1 w/v and 0.2 w/v) on the mechanical properties (TS, EB, and YM) of pectin-based films. The values that are significantly different from the control are indicated by "a" ($p < 0.05$), while the values indicated by "b" are significantly different ($p < 0.05$) from the films obtained with the same material at a different concentration.

3.3. Antioxidant Activity

The DPPH radical scavenging abilities of the obtained films were evaluated, and the results were compared with the control film that was prepared from only pectin and glycerol. The results are reported in Figure 3. According to many studies, PEC films do not have antioxidant activity naturally; in this context, OLE and GLE have previously been recognized for their antioxidant activity, and they have higher antioxidant activities compared to other materials [42,48,49,62,64]. The pure OLE and GLE were evaluated at two different concentrations (150 and 300 ppm), and the results were 67 ± 1.8% and 76 ± 1.8%, respectively, for OLE, and 61 ± 1.9% and 82 ± 1.2%, respectively, for GLE. The results indicated that when the concentration of pure OLE or GLE increased, the antioxidant activity increased, and GLE at 300 ppm showed the highest activity, reaching 82%, as compared to 76% for OLE. Moreover, the obtained films were evaluated to study the effects of the addition of OLE or GLE to the pectin film-forming solution on the film's antioxidant activity. The obtained results indicated that by increasing the proportion of OLE or GLE from 0.1 to 0.2% *w/v*, the DPPH radical scavenging abilities of the films were significantly improved compared to the control film. The addition of 0.2% (*w/v*) OLE to the films showed significantly higher activity compared to the control and to the 0.1% *w/v* OLE. Moreover, the incorporation of GLE into the films indicated that these films had higher antioxidant activity compared with the OLE at the same concentration. This increased activity may be attributed to the presence of quercetin, which is classified as the most potentially powerful antioxidant compound [48,49]. Similar results were found by Albertos et al. [64] when they evaluated the effects of the addition of different concentrations of OLE to gelatin-based films to enhance the quality of cold smoked salmon, and they reported that increasing the OLE concentrations also increased the antioxidant activity. The addition of OLE and OLE to packaging materials can significantly increase their antioxidant capacity, which can help to protect food from oxidation and deterioration.

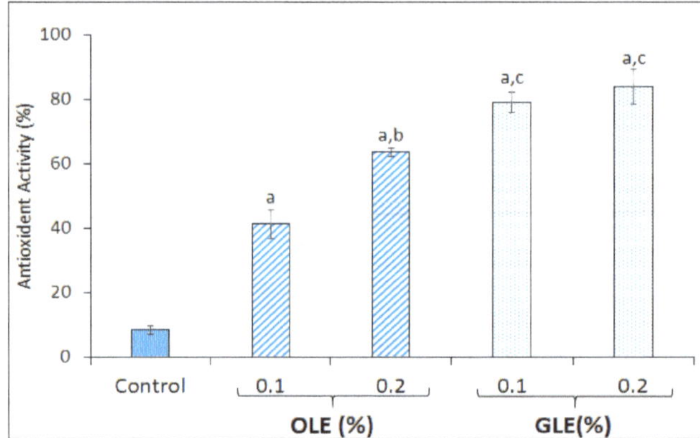

Figure 3. Effects of using OLE and GLE at different concentrations (0.1 *w/v* and 0.2 *w/v*) on the antioxidant activity of pectin-based films. The values significantly different from the control are indicated by "a" (*p* < 0.05); the values indicated by "b" are significantly different from the others films obtained with the same material (OLE); the values indicated by "c" are significantly different from the films made of OLE or GLE at a concentration of 0.1% or 0.2% (*p* < 0.05).

3.4. Water Content and Water Uptake

The effects of the incorporation of different concentrations of OLE and GLE on the water content and uptake of pectin-based films were measured, and the results are reported in Figure 4. The results show that OLE and GLE do not have a significant influence on the film's moisture content at any concentration. However, the moisture uptake of the pectin films modified with GLE increased significantly compared to the control film prepared from pectin alone or the pectin-based films containing OLE. The highest moisture uptake was found in the film that contained 0.2% GLE, reaching $20.20 \pm 1.80\%$, while the control film had about $13.90 \pm 0.88\%$ moisture uptake. Luo et al. [62] prepared sodium alginate films incorporating GLE and concluded that films containing GLE tended to have lower moisture content than the control films. Similar results were found by adding grape juice to protein-based films, where the moisture content did not change with increasing grape juice concentration [8].

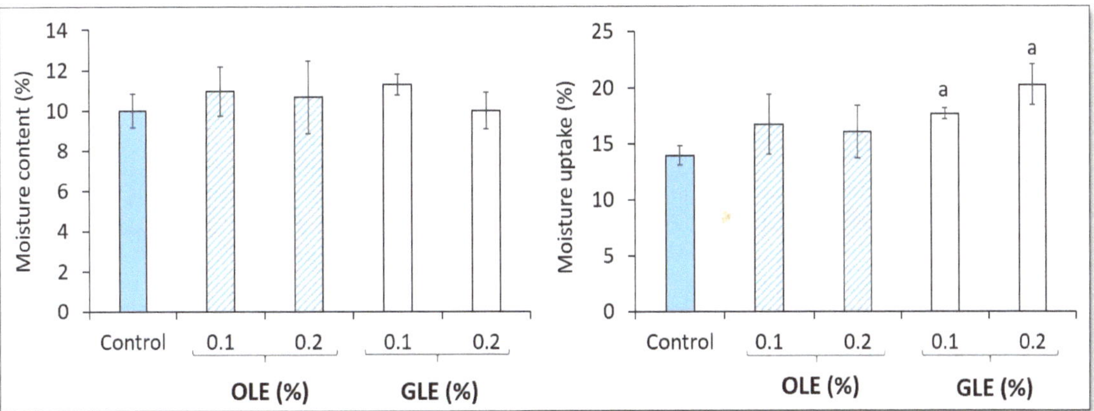

Figure 4. Effects of using OLE and GLE at different concentrations (0.1 w/v and 0.2 w/v) on the water content and water uptake of pectin-based films. The values that are significantly different from the control are indicated by "a" ($p < 0.05$).

3.5. Effects of OLE and GLE on the Color of Pectin-Based Films

The effects of adding different concentrations of OLE and GLE on the color of pectin-based films are shown in Table 1. The L* values of the films containing OLE (0.1 and 0.2% w/v) were significantly increased compared to the control, indicating that the films were lighter. In contrast, the L* values of the films containing GLE decreased significantly compared to the control and the films containing OLE, indicating that these films were darker. The a* value, which indicates greenness, was significantly changed in the films modified with OLE and GLE.

Table 1. Effects of different concentrations of OLE and GLE on pectin-based films' color.

Films	L*	a*	b*
PEC (control)	83.44 ± 0.09	-3.09 ± 0.04	17.41 ± 0.30
PEC + OLE 0.1%	85.02 ± 0.05 [A]	-1.50 ± 0.01 [A]	11.13 ± 0.17 [A]
PEC + OLE 0.2%	86.92 ± 0.14 [A,B]	0.72 ± 0.04 [A,B]	3.36 ± 0.29 [A,B]
PEC + GLE 0.1%	79.83 ± 0.91 [A,D]	-2.01 ± 0.03 [A,D]	21.76 ± 1.80 [A,D]
PEC + GLE 0.2%	74.89 ± 0.38 [A,C,D]	-3.08 ± 0.11 [C,D]	34.50 ± 0.43 [A,C,D]

The values that were significantly different from the controls are indicated by "A" ($p < 0.05$), while the values indicated by "B" were significantly different from the films obtained with OLE ($p < 0.05$), the values indicated by "C" were significantly different from the films obtained with GLE ($p < 0.05$), and the values indicated by "D" were significantly different from the films made of OLE or GLE at a concentration of 0.1% or 0.2% ($p < 0.05$). Further experimental details are given in the text.

The films containing OLE were greener than the control films, and the films containing GLE were even greener. However, the b* value, which indicates yellowness, decreased significantly when the concentration of OLE was increased. This indicates that the films became less yellow. In contrast, the b* value was significantly increased when GLE was incorporated into the pectin films. This indicates that the films became more yellow. In conclusion, the addition of GLE to the films resulted in a darker, more greenish, and more yellowish color. These findings explain the opacity results reported in Figure 1C. The obtained results are consistent with the results achieved by [62,63].

3.6. Permeability Properties

Table 2 reports the effects of the incorporation of OLE or GLE at different concentrations on the gas (CO_2 and O_2) and water vapor (WV) permeability of the pectin-based films. The permeability of the films that contained either OLE or GLE was higher toward all gases and WV compared to the control film (pectin alone). Gradually, the CO_2 permeability significantly increased when the OLE concentration increased. On the other hand, when increasing the concentration of OLE or GLE, the WV permeability of the films remained almost constant. The obtained results show a lower permeability toward CO_2 compared to Viscofan NDX—the commercial casing material used for processed meat products—and a similar value of O_2 permeability but a higher WV permeability. The WV permeability of films is a function of both the molecular diffusion coefficient and the water solubility of the film material [65]. The obtained results are consistent with those previously obtained by [64], who concluded that the addition of 3.75 and 5.63% OLE to gelatin-based films increased the WV permeability significantly, which they explained as being due to the increasing film thickness and the water solubility factor. However, the obtained results could also be explained by the presence of phenolic compounds that may accumulate in the polymer matrix and form voids, which can lead to higher WV permeability values compared to films that do not contain phenolics [62,66].

Table 2. Gas and water vapor (WV) permeability of 1% PEC films prepared in the presence of different concentrations of OLE or GLE with 30% GLY.

Films	CO_2	O_2	WV
	$cm^3 \cdot mm \cdot m^{-2} \cdot day^{-1} \cdot kPa^{-1}$		$g \cdot mm \cdot m^{-2} \cdot day^{-1} \cdot kPa^{-1}$
PEC *	0.08 ± 0.03	0.01 ± 0.00	0.12 ± 0.01
PEC + OLE 0.1%	0.15 ± 0.02 [A]	0.01 ± 0.00	3.87 ± 0.76 [A]
PEC + OLE 0.2%	0.35 ± 0.01 [A,B]	0.05 ± 0.00 [A,B]	4.64 ± 0.02 [A]
PEC + GLE 0.1%	0.31 ± 0.05 [A]	0.05 ± 0.00 [A]	4.64 ± 0.11 [A]
PEC + GLE 0.2%	0.37 ± 0.01 [A,B]	0.04 ± 0.00 [A]	3.45 ± 0.79 [A]
Viscofan (NDX) **	3.71 ± 0.16	0.03 ± 0.01	0.08 ± 0.01

* Data from Al-Asmar et al. [19]; ** data from Porta et al. [67]. The values that are significantly different from the control are indicated by "[A]" ($p < 0.05$); the values indicated by "[B]" are significantly different ($p < 0.05$) from the films obtained with the same material at different concentrations.

3.7. Film Application as a Soluble Sachet for Chicken Stock Powder

One of the important issues related to this type of research is the application of the obtained films that are prepared based on pectin—a well-known material made from citrus peels. OLE and GLE are both natural compounds that have antioxidant and antimicrobial properties. When these extracts are incorporated into pectin films, they improve the film's antioxidant and promising barrier properties, which will potentially extend its shelf-life. This is an active area of research, and more work is needed to understand it fully.

In this study, the obtained materials were kept for up to one year at room temperature, and the films' properties were evaluated the obtained results were extremely promising. Therefore, pectin films that contain OLE or GLE are promising for use as soluble sachets for chicken stock powder. A soluble sachet is a type of packaging that is made from a material that dissolves in water. This allows the food product to be easily poured out of the sachet, without the need for a separate container. Soluble sachets are a convenient and environmentally friendly alternative to traditional packaging materials such as plastic and foil. Pectin films containing OLE or GLE are sustainable and innovative materials that can be used as soluble bags for chicken stock powder (Figure 5). These films have the potential to improve the quality and shelf-life of chicken stock powder while also reducing environmental impacts. As shown in Figure 5, the pectin sachets were prepared and heat-sealed using the kitchen sealer machine, and chicken stock powder was transferred from commercial plastic foil to our sachets and then added to boiling water to demonstrate their solubility. The sachets were completely dissolved within a few seconds in boiling water, and the chicken stock was ready to use. As a result, pectin films that contain GLE or OLE are promising and warrant further research.

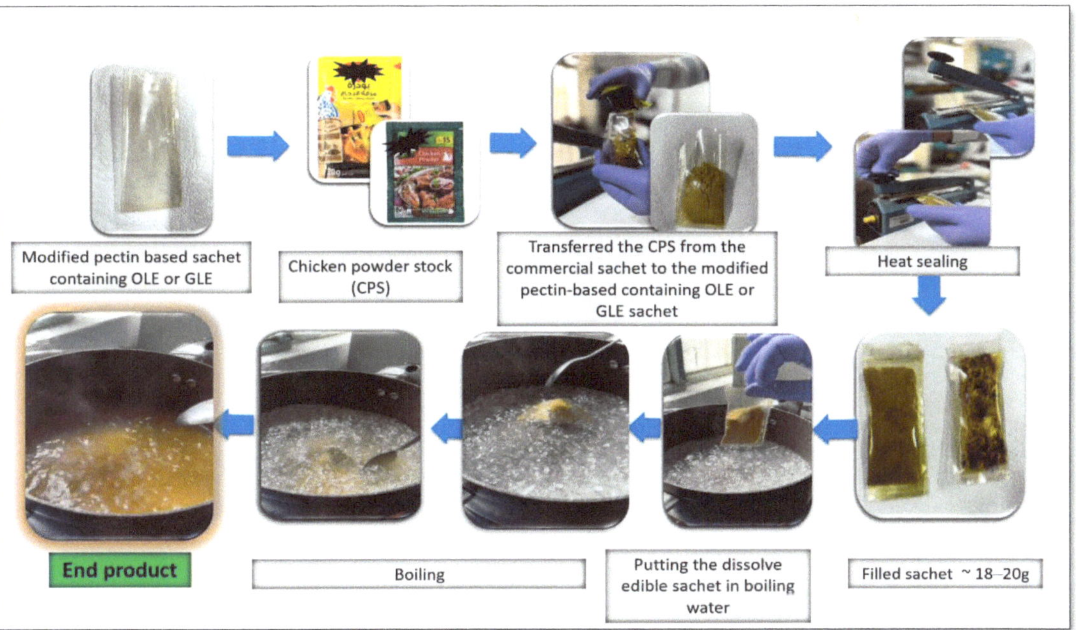

Figure 5. Scheme demonstrating the preparation and use of the pectin sachets containing OLE or GLE filled with chicken stock powder and dissolved in boiling water.

4. Conclusions

In conclusion, natural antioxidants and antimicrobials are becoming increasingly important today due to the worldwide strategy to reduce the use of synthetic additives in our foods. In this work, we successfully functionalized pectin films by incorporating OLE or GLE. The films produced in this study showed lower water vapor permeability than the commercial casing used for processed meat. Additionally, the results indicated a higher antioxidant level compared to the pectin film obtained without OLE or GLE. This helps to reduce the oxidation reactions that can occur during food distribution and storage. This suggests that the obtained films had a significant impact on extending the food's shelf-life. Finally, the films were used to produce soluble sachets filled with chicken stock powder, which was used to prepare soup.

Author Contributions: Conceptualization, M.S., F.A.-R. and A.A.-A.; methodology, F.A.-R., M.F., A.A.-A., and D.Y.; software, M.S.; validation, L.M., M.S. and F.A.-R.; formal analysis, D.Y., A.A.-A. and M.S.; investigation, M.S. and F.A.-R.; resources, F.A.-R., M.S. and L.M.; data curation, M.S. and M.F.; writing—original draft preparation, M.S. and L.M.; writing—review and editing, F.A.-R., A.A.-A., D.Y. and M.F.; visualization, M.S., D.Y., and A.A.-A.; supervision, M.S. and L.M.; project administration, A.A.-A.; funding acquisition, M.S. and L.M. All authors have read and agreed to the published version of the manuscript.

Funding: This research received no external funding.

Institutional Review Board Statement: Not applicable.

Informed Consent Statement: Not applicable.

Data Availability Statement: The original data of this work are available upon reasonable request.

Acknowledgments: We would like to thank the lab instructors at the Faculty of Agriculture and Vet Medicine at An-Najah National University for their help and support.

Conflicts of Interest: The authors declare no conflict of interest.

References

1. Porta, R.; Sabbah, M.; Di Pierro, P. Bio-based materials for packaging. *Int. J. Mol. Sci.* **2022**, *23*, 3611. [CrossRef] [PubMed]
2. Sing, N.; Ogunseitan, O.A.; Wong, M.H.; Tang, Y. Sustainable materials alternative to petrochemical plastics pollution: A review analysis. *Sustain. Horiz.* **2022**, *2*, 100016. [CrossRef]
3. Bauer, F.; Nielsen, T.D.; Nilsson, L.J.; Palm, E.; Ericsson, K.; Frane, A.; Cullen, J. Plastics and climate change—Breaking carbon lock-ins through three mitigation pathways. *One Earth* **2022**, *5*, 361–376. [CrossRef]
4. Plastics Europe. *Plastics—The Facts 2020*; Plastics Europe: Brussels, Belgium, 2020.
5. Dell'Olmo, E.; Gaglione, R.; Sabbah, M.; Schibeci, M.; Cesaro, A.; Di Girolamo, R.; Arciello, A. Host defense peptides identified in human apolipoprotein B as novel food biopreservatives and active coating components. *Food Microbiol.* **2021**, *99*, 103804. [CrossRef] [PubMed]
6. Gerschenson, L.N.; Fissore, E.N.; Rojas, A.M.; Idrovo Encalada, A.M.; Zukowski, E.F.; Higuera Coelho, R.A. Pectins obtained by ultrasound from agroindustrial by-products. *Food Hydrocoll.* **2021**, *118*, 106799. [CrossRef]
7. Rigueto, C.V.T.; Rosseto, M.; Alessandretti, I.; de Oliveira, R.; Wohlmuth, D.A.R.; Menezes, J.F.; Loss, R.A.; Dettmer, A.; Pizzutti, I.R. Gelatin films from wastes: A review of production, characterization, and application trends in food preservation and agriculture. *Food Res. Int.* **2022**, *162*, 112114. [CrossRef]
8. Yaseen, D.; Sabbah, M.; Al-Asmar, A.; Altamimi, M.; Famiglietti, M.; Giosafatto, C.V.L.; Mariniello, L. Functionality of films from *Nigella sativa* defatted seed cake proteins plasticized with grape juice: Use in wrapping sweet cherries. *Coatings* **2021**, *11*, 1383. [CrossRef]
9. Villanueva, V.; Valdés, F.; Zúñiga, R.N.; Villamizar-Sarmiento, M.G.; Soto-Bustamante, E.; Romero-Hasler, P.; Riveros, A.L.; Tapia, J.; Lisoni, J.; Oyarzun-Ampuero, F.; et al. Development of biodegradable and vermicompostable films based on alginate and waste eggshells. *Food Hydrocoll.* **2023**, *142*, 108813. [CrossRef]
10. Lahiri, A.; Daniel, S.; Kanthapazham, R.; Vanaraj, R.; Thambidurai, A.; Peter, L.S. A critical review on food waste management for the production of materials and biofuel. *J. Hazard. Mater. Adv.* **2023**, *10*, 100266. [CrossRef]
11. Gerna, S.; D'Incecco, P.; Limbo, S.; Sindaco, M.; Pellegrino, L. Strategies for Exploiting Milk Protein Properties in Making Films and Coatings for Food Packaging: A Review. *Foods* **2023**, *12*, 1271. [CrossRef]
12. Righetti, G.I.C.; Nasti, R.; Beretta, G.; Levi, M.; Turri, S.; Suriano, R. Unveiling the hidden properties of tomato peels: Cutin ester derivatives as bio-based plasticizers for polylactic acid. *Polymers* **2023**, *15*, 1848. [CrossRef]
13. Cai, Z.; Haque, A.N.M.A.; Dhandapani, R.; Naebe, M. Sustainable Cotton Gin Waste/Polycaprolactone Bio-Plastic with Adjustable Biodegradation Rate: Scale-Up Production through Compression Moulding. *Polymers* **2023**, *15*, 1992. [CrossRef] [PubMed]
14. Restaino, O.F.; Hejazi, S.; Zannini, D.; Giosafatto, C.V.L.; Di Pierro, P.; Cassese, E.; D'ambrosio, S.; Santagata, G.; Schiraldi, C.; Porta, R. Exploiting Potential Biotechnological Applications of Poly-γ-glutamic Acid Low Molecular Weight Fractions Obtained by Membrane-Based Ultra-Filtration. *Polymers* **2022**, *14*, 1190. [CrossRef]
15. Avila, L.B.; Schnorr, C.; Silva, L.F.O.; Morais, M.M.; Moraes, C.C.; da Rosa, G.S.; Dotto, G.L.; Lima, É.C.; Naushad, M. Trends in Bioactive Multilayer Films: Perspectives in the Use of Polysaccharides, Proteins, and Carbohydrates with Natural Additives for Application in Food Packaging. *Foods* **2023**, *12*, 1692. [CrossRef] [PubMed]
16. Restaino, O.F.; Giosafatto, C.V.L.; Mirpoor, S.F.; Cammarota, M.; Hejazi, S.; Mariniello, L.; Schiraldi, C.; Porta, R. Sustainable exploitation of posidonia oceanica sea Balls (*Egagropili*): A Review. *Int. J. Mol. Sci.* **2023**, *24*, 7301. [CrossRef]
17. Al-Asmar, A.; Giosafatto, C.V.L.; Sabbah, M.; Mariniello, L. Hydrocolloid-based coatings with nanoparticles and transglutaminase crosslinker as innovative strategy to produce healthier fried kobbah. *Foods* **2020**, *9*, 698. [CrossRef] [PubMed]
18. Giosafatto, C.V.L.; Fusco, A.; Al-Asmar, A.; Mariniello, L. Microbial Transglutaminase as a tool to improve the features of hydrocolloid-based bioplastics. *Int. J. Mol. Sci.* **2020**, *21*, 3656. [CrossRef]
19. Al-Asmar, A.; Giosafatto, C.V.L.; Sabbah, M.; Sanchez, A.; Villalonga Santana, R.; Mariniello, L. Effect of mesoporous silica nanoparticles on the physicochemical properties of pectin packaging material for strawberry wrapping. *Nanomaterials* **2020**, *10*, 52. [CrossRef]
20. Góral, D.; Góral-Kowalczyk, M. Application of metal nanoparticles for production of self-sterilizing coatings. *Coatings* **2022**, *12*, 480. [CrossRef]
21. Krivorotova, T.; Cirkovas, A.; Maciulyte, S.; Staneviciene, R.; Budriene, S.; Serviene, E.; Sereikaite, J. Nisin-loaded pectin nanoparticles for food preservation. *Food Hydrocoll.* **2016**, *54*, 49–56. [CrossRef]
22. Valdés, A.; Burgos, N.; Jiménez, A.; Garrigós, M.C. Natural pectin polysaccharides as edible coatings. *Coatings* **2015**, *5*, 865–886. [CrossRef]

23. Sabbah, M.; Di Pierro, P.; Cammarota, M.; Dell'Olmo, E.; Arciello, A.; Porta, R. Development and properties of new chitosan-based films plasticized with spermidine and/or glycerol. *Food Hydrocoll.* **2019**, *87*, 245–252. [CrossRef]

24. Porta, R.; Di Pierro, P.; Sabbah, M.; Regalado-Gonzales, C.; Mariniello, L.; Kadivar, M.; Arabestani, A. Blend films of pectin and bitter vetch (*Vicia ervilia*) proteins: Properties and effect of transglutaminase. *Innov. Food Sci. Emerg. Technol.* **2016**, *36*, 245–251. [CrossRef]

25. Sabbah, M.; Di Pierro, P.; Giosafatto, C.V.L.; Esposito, M.; Mariniello, L.; Regalado-Gonzales, C.; Porta, R. Plasticizing effects of polyamines in protein-based films. *Int. J. Mol. Sci.* **2017**, *18*, 1026. [CrossRef]

26. Mierczyńska, J.; Cybulska, J.; Pieczywek, P.; Zdunek, A. Effect of storage on rheology of water-soluble, chelate-soluble and diluted alkali-soluble pectin in carrot cell walls. *Food Bioprocess Technol.* **2015**, *8*, 171–180. [CrossRef]

27. Chandel, V.; Biswas, D.; Roy, S.; Vaidya, D.; Verma, A.; Gupta, A. Current advancements in pectin: Extraction, properties and multifunctional applications. *Foods* **2022**, *11*, 2683. [CrossRef] [PubMed]

28. Giosafatto, C.V.L.; Sabbah, M.; Al-Asmar, A.; Esposito, M.; Sanchez, A.; Villalonga Santana, R.; Cammarota, M.; Mariniello, L.; Di Pierro, P.; Porta, R. Effect of mesoporous silica nanoparticles on glycerol-plasticized anionic and cationic polysaccharide edible films. *Coatings* **2019**, *9*, 172. [CrossRef]

29. Sengar, A.S.; Rawson, A.; Muthiah, M.; Kumar Kalakandan, S. Comparison of different ultrasound assisted extraction techniques for pectin from tomato processing waste. *Ultrason. Sonochem.* **2019**, *61*, 104812. [CrossRef]

30. Torres-León, C.; Ramírez, N.; Londoño, L.; Martinez, G.A.; Díaz, R.; Navarro, V.; Alvarez-Pérez, O.B.; Picazo, B.; Villarreal, M.; Ascacio, J.; et al. Food Waste and Byproducts: An Opportunity to Minimize Malnutrition and Hunger in Developing Countries. *Front. Sustain. Food Syst.* **2018**, *2*, 52. [CrossRef]

31. Voragen, A.G.J.; Coenen, G.J.; Verhoef, R.P.; Schols, H.A. Pectin, a versatile polysaccharide present in plant cell walls. *Struct. Chem.* **2009**, *20*, 263. [CrossRef]

32. Tanhatan Naseri, A.; Thibault, J.-F.; Ralet-Renard, M.-C.; Tanhatan Nasseri, A.; Ralet, M.-C. Citrus Pectin: Structure and Application in Acid Dairy Drinks. In *Tree and Forestry Science and Biotechnology*; Global Science Books: Carrollton, GA, USA, 2008.

33. Brouns, F.; Theuwissen, E.; Adam, A.; Bell, M.; Berger, A.; Mensink, R.P. Cholesterol-lowering properties of different pectin types in mildly hyper-cholesterolemic men and women. *Eur. J. Clin. Nutr.* **2012**, *66*, 591–599. [CrossRef]

34. Nastasi, J.R.; Kontogiorgos, V.; Daygon, V.D.; Fitzgerald, M.A. Pectin-based films and coatings with plant extracts as natural preservatives: A systematic review. *Trends Food Sci. Technol.* **2022**, *120*, 193–211. [CrossRef]

35. Wu, W.; Wu, Y.; Lin, Y.; Shao, P. Facile fabrication of multifunctional citrus pectin aerogel fortified with cellulose nanofiber as controlled packaging of edible fungi. *Food Chem.* **2022**, *374*, 131763. [CrossRef] [PubMed]

36. Moradi, L.T.; Sharifan, A.; Larijani, K. The Effect of multilayered chitosan–pectin–mentha piperita and lemon essential oil on oxidation effects and quality of rainbow trout fillet (*Oncorhynchus Mykiss*) during refrigeration at 4 ± 1 °C storage. *Iran. J. Fish. Sci.* **2020**, *19*, 2544–2559.

37. Sumonsiri, N.; Danpongprasert, W.; Thaidech, K. Comparison of sweet orange (*Citrus sinensis*) and lemon (*Citrus limonum*) essential oils on qualities of fresh-cut apples during storage. *Chem. Chem. Eng. Biotechnol. Food Ind.* **2020**, *21*, 47–57.

38. Xiong, Y.; Li, S.; Warner, R.D.; Fang, Z. Effect of oregano essential oil and resveratrol nanoemulsion loaded pectin edible coating on the preservation of pork loin in modified atmosphere packaging. *Food Control.* **2020**, *114*, 107226. [CrossRef]

39. Nisar, T.; Wang, Z.; Yang, X.; Tian, Y.; Iqbal, M.; Guo, Y. Characterization of citrus pectin films integrated with clove bud essential oil: Physical, thermal, barrier, antioxidant and antibacterial properties. *Int. J. Biol. Macromol.* **2018**, *106*, 670–680. [CrossRef] [PubMed]

40. Liu, H.; Li, J.; Zhu, D.; Wang, Y.; Zhao, Y.; Li, J. Preparation of Soy Protein Isolate (SPI)-Pectin Complex Film Containing Cinnamon Oil and Its Effects on Microbial Growth of Dehydrated Soybean Curd (Dry Tofu). *J. Food Process. Preserv.* **2014**, *38*, 1371–1376. [CrossRef]

41. Rodsamran, P.; Sothornvit, R. Lime peel pectin integrated with coconut water and lime peel extract as a new bioactive film sachet to retard soybean oil oxidation. *Food Hydrocoll.* **2019**, *97*, 105173. [CrossRef]

42. Şahin, S.; Bilgin, M. Olive tree (*Olea europaea* L.) leaf as a waste by-product of table olive and olive oil industry: A review. *J. Sci. Food Agric.* **2018**, *98*, 1271–1279. [CrossRef]

43. Famiglietti, M.; Savastano, A.; Gaglione, R.; Arciello, A.; Naviglio, D.; Mariniello, L. Edible films made of dried olive leaf extract and chitosan: Characterization and applications. *Foods* **2022**, *11*, 2078. [CrossRef]

44. Hayes, J.E.; Stepanyan, V.; Allen, P.; O'Grady, M.N.; Kerry, J.P. Evaluation of the effects of selected phytochemicals on quality indices and sensorial properties of raw and cooked pork stored in different packaging systems. *Meat Sci.* **2010**, *85*, 289–296. [CrossRef]

45. Hayes, J.E.; Stepanyan, V.; Allen, P.; O'Grady, M.N.; Kerry, J.P. Evaluation of the effects of selected plant-derived nutraceuticals on the quality and shelf-life stability of raw and cooked pork sausages. *Meat Sci.* **2011**, *44*, 164–172. [CrossRef]

46. Nunes, M.A.; Pimentel, F.B.; Costa, A.S.; Alves, R.C.; Oliveira, M.B.P. Olive by-products for functional and food applications: Challenging opportunities to face environmental constraints. *Innov. Food Sci. Emerg. Technol.* **2016**, *35*, 139–148. [CrossRef]

47. Elsayed, N.; Hasanin, M.S.; Abdelraof, M. Utilization of olive leaves extract coating incorporated with zinc/selenium oxide nanocomposite to improve the postharvest quality of green beans pods. *Bioact. Carbohydr. Diet. Fibre* **2022**, *28*, 100333. [CrossRef]

48. Naseer, S.; Hussain, S.; Naeem, N.; Pervaiz, M.; Rahman, M. The phytochemistry and medicinal value of *Psidium guajava* (guava). *Clin. Phytosci.* **2018**, *4*, 32. [CrossRef]

49. Ruksiriwanich, W.; Khantham, C.; Muangsanguan, A.; Phimolsiripol, Y.; Barba, F.J.; Sringarm, K.; Rachtanapun, P.; Jantanasakulwong, K.; Jantrawut, P.; Chittasupho, C.; et al. Guava (*Psidium guajava L.*) leaf extract as bioactive substances for anti-androgen and antioxidant Activities. *Plants* **2022**, *11*, 3514. [CrossRef] [PubMed]

50. Nantitanon, W.; Yotsawimonwat, S.; Okonogi, S. Factors influencing antioxidant activities and total phenolic content of guava leaf extract. *LWT Food Sci. Technol.* **2010**, *43*, 1095–1103. [CrossRef]

51. Annegowda, H.V.; Anwar, L.N.; Mordi, M.N.; Ramanathan, S.; Mansor, S.M. Influence of sonication on the phenolic content and antioxidant activity of *Terminalia catappa L.* leaves. *Pharm. Res.* **2010**, *2*, 368–373. [CrossRef]

52. Tonyali, B.; Cikrikci, S.; Oztop, M.H. Physicochemical and microstructural characterization of gum tragacanth added whey protein based films. *Food Res. Int.* **2018**, *105*, 1–9. [CrossRef] [PubMed]

53. Galus, S.; Lenart, A. Development and characterization of composite edible films based on sodium alginate and pectin. *J. Food Eng.* **2013**, *115*, 459–465. [CrossRef]

54. Papadakis, S.E.; Abdul-Malek, S.; Kamdem, R.E.; Yam, K.L. A versatile and inexpensive technique for measuring colour of foods. *Food Technol.* **2000**, *54*, 48–51.

55. *ASTM F1249-13*; Standard Test Method for Water Vapor Transmission Rate through Plastic Film and Sheeting Using a Modulated Infrared Sensor. ASTM International: West Conshohocken, PA, USA, 2013.

56. *ASTM D3985-05*; Standard Test Method for Oxygen Gas Transmission Rate through Plastic Film and Sheeting Using a Colorimetric Sensor. ASTM International: West Conshohocken, PA, USA, 2010.

57. *ASTM F2476-05*; Standard Test Method for Carbon Dioxide Gas Transmission Rate through Barrier Materials Using an Infrared Detector. ASTM International: West Conshohocken, PA, USA, 2005.

58. Chou, M.; Osako, K.; Lee, T.; Wang, M.; Lu, W.; Wu, W.; Huang, P.; Li, P.; Ho, J. Characterization and antibacterial properties of fish skin gelatin/guava leaf extract bio-composited films incorporated with catechin. *LWT Food Sci. Technol.* **2023**, *178*, 114568. [CrossRef]

59. Du, H.; Liu, C.; Unsalan, O.; Altunayar-Unsalan, C.; Xiong, S.; Manyande, A.; Chen, H. Development and characterization of fish myofibrillar protein/chitosan/ rosemary extract composite edible films and the improvement of lipid oxidation stability during the grass carp fillets storage. *Int. J. Biol. Macromol.* **2021**, *184*, 463–475. [CrossRef]

60. Mondal, K.; Bhattacharjee, S.K.; Mudenur, C.; Ghosh, T.; Goud, V.V.; Katiyar, V. Development of antioxidant-rich edible active films and coatings incorporated with de-oiled ethanolic green algae extract: A candidate for prolonging the shelf life of fresh produce. *RSC Adv.* **2022**, *12*, 13295–13313. [CrossRef] [PubMed]

61. Tongnuanchan, P.; Benjakul, S.; Prodpran, T. Properties and antioxidant activity of fish skin gelatin film incorporated with citrus essential oils. *Food Chem.* **2012**, *134*, 1571–1579. [CrossRef]

62. Luo, Y.; Liu, H.; Yang, S.; Zeng, J.; Wu, Z. Sodium alginate-based green packaging films functionalized by guava leaf extracts and their bioactivities. *Materials* **2019**, *12*, 2923. [CrossRef] [PubMed]

63. Dou, L.; Li, B.; Zhang, K.; Chu, X.; Hou, H. Physical properties and antioxidant activity of gelatin-sodium alginate edible films with tea polyphenols. *Int. J. Biol. Macromol.* **2018**, *118*, 1377–1383. [CrossRef]

64. Albertos, I.; Avena-Bustillos, R.J.; Martín-Diana, A.B.; Du, W.; Rico, D.; McHugh, T.H. Antimicrobial olive leaf gelatin films for enhancing the quality of cold-smoked Salmon. *Food Packag. Shelf Life* **2017**, *13*, 49–55. [CrossRef]

65. McHugh, T.H.; Avena-Bustillos, R.J.; Krochta, J.M. Hydrophilic edible films: Modified procedure for water vapor permeability and explanation of thickness effects. *J. Food Sci.* **1993**, *58*, 899–903. [CrossRef]

66. Wu, H.; Lei, Y.; Zhu, R.; Zhao, M.; Lu, J.; Xiao, D.; Jiao, C.; Zhang, Z.; Shen, G.; Li, S. Preparation and characterization of bioactive edible packaging films based on pomelo peel flours incorporating tea polyphenol. *Food Hydrocoll.* **2019**, *90*, 41–49. [CrossRef]

67. Porta, R.; Di Pierro, P.; Roviello, V.; Sabbah, M. Tuning the functional properties of bitter vetch (*Vicia ervilia*) protein films grafted with spermidine. *Int. J. Mol. Sci.* **2017**, *18*, 2658. [CrossRef] [PubMed]

Article

Production and Preliminary Characterization of Linseed Mucilage-Based Films Loaded with Cardamom (*Elettaria cardamomum*) and Copaiba (*Copaifera officinalis*)

Mayra Z. Treviño-Garza [1], Ana Karen Saldívar-Vázquez [2], Sonia Martha López-Villarreal [2], María del Refugio Lara-Banda [3], Joel Horacio Elizondo-Luevano [4], Abelardo Chávez-Montes [4], Juan Gabriel Báez-González [1,*] and Osvelia Esmeralda Rodríguez-Luis [2,*]

[1] Departamento de Alimentos, Facultad de Ciencias Biológicas, Universidad Autónoma de Nuevo León (UANL), San Nicolás de los Garza 66455, NL, Mexico; mayra.trevinogrz@uanl.edu.mx

[2] Departamento de Microbiología, Facultad de Odontología, Universidad Autónoma de Nuevo León (UANL), Monterrey 64460, NL, Mexico; ana.saldivarvz@uanl.edu.mx (A.K.S.-V.); sonia.lopezvl@uanl.edu.mx (S.M.L.-V.)

[3] Centro de Investigación e Innovación en Ingeniería Aeronáutica, Facultad de Ingeniería Mecánica y Eléctrica, Universidad Autónoma de Nuevo León (UANL), San Nicolás de los Garza 66455, NL, Mexico; maria.laraba@uanl.edu.mx

[4] Departamento de Química, Facultad de Ciencias Biológicas, Universidad Autónoma de Nuevo León (UANL), San Nicolás de los Garza 66455, NL, Mexico; joel.elizondolv@uanl.edu.mx (J.H.E.-L.); abelardo.chavezmn@uanl.edu.mx (A.C.-M.)

* Correspondence: juan.baezgn@uanl.edu.mx (J.G.B.-G.); osvelia.rodriguezls@uanl.edu.mx (O.E.R.-L.); Tel.: +52-8183294000 (ext. 3654) (J.G.B.-G.); +52-8183294230 (ext. 3117) (O.E.R.-L.)

Citation: Treviño-Garza, M.Z.; Saldívar-Vázquez, A.K.; López-Villarreal, S.M.; Lara-Banda, M.d.R.; Elizondo-Luevano, J.H.; Chávez-Montes, A.; Báez-González, J.G.; Rodríguez-Luis, O.E. Production and Preliminary Characterization of Linseed Mucilage-Based Films Loaded with Cardamom (*Elettaria cardamomum*) and Copaiba (*Copaifera officinalis*). *Coatings* **2023**, *13*, 1574. https://doi.org/10.3390/coatings13091574

Academic Editors: Jun Mei, Jing Xie and Biao Zhang

Received: 17 August 2023
Revised: 4 September 2023
Accepted: 6 September 2023
Published: 9 September 2023

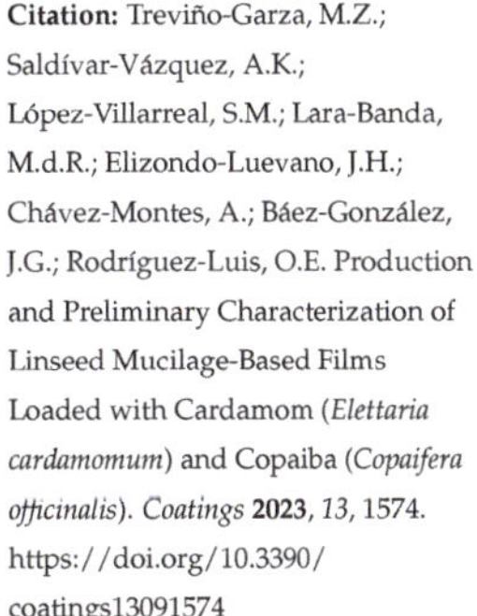

Abstract: In this research, developed linseed mucilage (M)-based films loaded with *E. cardamom* (MCA), *C. officinalis* (MCO), and co-loaded with both compounds (MCACO) were evaluated. The incorporation of the active compounds modified the color (redness–greenness, and yellowness); however, the thickness remained constant in all treatments (0.0042–0.0052 mm). In addition, the solubilization time of the films (in artificial saliva) to release the active compounds fluctuates between 9 and 12 min. Furthermore, the incorporation of bioactive compounds increased the total phenolic content and antioxidant activity (DPPH and ABTS, respectively), mainly in MCA (inhibition of 81.99 and 95.80%, respectively) and MCACO (inhibition of 47.15% and 39.73%, respectively). In addition, the incorporation of these compounds also decreased the hardness (39.50%–70.81%), deformation (49.16%–78.30%), and fracturability (39.58%–82.95%). On the other hand, it did not modify the adhesiveness, except in MCO. Moreover, SEM micrographs showed a more homogeneous structure in the MCO films among the films that contained CA in the formulation (heterogeneous structure with the presence of protuberances). Finally, due to the previously reported pharmacological properties of *E. cardamomun* and *C. officinalis*, the films developed in this study could have an application as a wound dressing in dentistry.

Keywords: films; linseed mucilage; cardamom; copaiba; physicochemical properties; antioxidant activity; texture analysis; dentistry

1. Introduction

Nowadays, there is a growing global trend towards developing new natural products, such as therapeutic options for the prevention and treatment of various diseases [1,2]. Due to this, numerous researchers have determined the properties of multiple natural products with active properties of interest to the medical area [3]. In addition, recent research has focused on studying phytotherapeutic products of interest to dentistry [4–6].

Cardamom (*Elettaria cardamomum*; CA) is a perennial, aromatic, herbaceous plant belonging to the *Zingiberaceae* family, whose use is culinary, domestic, and medicinal. It is a species native to southwest Asia (India, Sri Lanka, Malaysia, and Indonesia), Tanzania, and Guatemala, which has been used since ancient times due to its wide variety of antimicrobial, antifungal, antioxidant, analgesic, anti-inflammatory, and anticancer properties associated with the presence of components such as phenols, tannins, terpenoids, flavonoids, and sterols, among others [7–9]. Further, in dentistry, it has been reported that CA possesses anti-bad breath, anticaries, antiseptic, and antimicrobial properties [10]. Multiple studies have shown that CA has been effective against microorganisms that cause infections and dental cavities [11], such as *Streptococcus mutans, Candida albicans*, and *Lactobacillus casei* [12–15]. Additionally, it has been shown that the use of CA in combination with other extracts (black pepper/black cumin/cardamom and black pepper/black cumin/cardamom/cinnamon) has a high antibacterial susceptibility against microorganisms from oral isolates [16]. Finally, some reports indicated the potential therapeutic benefits of CA for periodontal infections [17].

Moreover, *C. officinalis* is a plant belonging to the *Fabaceae* family, distributed mainly in Latin America (Brazil, Bolivia, Colombia, Peru, and Venezuela) and West Africa [18–20]. The main uses of CO include medical applications, nutrition, cosmetics, fuels, and wood, among others [1]. CO has been widely used in folk medicine through topical and oral administration due to its antimicrobial, antifungal, antiseptic, anti-inflammatory, and antioxidant properties, among others [3,6,20]. Regarding dental applications, various studies have demonstrated the antibacterial activity of CO against *Streptococcus* spp., *S. mutans* [21–23], and other oral pathogens [4]. Likewise, it has been recognized that CO has anti-inflammatory and healing activity in the oral cavity [6] and for acting as a dentin biomodifier [19]. Finally, some studies in animal models report that CO oleoresin is a safe and effective alternative therapy for inflammation and tissue repair in oral wounds [24].

On the other hand, biopolymer-based delivery systems (e.g., nano-composites, microcapsules, emulsions, hydrogels, films, or membranes, among others) have been widely used for the incorporation of bioactive compounds [6,9,20,25–28]. Additionally, membranes or biodegradable films have been designed as controlled-release systems (e.g., would dressing) for various applications in the medical area [6,20,27,29,30]. A biodegradable film is a multi-component system made from polymers (e.g., polysaccharides, lipids, and proteins), which contain in their polymer matrix a plasticizing agent (e.g., glycerol, polyethylene glycol, and sorbitol, among others) and some active compounds of interest. The most widely used polymers for the development of these biomaterials are polysaccharides, such as cellulose and its derivatives, chitosan, pectin, alginate, and recently the use of plant mucilages, among others [31].

Linseed mucilage (*Linum usitatissimum*) is a natural polysaccharide composed of an acidic fraction of pectic-like material (L-galactose, L-fucose, L-rhamnose, and D-galacturonic acid) and a neutral fraction (arabinoxylan, D-galactose, L-arabinose, and D-xylose) [32]. This polymer has been used to produce biodegradable films in various applications, either as an individual polymer [33] or in combination with other polymers, such as chitosan [34,35], pectin [36,37], and polyvinyl alcohol [38]. Likewise, this mucilage has been used to produce active biodegradable films incorporated with carvacrol [39] and *Hamamelis virginiana* extract [31].

Finally, various studies have focused on the production of membranes or active films based on chitosan [20,27,30,40], poly (L-co-D, L lactic acid), and poly (lactic acid)/poly (vinyl pyrrolidone) [6] incorporated with CO for applications in the medical area, mainly as wound dressing. Furthermore, films based on mung bean protein-apple pectin [41] and soy protein isolate loaded with CA [42] have been developed, mainly for applications in the food area.

However, as far as we know, there are no reports of linseed mucilage-based films incorporated with CO and CA for dental applications. Therefore, in this context, this research aimed to produce and characterize films based on linseed mucilage (M) loaded with *E. cardamomun* (MCA), *C. officinalis* (MCO), and co-loaded with both compounds (MCACO) for future application in the dentistry area.

2. Materials and Methods

2.1. Vegetal Material

Linseed was purchased at a local supermarket (Mty NL, Mexico). On the other hand, the seeds of *E. cardamomum* (SKU: 209740-01) were purchased from Starwest Botanicals (Sacramento, CA, USA). Finally, *C. officinalis* essential oil was purchased from Young Living Essential Oils, LC Company (Lehi, UT, USA).

2.2. Obtaining the Extract of E. cardamomum

The extract was obtained by cold maceration without stirring [43]. Vegetable material (100 g) was placed in a flask containing 400 mL of absolute ethanol (99.5%, CTR Scientific, Mty NL, Mexico; extraction solvent). The sample was left to rest for a period of 24 h, filtered (Whatman™ qualitative filter paper, grade 1), and then placed in a rotary evaporator (IKA RV10, Hayward, CA, USA). Finally, the yield of the extract (% Yield = [final weight of dry extract/initial weight of the plant] × 100) was determined, and the product was stored in a dark container under refrigerated conditions (4.0 ± 2.0 °C) until its later use.

2.3. Linseed Mucilage Extraction

The linseed mucilage was extracted according to the methodology reported by Treviño-Garza et al. [31]. Briefly, the flaxseeds (300 g) were placed in distilled water (1000 mL) and kept under constant agitation (250 rpm, 25 ± 2 °C for 2 h). Next, the linseeds were removed with a strainer. Subsequently, ethanol (96%, 2000 mL; CTR Scientific, Mty N.L., Mexico) was added to the resulting aqueous suspension for mucilage precipitation. Finally, the precipitated mucilage was recovered with the help of a strainer and subsequently dried (55 ± 2 °C for 24 h), pulverized, and stored until later use.

2.4. Phytochemical Tests of Plant Material

The phytochemical profile of plant material was carried out according to the methodology reported by Guillén-Meléndez et al. [44] and Rodríguez-Garza et al. [45]. In these determinations, the presence (+) or absence (−) of compound groups was analyzed using the following tests: Lieberman–Buchard (sterols, triterpenes), Shinoda (flavonoids), Baljet (sesquiterpene-lactones), sulfuric acid (quinones), ferric chloride (tannins), potassium permanganate (unsaturations), 2,4-dinitrophenylhydrazine (carbonyl group), Dragendorff (alkaloids), sodium hydroxide (coumarins, lactones), Molish (carbohydrates), and foam test (saponins).

2.5. Production and Preliminary Characterization of Films Based on Linseed Mucilage Loaded with CA and CO

Four film-forming solutions based on linseed mucilage and loaded with *E. cardamom* (MCA), *C. officinalis* (MCO), and co-loaded with both compounds (MCACO) were designed. Glycerol (99.50% purity, CTR Scientific, Mty N.L., Mexico) was used as a plasticizer and tween 20 (polyoxyethylene-20-sorbitan monolaurate; Sigma Aldrich, St. Louis, MO, USA) as a surfactant agent. The formulations were prepared with the concentrations indicated in Table 1 in distilled water and by constant mechanical stirring (500 rpm, 25 ± 2 °C) until complete dissolution of all the components was achieved. Subsequently, the films were prepared by the casting method; the film-forming solutions (~10 mL) were placed in plastic boxes (60 mm × 15 mm) to be later subjected to a drying process (55 ± 2 °C for 24 h). Finally, the films were manually retrieved and characterized, as indicated in the following sections.

Table 1. Film-forming solutions based on linseed mucilage (M) and incorporated with *E. cardamomum* (CA) and *C. officinalis* (CO).

Formulation	Linseed Mucilage (%)	Glycerol (%)	Tween 20 (%)	*E. cardamomum* (%)	*C. officinalis* (%)
M	3.0	0.5	1.0	-	-
MCA	3.0	0.5	1.0	2.0	0.0
MCO	3.0	0.5	1.0	-	2.0
MCACO	3.0	0.5	1.0	1.0	1.0

2.5.1. Color Determination

The color analysis of the films was performed using a colorimeter (Hunterlab model, Colorflex® EZ, Reston, VA, USA), based on the CIE L*a*b* color space (CIELAB; L* coordinate or lightness, 0 black—100 white; coordinate a*, (−) green and (+) red; coordinate b*, (−) blue and (+) yellow) [27,42].

2.5.2. Thickness Measurement

Thickness determinations were carried out using a digital micrometer (Quickmike Model Mitutoyo, Kawasaki, Japan); measurements were made at three random positions on each film, and the results were expressed in millimeters (mm) [40,42].

2.5.3. Solubility Tests in Artificial Saliva

For the solubility analysis, the films were cut into discs (20 mm in diameter) and placed in beakers containing 50 mL of artificial saliva (Viarden brand, $25 \pm 2\ ^{\circ}$C). Subsequently, the solubilization time of the complete films (minutes, min) was determined [31].

2.5.4. Antioxidant Activity (DPPH, 1,1-Diphenyl-2-picrylhydrazyl, and ABTS, 2,2′-Azino-bis-3-ethylbenzothiazoline-6-sulfonic Acid) and Total Phenolic Content

The antioxidant activity was calculated by using the DPPH and ABTS free radical (Sigma Aldrich, St. Louis, MO, USA) methods based on previous studies with some modifications [31,46,47]. For sample preparation, films (1 cm × 1 cm) were cut, weighed, and placed in 1 mL of distilled water and vortexed until complete solubilization. Later, the samples were centrifuged (Spectrafuge 6C Labnet International, Inc., Edison, NJ, USA) at 6500 rpm for 10 min to recover the supernatant.

For the DPPH method, the sample (0.75 mL) was placed in conical tubes containing 2.25 mL of the DPPH ethanolic solution (0.039 mg/mL, absorbance = 1.0 ± 0.005). The samples were incubated at 25 °C in the dark for a period of 90 min to allow for the reaction. Subsequently, the samples were centrifuged (Spectrafuge 6C, Labnet International, Inc., 6500 rpm for 5–10 min), and the absorbance was determined with a UV-VIS spectrophotometer (Genesys 5, Thermo Spectronic, Rochester, NY, USA) at 517 nm. A calibration curve was made ($y = -0.0052x + 0.7161$, $R^2 = 0.99$) using Trolox (Sigma Aldrich, St. Louis, MO, USA) as a standard.

For the ABTS method, the sample (0.30 mL) was added to conical tubes containing 2.70 mL of the ABTS radical solution (ABTS, 7.00 mM, and potassium persulfate, 2.45 mM, 1:1 ratio, absorbance = 0.7 ± 0.005). The samples were incubated at 25 °C in the dark for a period of 7 min to allow for the reaction. Subsequently, the samples were centrifuged (Spectrafuge 6C, Labnet International, Inc., 6500 rpm for 5–10 min), and the absorbance was determined with a UV-Vis spectrophotometer (Genesys 5, Thermo Spectronic, Rochester, NY, USA) at 734 nm (ABTS). A calibration curve was made ($y = -0.0024x + 0.577$, $R^2 = 0.99$) using Trolox (Sigma Aldrich, St. Louis, MO, USA) as a standard.

Finally, the results were expressed as µM Trolox equivalents (TE)/g of film for both methods. Additionally, the antioxidant activity was also determined by the percentage of inhibition, according to the following equation: inhibition scavenging activity (%) =

(absorbance of the radical solution − absorbance of the sample/absorbance of the radical solution) × 100.

The total phenolic content of the films was calculated using the Folin–Ciocalteu (Sigma Aldrich, St. Louis, MO, USA) technique with some modifications [48,49]. The samples (1.6 mL) were placed in conical polypropylene tubes (15 mL). Subsequently, the Folin–Ciocalteu reagent (0.1 mL) and sodium carbonate (20% w/v, 0.3 mL, CTR Scientific, Mty N.L., Mexico) were added, and the solutions were homogenized in a vortex (Mixer Labnet International, Inc., Edison, NJ, USA). The samples were kept at rest in dark conditions (90 min at 25 °C) and then were centrifuged (Spectrafuge 6C Labnet International, Inc., Edison, NJ, USA; 6500 rpm for 10 min) and analyzed at 760 nm in a UV-Vis spectrophotometer (Genesys 5, Thermo Spectronic, Rochester, NY, USA). Finally, a calibration curve was made using gallic acid (Sigma Aldrich, St. Louis, MO, USA) as a standard (y = 0.07x − 0.2533 R^2 = 0.9958), and the total phenolic content was expressed as µg of gallic acid equivalents (GAE)/g of film.

2.5.5. Texture Profile Analysis (APT)

The texture analysis of the films was carried out with the help of a texturometer (Brookfield, CT3, Middleboro, MA, USA). The films (60 mm in diameter) were placed in the equipment and analyzed with the TA44 probe (cylinder 4 mm diameter), with an activation load of 0.070 N, a test speed of 0.50 mm/s, and a load range of 1000 g. The parameters evaluated were hardness (N; newtons), percentage deformation (%), adhesiveness (mJ; millijoules), and fracturability (N; newtons).

2.5.6. Field Emission Scanning Electron Microscopy (FE-SEM)

Film microscopy (10 × 10 mm) was performed with a field emission scanning electron microscope (Zeiss Sigma 300 VP, Jena, Germany) under high vacuum conditions, with a secondary electron detector considering a voltage of 5 kV, and at a working distance of 6 mm. Prior to analysis, the films were coated with gold (sputtering) to obtain better imaging conditions. Finally, micrographs were taken at 200× and 1500× magnification.

2.6. Statistical Analysis

The results of color, thickness, solubility, antioxidant activity, total phenols, and texture profile were analyzed by analysis of variance (ANOVA) and Tukey's test, with a significance level of $p \leq 0.05$, using the SPSS software (IBM version 22, SPSS Inc., Chicago, IL, USA).

3. Results and Discussion

3.1. Phytochemical Analysis of Plant Extracts

In the first place, the yield of the CA extract was 5.80%, in agreement with that reported by Cárdenas-Garza et al. [2]. Moreover, the phytochemical analysis showed that the plant material contains a complex mixture of various components [45]. Table 2 shows that *E. cardamomum* and linseed mucilage were positive for eight of the phytochemical tests, except for quinones, coumarins, and saponins. Additionally, the chemical composition of the cardamom extract (α-terpinyl acetate, mainly) has been reported in previous studies by our working group [2]. In addition, *C. officinalis* was positive only for six tests performed (sterols and triterpenes, flavonoids, unsaturations, carbonyl group, alkaloids, and carbohydrates). These results agree with what was reported by Hanaa et al. [50] and Yasmeen et al. [51], who have found some of these components in the linseed. In addition, the presence of some of these phytochemical compounds has also been reported in the essential oil of *C. officinalis* [52] and natural extracts of *E. cardamomum* [2,53,54]. Finally, this type of phytochemical compound has been characterized by having multiple pharmacological properties, among which stand out anti-inflammatory, anticancer, antiseptic, antimicrobial, and antioxidant activities, among others [2,4]. Likewise, due to its composition, copaiba oil has been characterized by reducing the formation of dental biofilm [21], thus counteracting some oral pathogens [4], presenting an anti-inflammatory effect and healing in the oral cavity [55], and acting as a dentin biomodifier [19]. On the other hand, cardamom is efficient in reducing the levels of microbial viability in the dental biofilm [15], having a potential therapeutic effect on periodontal infections [17], and improving oral hygiene [16].

Table 2. Partial phytochemical screening of plant extracts.

Tests	Phytoconstituents	Reaction	CA	CO	M
Liebermann-Burchard	Sterols, triterpenes	Reddish-brown ring	+	+	+
Shinoda	Flavonoids	Reep red	+	+	+
Baljet	Sesquiterpene-Lactones	Orange color	+	−	+
Sulfuric acid (H_2SO_4)	Quinones	Red color	−	−	−
Ferric chloride ($FeCl_3$)	Phenols, tannins	Green color	+	−	+
Potassium permanganate ($KMnO_4$)	Unsaturations	Brown precipitate	+	+	+
DNPH (2,4-dinitrophenylhydrazine)	Carbonyl group	Orange color	+	+	+
Dragendorff	Alkaloids	Orange precipitate	+	+	+
Sodium hydroxide (NaOH)	Coumarins, lactones	Yellow color	−	−	−
Molish	Carbohydrates	Purple ring formation	+	+	+
Foam	Saponins	Presence of stable foam	−	−	−

3.2. Development and Partial Characterization of Films Based on Linseed Mucilage Loaded with CA and CO

3.2.1. Film Production

The control films and those incorporated with the extracts (MCA, MCO, MCACO) were successfully produced (Table 3). The films obtained (60 mm in diameter) were thin and slightly flexible, with an opaque appearance, a light brown hue (control and MCO), and a more intense brown color (MCA and MCOCA). These results are similar to those reported in previous studies [30,31,42].

Table 3. Macroscopic characteristics of films based on linseed mucilage (M) and incorporated with *E. cardamomum* (CA) and *C. officinalis* (CO).

Formulation	Representative Image	Homogeneity	Appearance
MCA		Intermediate	Opaque, dark brown, and slightly flexible
MCO		Intermediate	Opaque, slightly brown, and slightly flexible
MCACO		Low	Opaque, light brown, and slightly flexible
Control		Intermediate	Opaque, slightly brown, and slightly flexible

3.2.2. Physicochemical Properties

Color and Thickness

The color properties affect the visual appearance and could influence the patient's perception of the product. Regarding color, no significant difference ($p > 0.05$) was found in the luminosity values between the different treatments, whose values fluctuated between 25.33 and 32.93 (Figure 1). This finding agrees with the literature since Hajirostamloo et al. [42] reported that the L* values remain constant at a concentration of 1% CO but increase at a concentration of 5% in protein isolate films containing cardamom essential oil. Likewise, Rodrigues et al. [56] indicated that the addition of CO oil (10%) does not affect the luminosity of starch-based films incorporated with copaiba oil. On the contrary, for the values of a* and b*, a significant difference ($p < 0.05$) was found between the treatments; the values in both coordinates were highest for MCA (a* = 1.60 ± 0.98 and b* = 5.93 ± 1.25) and MCOCA (a* = 1.14 ± 0.19 and b* = 6.27 ± 0.76), while the lowest values were for MCO (a* = −0.40 ± 0.03 and b* = 1.05 ± 0.88) and the control (a* = −0.31 ± 0.02 and b* = 1.31 ± 0.68) (Figure 1). According to what was reported by Hajirostamloo et al. [42], the incorporation of CO into the films affected the optical attributes, increasing the values of a* and b* coordinates, as observed in the MCO and MCACO treatments. On the other hand, Rodrigues et al. [56] reported that the color of the films was directly related to the color of the polymer and the oil or extract incorporated into the formulation. In addition, they also reported that the CO oil provided a slightly yellowish coloration to the films, which was more accentuated as the concentration of CO in the polymer matrix increased. Therefore, it is likely that, since the color of CO is similar to that of linseed mucilage, the MCO and control films are very similar in the a* and b* coordinate values.

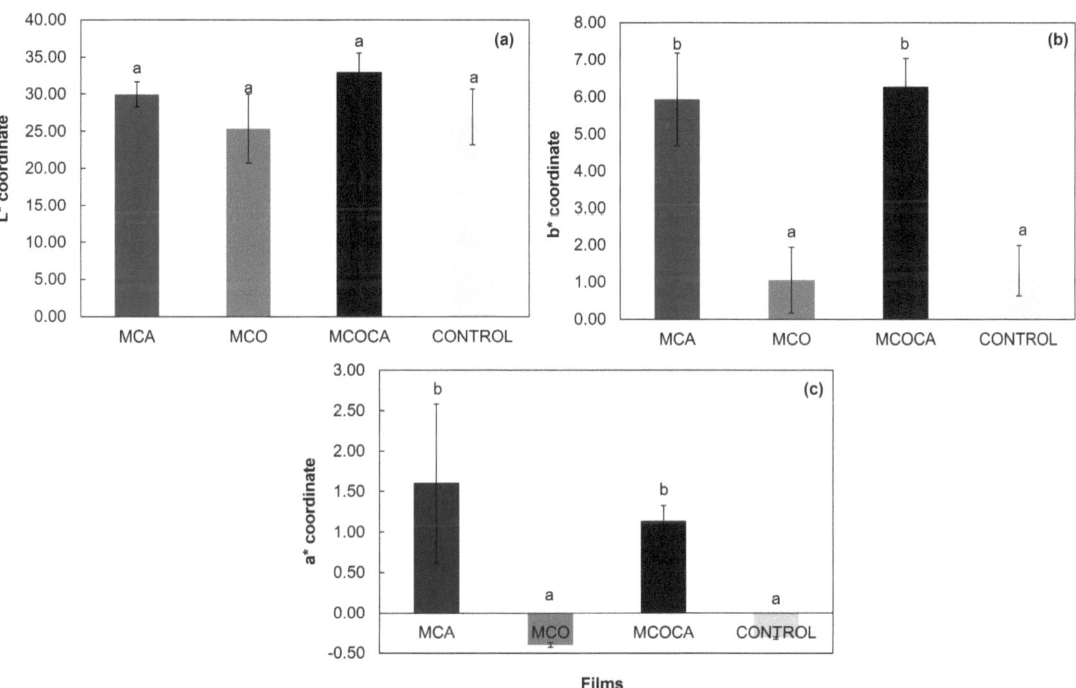

Figure 1. Color properties (**a**) luminosity, (**b**) blueness (−)—yellowness (+), and (**c**) redness (−)—greenness (+) of films based on linseed mucilage (M) and incorporated with *E. cardamomum* (CA) and *C. officinalis* (CO). Different letters (a,b) indicate significant differences ($p < 0.05$) between treatments.

The thickness is an important parameter since thin films tend to be less uncomfortable for the patient. In the thickness parameter, no significant difference ($p < 0.05$) was found between the treatments, and the values fluctuated between 0.0043 and 0.0052 mm (Figure 2). These results agree with those found by Herrera Brandelero et al. [57] and Rodrigues et al. [56], who reported that the incorporation of CO oil did not modify the thickness of films made from starch/PVOH/alginate and starch, respectively. Moreover, Hajirostamloo et al. [42] reported an increase in the thickness of the films when incorporating CA; this parameter increases with the increase in the amount of extract (5–20%) in a polymeric matrix of soy protein isolate.

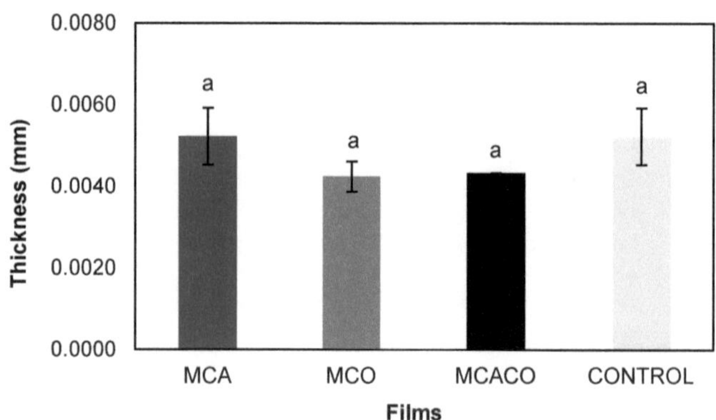

Figure 2. Thickness of films based on linseed mucilage (M) and incorporated with *E. cardamomum* (CA) and *C. officinalis* (CO). Letters (a) indicate no significant differences ($p > 0.05$) between treatments.

Solubility Tests in Artificial Saliva

In the case of the solubility of the films, although the values were higher for MCO (12.00 ± 2.65 min) and MCA (9.00 ± 1.73 min), no significant difference ($p < 0.05$) was found between the treatments (MCOCA $= 12.33 \pm 1.15$ and control $= 8.67 \pm 0.58$), indicating that the incorporation of the extracts into the polymeric matrix does not modify this property (Figure 3).

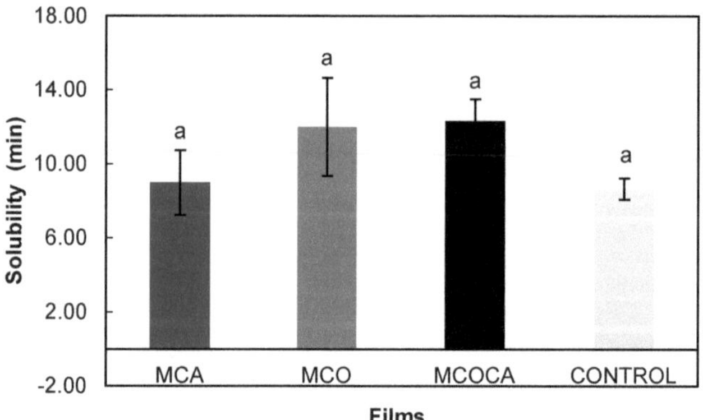

Figure 3. Solubility in artificial saliva of films based on linseed mucilage (M) and incorporated with *E. cardamomum* (CA) and *C. officinalis* (CO). Letters (a) indicate no significant differences ($p > 0.05$) between treatments.

According to what was reported by Puligundla and Lim [32], linseed mucilage shows a high degree of swelling and solubility in aqueous solutions. The water absorption leads to swelling and solubilization of the films, allowing the complete release of the active compounds incorporated within the polymeric matrix towards the artificial saliva. This behavior is directly associated with the hydrophilic nature of the polysaccharide and with the slightly branched structure of the linseed mucilage, which leads to the film dissolving and losing its structure over time [37]. In addition, the difference in the solubility time of the MCO and MCACO films could be related to the low solubility of the CO oil in water (the main component of artificial saliva) [3,26,30]. Likewise, Rodrigues et al. [56] reported that the addition of CO oil decreased the hydrophilic character of the films, leading to lower solubility. This behavior can be attributed to the fact that the interaction between the hydroxyl groups of the polymeric matrix and the oil components makes the -OH groups less available and, consequently, the solubility time of the films in artificial saliva increases.

Antioxidant Activity (DPPH and ABTS) and Total Phenolic Content

Regarding the antioxidant activity of the films, significantly higher values ($p < 0.05$) were for MCA (DPPH = 5781.38 ± 92.10 µm TE/g, 81.99 ± 3.56% inhibition and ABTS = 11,855.82 ± 1159.46 µm TE/g, 95.80 ± 3.98% inhibition), followed by MCACO treatment (DPPH = 4345.29 ± 150.87 µm TE/g, 47.15 ± 1.46% inhibition and ABTS = 5598.14 ± 1051.41, 39.73 ± 7.77% inhibition). The lowest values were for MCO (DPPH = 1740.18 ± 236.51 µm TE/g, 21.78 ± 4.83% inhibition and ABTS = 5174.77 ± 1487.98 µm TE/g, 35.38 ± 6.22% inhibition) and the control (DPPH = 2132. 39 ± 229.87 µm TE/g, 23.85 ± 2.57% inhibition and ABTS = 4615.74 ± 154.00 µm TE/g, 34.04 ± 1.66% inhibition) (Figure 4).

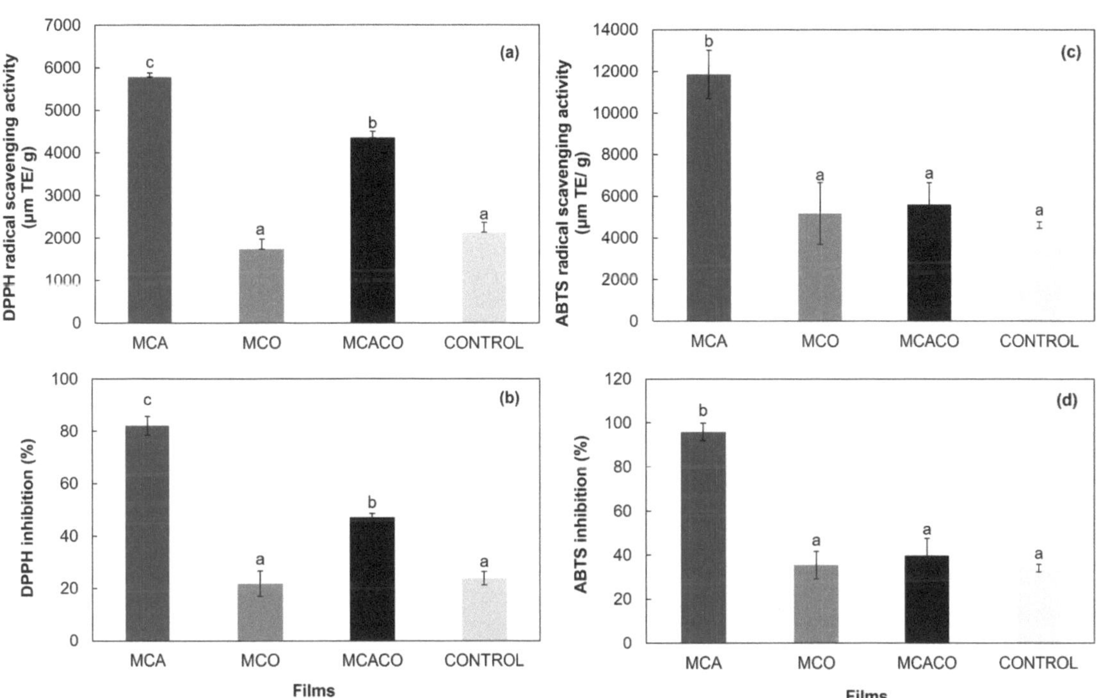

Figure 4. (**a**,**b**) DPPH and (**c**,**d**) ABTS radical scavenging activity and inhibition (%) of films based on linseed mucilage (M) and incorporated with *E. cardamomum* (CA) and *C. officinalis* (CO). Different letters (a,b,c) indicate significant differences ($p < 0.05$) between treatments.

These results agree with what was found by Hajirostamloo et al. [42], who reported that the incorporation of CA significantly increased the antioxidant activity of the films, which is accentuated by the increase in the concentration of CA in the polymeric matrix (% of inhibition of 9.10–63.61% with concentrations of 1–20% of CA, respectively), as we found in the MCA and MCACO films (Figure 4). Furthermore, the antioxidant activity provided by CO has also been previously reported in other applications [58], similar to that found in MCO films. Finally, the antioxidant activity reported in the control films is higher than that reported in other films made from linseed mucilage [31]; this behavior could be related to the polymer extraction processes, as well as the variety of seeds used for its recovery.

In the case of the total phenolic content (Figure 5), the highest values were for MCA (8076.49 ± 297.66 µg GAE/g), followed by MCACO (3985.25 ± 213.21 µg GAE/g), while the lowest values were for MCO and the control (3270.18 ± 243.99 and 2831.16 ± 392.69 µg GAE/g, respectively). These results agree with what was found in the DPPH and ABTS antioxidant activity analyses (Figure 4), which reflect a higher activity in MCA and MCACO, which can be associated with a higher total phenolic content in CA. In general, the antioxidant properties of the films developed in this study can be attributed to the content of antioxidant compounds present in the extract, oil, and linseed mucilage, such as sterols, triterpenes, flavonoids, tannins, phenols, and alkaloids, among others (Table 1), which have been reported in previous studies [2,18,52,59,60]. In addition, the mechanism for the action of these compounds is associated with the donation of a hydrogen atom and/or electrons to free radical chemical species to prevent cell damage by reducing oxidative stress, reactive oxygen species (ROS) and nitrogen (RON), oxidation of lipids, proteins, and DNA [58,61].

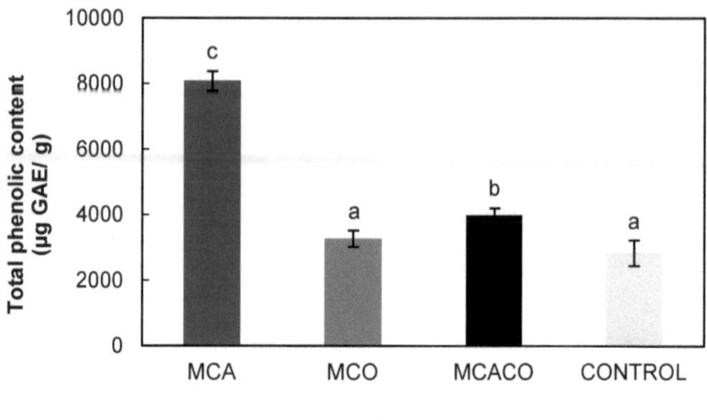

Figure 5. Total phenolic content of films based on linseed mucilage (M) and incorporated with *E. cardamomum* (CA) and *C. officinalis* (CO). Different letters (a,b,c) indicate significant differences ($p < 0.05$) between treatments.

3.3. Texture Profile Analysis (TPA)

Hardness, Deformation, Adhesiveness and Fracturability

The textural properties of the films provide information to understand the behavior of materials. Regarding the texture analysis of the films, it was observed that the incorporation of CA or CO decreased ($p < 0.05$) the hardness of the films (Figure 6a). Hardness refers to the maximum force required to compress the films between the molars, the tongue, or the palate [62]. The control film presented the highest hardness value (8.86 ± 1.54 N), and the MCA film had the lowest value (2.59 ± 0.35 N). In addition, the MCOCA and MCO films presented similar values, fluctuating between 4.78 and 5.35 N. According to the report by [31], the incorporation of extracts and essential oils into the films provides them a plasticizing effect since they reduce the interchain interaction forces of the polymer, increasing the mobility of the chains and providing greater flexibility. Consequently, the film has a less rigid and more malleable structure; this behavior can be attributed to the presence of hydroxyl groups (-OH) that form H bonds with the mucilage, increasing the mobility and decreasing the hardness of the polymeric matrix.

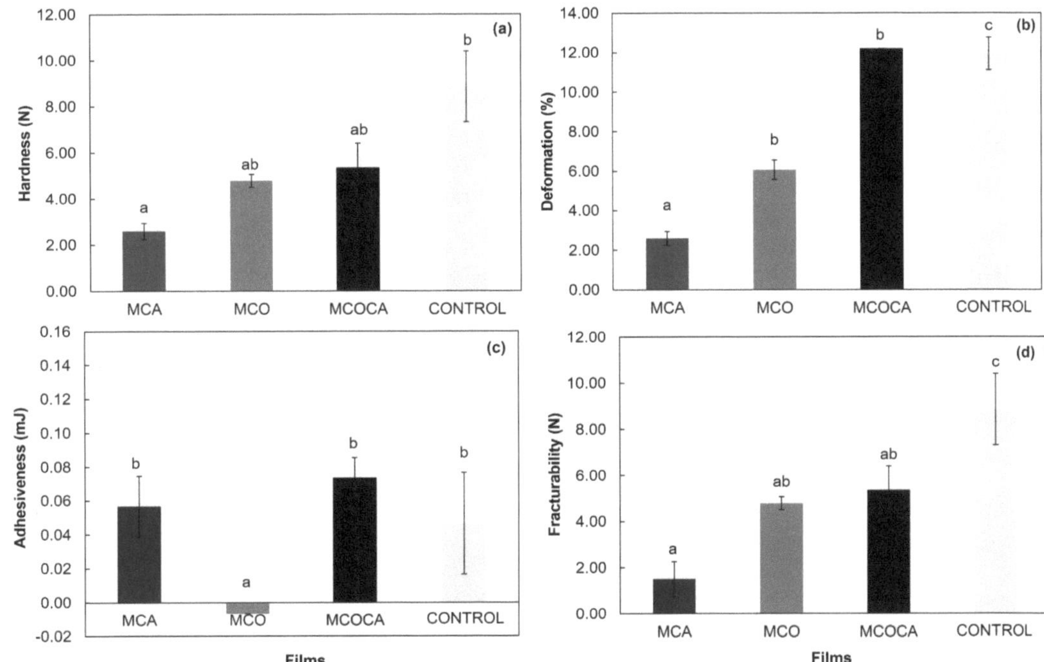

Figure 6. Textural properties: (**a**) hardness, (**b**) deformation, (**c**) adhesiveness and (**d**) fracturability of films based on linseed mucilage (M) and incorporated with *E. cardamomum* (CA) and *C. officinalis* (CO). Different letters (a,b,c) indicate significant differences ($p < 0.05$) between treatments.

Furthermore, it was found that the incorporation of CA and CO significantly decreased ($p < 0.05$) the deformation of the films (Figure 6b). In general, the deformation of the film occurs when the applied compressive stress changes the internal structure of the material, which prevents it from returning to its structural or dimensional original. The highest deformation values were found in MCACO and control treatments ($11.93 \pm 0.83\%$ and $12.20 \pm 0.00\%$, respectively). The lowest deformation values were obtained for MCA and MCO ($2.59 \pm 0.35\%$ and $6.07 \pm 0.49\%$, respectively). The improvement in this property could be related to the ability of the essential oil or extract to induce the reorganization of the polymeric network in the film matrix, which leads to the formation of films with a more stable structure [63].

The adhesiveness parameter refers to the work required to overcome the attractive force between the film and a surface [62]. The lowest value ($p < 0.05$) was obtained for the MCO treatment (-0.01 ± 0.00 mJ). The control, MCA, and MCO films presented similar values that fluctuated between 0.05 and 0.07 mJ (Figure 6c). In general, most of the films (MCA, MCOCA, and control) showed very similar adhesiveness values; the low values found for MCO could be related to the characteristics of the CO oil, which influence the final properties of the film.

Finally, fracturability refers to the force necessary to fracture the film [62]. In this parameter, it was found that the incorporation of the extracts decreased the fracturability of the films. The highest values were for the control (8.86 ± 1.54 N), followed by MCOCA (5.35 ± 1.05 N), MCO (4.78 ± 0.28 N), and MCA (1.51 ± 0.75) (Figure 6d). As previously mentioned, the incorporation of essential oils and extracts into the film matrix can cause a rearrangement of the polymer chains, as well as a plasticizing effect that provides a film with a more stable and fracture-resistant polymer structure [63].

3.4. Microscopic Analysis

The FE-SEM analysis is an essential tool for analyzing the films at the microscopic level. Regarding the microscopic analysis of the control and MCO, the films presented a more homogeneous microstructure, as well as a smooth (with few bulges) and continuous surface, similar to that reported by Karami et al. (chitosan-linseed mucilage films) [35], Debone et al. (chitosan/copaiba oleoresin films) [27], Norcino et al. (pectin films loaded with copaiba oil nanoemulsions) [36], and Pinto et al. (copaiba essential oil loaded-nanocapsules films) [40]. Otherwise, the MCA and MCACO films presented a more heterogeneous microstructure, a rough surface with the presence of discontinuous matter (bulges) in the polymer matrix associated with the presence of components of the CA extract; this effect has also been reported in protein isolate films containing cardamom essential oil [42] (Figure 7).

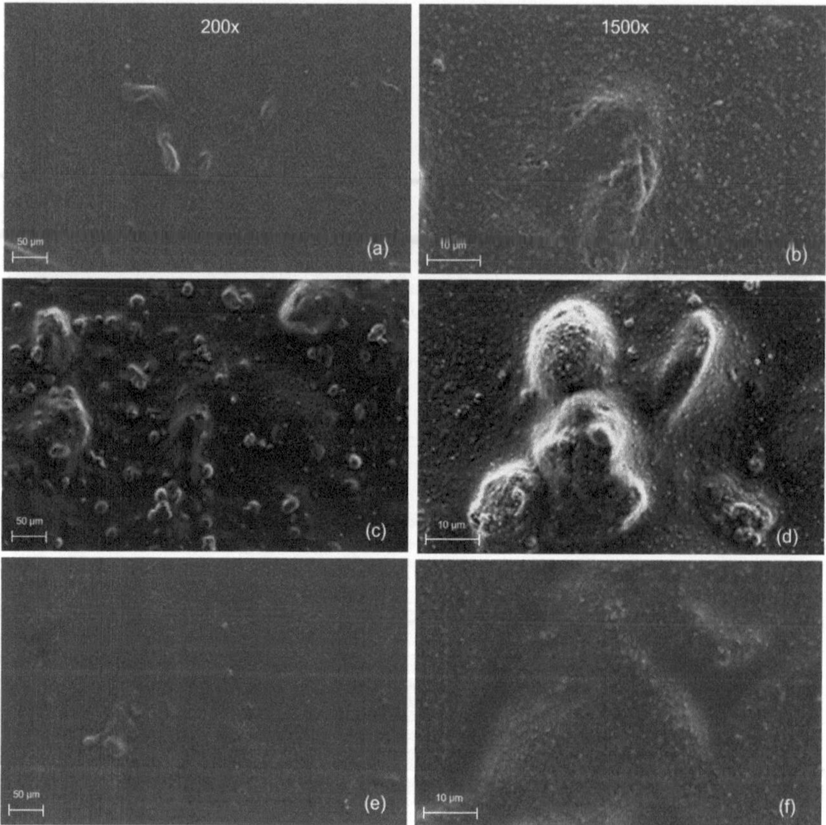

Figure 7. *Cont.*

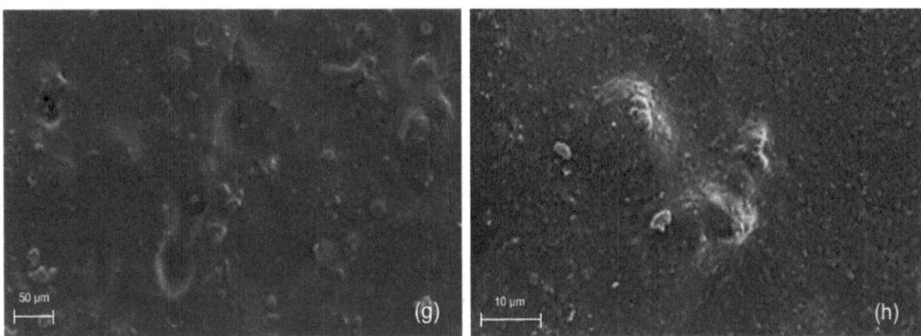

Figure 7. Surface micrographs (200× and 1500×) of films based on linseed mucilage (M) and incorporated with *E. cardamomum* (CA) and *C. officinalis* (CO); control (**a,b**), MCA (**c,d**), MCO (**e,f**), and MCOCA (**g,h**).

4. Conclusions

In their composition, *E. cardamomum* and *C. officinalis* compounds had evidence of the presence of phytochemicals, such as sterols, triterpenes, flavonoids, alkaloids, tannins, and phenols, among others. Moreover, it was possible to produce linseed mucilage-based films incorporated with *E. cardamomun*, *C. officinalis*, and co-loaded with both bioactive compounds. The addition of the active compounds in the films modified the color (redness–greenness and yellowness). However, the thickness parameter remained constant in all treatments (0.0042–0.0052 mm). In addition, the solubility time was higher for films containing CO (~12 min). Furthermore, the addition of the bioactive compounds increased the antioxidant activity and the total phenolic content, mainly in the films incorporated with CA. Regarding the texture analysis, the incorporation of the active ingredients decreased the hardness (39.50%–70.81%), deformation (49.16%–78.30%), and fracturability (39.58%–82.95%); however, it did not modify the adhesiveness, except in MCO. On the other hand, the FE-SEM micrographs showed a more homogeneous structure for the MCO films with respect to the films that contained CA in the formulation (heterogeneous structure with the presence of protuberances). Finally, due to the previously reported pharmacological properties of *E. cardamomun* and *C. officinalis* (anti-inflammatory, antimicrobial, antioxidant, among others), the films developed in this research could have a potential application in the area of dentistry as a wound dressing. However, it is necessary to continue with this research to evaluate the mechanical (tensile strength and elongation at break), antimicrobial, anti-inflammatory, and cytotoxic properties, as well as consider the development of in-vivo studies to broadly understand the properties of the films.

Author Contributions: Conceptualization, O.E.R.-L., M.Z.T.-G. and A.K.S.-V.; methodology, A.K.S.-V., M.Z.T.-G., J.G.B.-G., J.H.E.-L. and M.d.R.L.-B.; software, O.E.R.-L., J.G.B.-G. and M.Z.T.-G.; validation, O.E.R.-L., J.G.B.-G. and M.Z.T.-G.; formal analysis, O.E.R.-L., M.Z.T.-G., A.C.-M., S.M.L.-V., J.G.B.-G., J.H.E.-L. and M.d.R.L.-B.; investigation, A.K.S.-V., M.Z.T.-G., J.G.B.-G. and O.E.R.-L.; resources, O.E.R.-L., J.G.B.-G. and M.Z.T.-G.; data curation, O.E.R.-L., J.G.B.-G. and M.Z.T.-G.; writing—original draft preparation, M.Z.T.-G. and A.K.S.-V.; writing—review and editing, M.Z.T.-G., O.E.R.-L. and J.G.B.-G.; visualization, O.E.R.-L., A.K.S.-V. and M.Z.T.-G.; supervision, M.Z.T.-G., O.E.R.-L. and J.G.B.-G.; project administration, O.E.R.-L. and M.Z.T.-G.; funding acquisition, O.E.R.-L., M.Z.T.-G. and J.G.B.-G. All authors have read and agreed to the published version of the manuscript.

Funding: This research was funded by the Scientific and Technological Research Support Program of the Universidad Autónoma de Nuevo Leon (UANL), PAICYT CT1600-21 and CN1930-21.

Institutional Review Board Statement: Not applicable.

Informed Consent Statement: Not applicable.

Data Availability Statement: Not applicable.

Acknowledgments: To the CONAHCYT, National Council of Humanities, Sciences, and Technologies, for the support granted through scholarship 003989.

Conflicts of Interest: The authors declare no conflict of interest.

References

1. Da Trindade, R.; Da Silva, J.K.; Setzer, W.N. *Copaifera* of the Neotropics: A review of the phytochemistry and pharmacology. *Int. J. Mol. Sci.* **2018**, *19*, 1511. [CrossRef] [PubMed]
2. Cárdenas-Garza, G.R.; Elizondo-Luévano, J.H.; Bazaldúa-Rodríguez, A.F.; Chávez-Montes, A.; Pérez-Hernández, R.A.; Martínez-Delgado, A.J.; López-Villarreal, S.M.; Rodríguez-Rodríguez, J.; Sánchez-Casas, R.M.; Castillo-Velázquez, U.; et al. Benefits of cardamom (*Elettaria cardamomum* (L.) Maton) and turmeric (*Curcuma longa* L.) extracts for their applications as natural anti-inflammatory adjuvants. *Plants* **2021**, *10*, 1908. [CrossRef]
3. De Medeiros, M.L.; Xavier-Júnior, F.H.; Araújo-Filho, I.; Rêgo, A.C.; Veiga-Júnior, V.F.; Maciel, M.A. Copaiba oil for nanopharmaceutics and drug delivery. *Encycl. Nanosci. Nanotechnol.* **2019**, *27*, 165–189.
4. Diefenbach, A.L.; Muniz, F.W.M.G.; Oballe, H.J.R.; Rösing, C.K. Antimicrobial activity of copaiba oil (*Copaifera* ssp.) on oral pathogens: Systematic review. *Phytother. Res.* **2018**, *32*, 586–596. [CrossRef]
5. Ilyas, H.; Iqbal, M.J. Importance of some therapeutic plants and their antimicrobial activity to maintain oral health. *Biochem. Pharmacol. An. Open Access J.* **2021**, *10*, 08.
6. Pinto, M.F.; Quevedo, B.V.; Asami, J.; Komatsu, D.; Hausen, M.D.A.; Duek, E.A.D.R. Electrospun membrane based on poly(L-co-D,L lactic acid) and natural rubber containing copaiba oil designed as a dressing with antimicrobial properties. *Antibiotics* **2023**, *12*, 898. [CrossRef] [PubMed]
7. Noumi, E.; Snoussi, M.; Alreshidi, M.M.; Rekha, P.D.; Saptami, K.; Caputo, L.; De Martino, L.; Souza, L.F.; Msaada, K.; Mancini, E.; et al. Chemical and biological evaluation of essential oils from cardamom species. *Molecules* **2018**, *23*, 2818. [CrossRef] [PubMed]
8. Moulai-Hacene, F.; Boufadi, M.Y.; Keddari, S.; Homrani, A. chemical composition and antimicrobial properties of *Elettaria cardamomum* extract. *Pharmacogn. J.* **2020**, *12*, 1058–1063. [CrossRef]
9. Souza, A.G.; Ferreira, R.R.; de Oliveira, E.R.; Kato, M.M.; Mitra, S.K.; Rosa, D.D.S. Chemical stabilization behind cardamom pickering emulsion using nanocellulose. *Polysaccharides* **2022**, *3*, 200–216. [CrossRef]
10. Meenakshi, M.; Muralidharan, N.P. The role of cardamom oil in oral health. *IJLSR* **2015**, *1*, 322–325.
11. Aneja, K.R.; Joshi, R. Antimicrobial activity of *Amomum subulatum* and *Elettaria cardamomum* against dental caries causing microorganisms. *Ethnobotanical Leaflets* **2009**, *7*, 3.
12. Ibrahim, G.A.; Al–Obaidi, W.A. Effect of small cardamom extracts on Mutans streptococci in comparison to chlorhexidine gluconate and de-ionized water (in vitro study). *J. Bagh Coll. Dent.* **2013**, *25*, 104–108.
13. Dahigaonkar, K.; Yelpure, S.C.; Syed, F.N.; Wajid, A.F. Use of spices in treatment of dental infections. *World J. Pharm. Res.* **2018**, *7*, 360–371. [CrossRef]
14. Karimi, N.; Jabbari, V.; Nazemi, A.; Ganbarov, K.; Karimi, N.; Tanomand, A.; Karimi, S.; Abbasi, A.; Yousefi, B.; Khodadadi, E.; et al. Thymol, cardamom and *Lactobacillus plantarum* nanoparticles as a functional candy with high protection against *Streptococcus mutans* and tooth decay. *Microb. Pathog.* **2020**, *148*, 104481. [CrossRef]
15. Binimeliz, M.F.; Martins, M.L.; Filho, J.C.C.F.; Cabral, L.M.; Da Cruz, A.G.; Maia, L.C.; Fonseca-Gonçalves, A. Antimicrobial effect of a cardamom ethanolic extract on oral biofilm: An ex vivo study. *Nat. Oral. Care Dent. Ther.* **2020**, 121–131. [CrossRef]
16. Makky, E.A.; Ali, M.J.; Yusoff, M.M. Oral Hygiene Improvement using Combined Mouthwash with Plant extracts. *MATEC Web Conf.* **2018**, *150*, 06034. [CrossRef]
17. Souissi, M.; Azelmat, J.; Chaieb, K.; Grenier, D. Antibacterial and anti-inflammatory activities of cardamom (*Elettaria cardamomum*) extracts: Potential therapeutic benefits for periodontal infections. *Anaerobe* **2019**, *61*, 102089. [CrossRef]
18. Leandro, L.M.; De Sousa, V.F.; Barbosa, P.C.S.; Neves, J.K.O.; Da Silva, J.A.; Da Veiga-Junior, V.F. Chemistry and biological activities of terpenoids from copaiba (*Copaifera* spp.) Oleoresins. *Molecules* **2012**, *17*, 3866–3889. [CrossRef]
19. Bandeira, M.F.C.L.; Freitas, A.L.; Menezes, M.D.S.C.; Silva, J.D.S.; Sombra, G.A.D.; Araújo, E.A.M.; Toda, C.; Moreschi, A.R.C.; Conde, N.C.O. Adhesive resistance of a copaiba oil-based dentin biomodifier. *Braz. Oral. Res.* **2020**, *34*, e001. [CrossRef]
20. Genesi, B.P.; De Melo Barbosa, R.; Severino, P.; Rodas, A.C.; Yoshida, C.M.; Mathor, M.B.; Lopes, P.S.; Viseras, C.; Souto, E.B.; Da Silva, C.F. *Aloe vera* and copaiba oleoresin-loaded chitosan films for wound dressings: Microbial permeation, cytotoxicity, and in vivo proof of concept. *Int. J. Pharm.* **2023**, *634*, 122648. [CrossRef]
21. Simões, C.; Conde, N.; Venâncio, G.; Milério, P.; Fulgência, M.; Veiga, V. Antibacterial activity of copaiba oil gel on dental biofilm. *Open Dent. J.* **2016**, *10*, 188–195. [CrossRef]
22. Valadas, L.A.R.; Gurgel, M.F.; Mororó, J.M.; Da Cruz Fonseca, S.G.; Fonteles, C.S.R.; De Carvalho, C.B.M.; Fechine, F.V.; Neto, E.M.R.; De França Fonteles, M.M.; Chagas, F.O.; et al. Dose-response evaluation of a copaiba-containing varnish against *Streptococcus mutans* in vivo. *Saudi Pharm. J.* **2019**, *27*, 363–367. [CrossRef]
23. Valadas, L.A.R.; Lobo, P.L.D.; Fonseca, S.G.D.C.; Fechine, F.V.; Rodrigues Neto, E.M.; Fonteles, M.M.D.F.; Aguiar Trévia, L.R.D.; Vasconcelos, H.L.P.; Lima, S.M.D.S.; Lotif, M.A.L.; et al. Clinical and antimicrobial evaluation of *Copaifera langsdorffii* Desf. dental varnish in children: A clinical study. *Evid.-Based Complement. Altern. Med.* **2021**, *2021*, 6647849. [CrossRef]

24. Alvarenga, M.O.P.; Bittencourt, L.O.; Mendes, P.F.S.; Ribeiro, J.T.; Lameira, O.A.; Monteiro, M.C.; Barboza, C.A.G.; Martins, M.D.; Lima, R.R. Safety and effectiveness of copaiba oleoresin (*C. reticulata* Ducke) on inflammation and tissue repair of oral wounds in rats. *Int. J. Mol. Sci.* **2020**, *21*, 3568. [CrossRef]
25. Mehyar, G.F.; Al-Isamil, K.M.; Al-Ghizzawi, H.A.M.; Holley, R.A. Stability of cardamom (*Elettaria Cardamomum*) essential oil in microcapsules made of whey protein isolate, guar gum, and carrageenan. *J. Food Sci.* **2014**, *79*, C1939–C1949. [CrossRef]
26. Jamil, B.; Abbasi, R.; Abbasi, S.; Imran, M.; Khan, S.U.; Ihsan, A.; Javed, S.; Bokhari, H.; Imran, M. Encapsulation of cardamom essential oil in chitosan nano-composites: In-vitro efficacy on antibiotic-resistant bacterial pathogens and cytotoxicity studies. *Front. Microbiol.* **2016**, *7*, 1580. [CrossRef]
27. Debone, H.; Lopes, P.; Severino, P.; Yoshida, C.; Souto, E.B.; Silva, C. Chitosan/copaiba oleoresin films for would dressing application. *Int. J. Phar.* **2019**, *555*, 146–152. [CrossRef]
28. Dehghani, S.; Noshad, M.; Rastegarzadeh, S.; Hojjati, M.; Fazlara, A. Electrospun chia seed mucilage/PVA encapsulated with green cardamonmum essential oils: Antioxidant and antibacterial property. *Int. J. Biol. Macromol.* **2020**, *161*, 1–9. [CrossRef] [PubMed]
29. Bonan, R.F.; Bonan, P.R.; Batista, A.U.; Perez, D.E.; Castellano, L.R.; Oliveira, J.E.; Medeiros, E.S. Poly (lactic acid)/poly (vinyl pyrrolidone) membranes produced by solution blow spinning: Structure, thermal, spectroscopic, and microbial barrier properties. *J. Appl. Polym. Sci.* **2017**, *134*. [CrossRef]
30. Paranhos, S.B.; Ferreira, E.D.; Canelas, C.A.; Da Paz, S.P.; Passos, M.F.; Da Costa, C.E.; Da Silva, A.C.; Monteiro, S.N.; Candido, V.S. Chitosan membrane containing copaiba oil (*Copaifera* spp.) for skin wound treatment. *Polymers* **2021**, *14*, 35. [CrossRef]
31. Treviño-Garza, M.Z.; Yañez-Echeverría, S.; García, S.; Mora-Zúñiga, A.; Arévalo, N.K. Physico-mechanical, barrier and antimicrobial properties of linseed mucilage films incorporated with *H. virginiana* extract. *Rev. Mex. Ing. Quim.* **2019**, *19*, 983–996. [CrossRef]
32. Puligundla, P.; Lim, S. A Review of extraction techniques and food applications of flaxseed mucilage. *Foods* **2022**, *11*, 1677. [CrossRef] [PubMed]
33. Tee, Y.B.; Wong, J.; Tan, M.C.; Talib, R.A. Development of edible film from flaxseed mucilage. *BioResources* **2016**, *11*, 10286–10295. [CrossRef]
34. Salamanca, L.E.L.; Cabreta, L.E.P.; Narveaz, G.C.D.; de Alba, L.R.B. Linseed mucilage and chitosan composite films: Preparation, physical, mechanical, and microstructure properties. In Proceedings of the 11th International Congress on Engineering and Food, Athens, Greece, 22–26 May 2011.
35. Karami, N.; Kamkar, A.; Shahbazi, Y.; Misaghi, A. Effects of active chitosan-flaxseed mucilage-based films on the preservation of minced trout fillets: A comparison among aerobic, vacuum, and modified atmosphere packaging. *Pack. Technol. Sci.* **2020**, *33*, 469–484. [CrossRef]
36. Norcino, L.B.; Mendes, J.F.; Natarelli, C.V.L.; Manrich, A.; Oliveira, J.E.; Mattoso, L.H.C. Pectin films loaded with copaiba oil nanoemulsions for potential use as bio-based active packaging. *Food Hydrocoll.* **2020**, *106*, 105862. [CrossRef]
37. Bangar, S.P.; Singh, A.; Trif, M.; Kumar, M.; Kumar, P.; Kaur, R.; Kaur, N. Process parameter optimization and characterization for an edible film: Flaxseed concern. *Coatings* **2021**, *11*, 1106. [CrossRef]
38. De Paiva, P.H.; Correa, L.G.; Paulo, A.F.; Balan, G.C.; Ida, E.I.; Shirai, M.A. Film production with flaxseed mucilage and polyvinyl alcohol mixtures and evaluation of their properties. *J. Food Sci. Technol.* **2021**, *58*, 3030–3038. [CrossRef] [PubMed]
39. Fang, S.; Qiu, W.; Mei, J.; Xie, J. Effect of sonication on the properties of flaxseed gum films incorporated with carvacrol. *Int. J. Mol. Sci.* **2020**, *21*, 1637. [CrossRef] [PubMed]
40. Pinto, E.P.; Menezes, R.P.; Tavares, W.D.S.; Ferreira, A.M.; de Sousa, F.F.O.; Da Silva, G.A.; Zamora, R.R.; Araújo, R.S.; Souza, T.M. Copaiba essential oil loaded-nanocapsules film as a potential candidate for treating skin disorders: Preparation, characterization, and antibacterial properties. *Int. J. Pharm.* **2023**, *633*, 122608. [CrossRef]
41. Sani, K.I.; Alizadeh, M. Isolated mung bean protein-pectin nanocomposite film containing true cardamom extract microencapsulation/CeO$_2$ nanoparticles/graphite carbon quantum dots: Investigating fluorescence, photocatalytic and antimicrobial properties. *Food Packag. Shelf Life* **2022**, *33*, 100912. [CrossRef]
42. Hajirostamloo, B.; Molaveisi, M.; Jafarian Asl, P.; Rahman, M.M. Novel soy protein isolate film containing cardamom essential oil microcapsules: Study of physicochemical properties and its application in Iranian white cheese packaging. *J. Food Meas. Charact.* **2023**, *17*, 324–336. [CrossRef]
43. Castillo, N.E.T.; Teresa-Martínez, G.D.; Alonzo-Macías, M.; Téllez-Pérez, C.; Rodríguez-Rodríguez, J.; Sosa-Hernández, J.E.; Parra-Saldívar, R.; Melchor-Martínez, E.M.; Cardador-Martínez, A. Antioxidant activity and GC-MS profile of cardamom (*Elettaria cardamomum*) essential oil obtained by a combined extraction method-instant controlled pressure drop technology coupled with sonication. *Molecules* **2023**, *28*, 1093. [CrossRef] [PubMed]
44. Guillén-Meléndez, G.A.; Villa-Cedillo, S.A.; Pérez-Hernández, R.A.; Castillo-Velázquez, U.; Salas-Treviño, D.; Saucedo-Cárdenas, O.; Montes-de-Oca-Luna, R.; Gómez-Tristán, C.A.; Garza-Arredondo, A.J.; Zamora-Ávila, D.E.; et al. Cytotoxic effect in vitro of *Acalypha monostachya* extracts over human tumor cell lines. *Plants* **2021**, *10*, 2326. [CrossRef]
45. Rodríguez-Garza, N.E.; Quintanilla-Licea, R.; Romo-Sáenz, C.I.; Elizondo-Luevano, J.H.; Tamez-Guerra, P.; Rodríguez-Padilla, C.; Gomez-Flores, R. In vitro biological activity and lymphoma cell growth inhibition by selected mexican medicinal plants. *Life* **2023**, *13*, 958. [CrossRef] [PubMed]

46. Brand-Williams, W.; Cuvelier, M.E.; Berset, C. Use of a free radical method to evaluate antioxidant activity. *LWT—Food Sci. Technol.* **1995**, *28*, 25–30. [CrossRef]

47. Guzmán-Díaz, D.A.; Treviño-Garza, M.Z.; Rodríguez-Romero, B.A.; Gallardo-Rivera, C.T.; Amaya-Guerra, C.A.; Báez-González, J.G. Development and characterization of gelled double emulsions based on chia (Salvia hispanica L.) mucilage mixed with different biopolymers and loaded with green tea extract (Camellia sinensis). *Foods* **2019**, *8*, 677. [CrossRef]

48. Folin, O.; Ciocalteu, V. On tyrosine and tryptophane determinations in proteins. *J. Biol. Chem.* **1927**, *73*, 627–650. [CrossRef]

49. Drosou, C.; Kyriakopoulou, K.; Bimpilas, A.; Tsimogiannis, D.; Krokida, M.A. Comparative study on different extraction techniques to recover red grape pomace polyphenols from vinification by products. *Ind. Crops Prod.* **2015**, *75*, 141–149. [CrossRef]

50. Hanaa, M.H.; Ismail, H.A.; Mahmoud, M.E.; Ibrahim, H.M. Antioxidant activity and phytochemical analysis of flaxseeds (*Linum usitatisimum* L). *Minia J. Agric. Res. Develop.* **2017**, *37*, 129–140.

51. Yasmeen, M.; Nisar, S.; Tavallali, V.; Khalid, T. A review of phytochemicals and uses of flaxseed. *Int. J. Chem. Biochem. Sci.* **2018**, *13*, 70–75.

52. Arroyo, J.; Almora, Y.; Quino, M.; Martínez, J.; Condorhuamán, M.; Flores, M.; Bonilla, P. *Copaifera officinalis* oil cytoprotector and antisecretory effects in induced gastric lesions in rats. *An. Fac. Med.* **2009**, *70*, 89–96. [CrossRef]

53. Bano, S.; Ahmad, N.; Sharma, A.K. Phytochemical screening and evaluation of anti-microbial and anti-oxidant activity of *Elettaria cardamom* (Cardamom). *J. Appl. Nat. Sci.* **2016**, *8*, 1966–1970. [CrossRef]

54. Khatri, P.; Rani, A.; Hameed, S.; Chandra, S.; Chang, C.-M.; Pandey, R.P. Current understanding of the molecular basis of spices for the development of potential antimicrobial medicine. *Antibiotics* **2023**, *12*, 270. [CrossRef] [PubMed]

55. Dos Santos, M.A.C.; Alves, L.D.B.; Goldemberg, D.C.; De Melo, A.C.; Antunes, H.S. Anti-inflammatory and wound healing effect of Copaiba oleoresin on the oral cavity: A systematic review. *Heliyon* **2022**, *8*, e08993. [CrossRef]

56. Rodrigues, G.M.; Filgueiras, C.T.; Garcia, V.A.D.S.; Carvalho, R.A.; Velasco, J.I.; Fakhouri, F.M. Antimicrobial activity and GC-MS profile of copaiba oil for incorporation into *Xanthosoma mafaffa* Schott starch-based films. *Polymers* **2020**, *12*, 2883. [CrossRef]

57. Brandelero, R.P.H.; Brandelero, E.M.; Almeida, F.M.D. Biodegradable films of starch/PVOH/alginate in packaging systems for minimally processed lettuce (*Lactuca sativa* L.). *Cien Agrotec.* **2016**, *40*, 510–521. [CrossRef]

58. Monteschio, J.D.O.; De Vargas Junior, F.M.; Alves Da Silva, A.L.; Das Chagas, R.A.; Fernandes, T.; Leonardo, A.P.; Kaneko, I.N.; De Moraes Pinto, L.A.; Guerrero, A.; De Melo Filho, A.A.; et al. Effect of copaíba essential oil (*Copaifera officinalis* L.) as a natural preservative on the oxidation and shelf life of sheep burgers. *PLoS ONE* **2021**, *16*, e0248499. [CrossRef]

59. Kasote, D.M. Flaxseed phenolics as natural antioxidants. *Int. Food Res. J.* **2013**, *20*, 27–34.

60. Bouaziz, F.; Koubaa, M.; Barba, F.J.; Roohinejad, S.; Chaabouni, S.E. Antioxidant properties of water-soluble gum from flaxseed hulls. *Antioxidants* **2016**, *5*, 26. [CrossRef]

61. Arias, A.; Feijoo, G.; Moreira, M.T. Exploring the potential of antioxidants from fruits and vegetables and strategies for their recovery. *IFSET* **2022**, *77*, 102974. [CrossRef]

62. González, A.; Alvis, A.; Arrázola, G. Effect of edible coating on the properties of sweet potato slices (*Ipomoea Batatas* Lam) cooked by deep-fat frying: Part 1: Texture. *Inf. Technol.* **2015**, *26*, 95–102. [CrossRef]

63. Hashim, R.H.R.; Gunny, A.A.N.; Sam, S.T.; Kamaludin, N.H.I.; Shamsuddin, M.R. Effects of incorporation of essential oil into film for fruit packaging application. *IOP Conf. Ser. Earth Environ. Sci.* **2021**, *765*, 012020. [CrossRef]

Article

Fluorine-Free Compound Water- and Oil-Repellent: Preparation and Its Application in Molded Pulp

Xin Weng [1], Na Lin [1], Wenting Huang [1] and Minghua Liu [1,2,*]

[1] Fujian Provincial Engineering Research Center of Rural Waste Recycling Technology, College of Environment and Safety Engineering, Fuzhou University, Fuzhou 350116, China; 210620030@fzu.edu.cn (X.W.); bob3723@163.com (N.L.); 200620044@fzu.edu.cn (W.H.)

[2] College of Environmental and Biological Engineering, Putian University, No. 1133, Xueyuan Middle Street, Chengxiang District, Putian 351100, China

* Correspondence: mhliu2000@fzu.edu.cn; Tel.: +86-13305022089

Abstract: Molded pulp is considered an alternative to plastic packaging for its low cost, recyclability and non-pollution characteristics. However, the range of its applications has been limited by hydrophilicity and lipophilicity. Presented herein is a facile and straightforward method for the preparation biodegradable water- and oil-repellant for molded pulp. Sodium alginate-based oil repellent and PDMS-based water repellent were prepared by cross-linking and modification. The two were then mixed in various ratios to obtain compound water- and oil-repellent, which was applied to the molded pulp by dip-coating. The coated paper demonstrated excellent oil resistance (with a kit rating of 11/12) and outstanding water resistance (with a water contact angle of 121.9° and water absorption of 25.8%). This novel, eco-friendly, water- and oil-resistant molded pulp coating prepared from biodegradable and food-contactable materials is a potential candidate to replace petroleum-based coatings and has excellent possibilities to be manufactured on a large-scale intended for food and non-food contact applications.

Keywords: water resistant; oil resistant; molded pulp coating; fluorine-free; sustainable packaging; barrier property

Citation: Weng, X.; Lin, N.; Huang, W.; Liu, M. Fluorine-Free Compound Water- and Oil-Repellent: Preparation and Its Application in Molded Pulp. *Coatings* **2023**, *13*, 1257. https://doi.org/10.3390/coatings13071257

Academic Editor: Isabel Coelhoso

Received: 6 June 2023
Revised: 6 July 2023
Accepted: 12 July 2023
Published: 17 July 2023

1. Introduction

Packaging is vital in the supply chain system of procurement, production, transportation, sales and storage, which is closely related to human daily life [1]. As one of the four primary packaging materials, plastic products are universally utilized in the packaging industry due to their outstanding barrier properties, excellent mechanical properties and low prices [2,3]. However, plastic products are difficult to degrade in the natural environment [4], and their recycling is limited due to technical and economic factors. According to incomplete statistics, less than 3% of plastics are recycled in the world [5]. Therefore, there is an urgent need to find biodegradable packaging to replace plastic [6].

In this context, paper-based packaging stands out for its light weight, low cost and biodegradability [7–9]. Many kinds of paper-based packaging, including molded pulp, are typically made from recycled paperboard and/or newsprint [10]. Molded pulp products have been used extensively in the disposables market as the most promising alternative to plastic packaging [11]. However, molded pulp is made up of natural fibers with a porous and sparse structure causing it to be highly hygroscopic [12]. Compared with plastic materials, the molded pulp has a lower resistance to water vapor, water and grease [13].

Lamination and coating are the most commonly employed commercial approaches for providing water and oil repellency to paper products [14]. Low-density polyethylene (LDPE) liners are usually used in lamination treatment. However, the lining and paper fibers will be tightly bonded together after lamination, which is hard to separate and has a high recovery cost [15]. Therefore, laminated paper usually ends up in landfills as waste [16]. Currently, coating is the most common method to improve the waterproof and oil-resistant properties of paper. Perfluoroalkyl compounds are considered to be the most widely used water- and oil-repellents [17]. However, it has been found that perfluorooctane sulfonate (PFOS) and perfluorooctanoic acid (PFOA) monomers used in the preparation of perfluoroalkyl compounds are critical organic pollutants with high persistence and bioaccumulation in the environment, which can increase the risk of various diseases and cause harm to humans [18–21].

Recently, research on coatings has shifted to the development of renewable and biodegradable biopolymer materials that have great water- and oil-repellency [22–24]. Polylactic acid (PLA) is a new type of biological degradable material prepared from starch that is extracted from renewable plant resources, such as corn. The FDA has recognized it as a suitable packaging material for all food products [25,26]. However, it is hard to detach PLA from the paper substrate in the pulping process as with LDPE, and it can only be degraded under composting conditions [24].

In addition, starch [27,28], chitosan [29,30], nanocellulose [31,32] and protein [33] have successfully caught the attention of researchers and paper industries for the preparation of water- and oil-repellent [8]. Dhwani Kansal et al. [28] coated the kraft paper with starch and corn protein separately, and the coated paper demonstrated good tolerance to oil and water, resulting in a Cobb 60 value of 4.81 $g \cdot m^{-2}$ and a kit rating of 12/12. Research by Zhao et al. [34,35] reported that the kit rating of chitosan-coated paper was 8/12 at a coating load of 9.33 $g \cdot m^{-2}$. After chemical modification with polydimethylsiloxane and castor oil, the oil resistance of chitosan-coated paper was improved, with a kit rating of 12/12 and 10/12 when coated with 10.33 $g \cdot m^{-2}$ and 4.3 $g \cdot m^{-2}$ paper, respectively. Xie et al. [36] proposed to coat the cellophane base paper with a fluorine-free cellulose coating, and the coated paper exhibited good barrier properties and mechanical properties. In general, the majority studies in order to achieve great water or/and oil resistance usually increase the amount of coating, which will inevitably increase the production cost of coated paper. In addition, there is a lack of research on biomass-based water- and oil-repellents that can be applied to pulp molding.

In this work, we proposed a facile method to fabricate an environmentally, economical and water- and oil-resistant molded pulp coating mostly using sodium alginate (SA) and polydimethylsiloxane (PDMS). Extracted from brown algae, sodium alginate has strong film-forming properties. The film has a strong barrier effect on oil and water vapor. PDMS has a low surface energy and a strong hydrophobicity as an inert material while still acceptable for food contact applications. However, the surface energy of PDMS is not that low-enough to block the penetration of oil into the porous paper [34]. In this study, SA and Polyethylene glycol (PEG) were produced as oil repellent under esterification reaction. PDMS was modified to prepare the water repellent. Then the water repellent and oil repellent were mixed in different volume ratios to derive the compound water- and oil-repellent for the coating of molded pulp. The fabricated SP/VAPDMS coated paper possessed excellent oil and water resistance and was able to wash off the coating completely after a simple pulp-washing procedure, which facilitated its application in sustainable development.

2. Materials and Methods

2.1. Materials

Polyethylene glycol (PEG, Mn = 1000) and p-Toluenesulfonic acid monohydrate (p-TSA) with 98% purity were obtained from Aladdin Co., Ltd. (Shanghai, China). Food grade sodium alginate (SA) was purchased from Mingyue Group Co. (Shandong, China). hydroxyl-terminated polydimethylsiloxane (PDMS-OH, 40 mpa·s), vinyltrimethoxysilane (VTMS, >98%), (3-Aminopropyl)triethoxysilane (APTES, >98%) and Ethanol absolute were obtained from Aladdin Co., Ltd. (Shanghai, China). Except for sodium alginate, all chemicals referred to are analytical grade. Molded paper was obtained from Ali-benben paper products Co., Ltd. (Shandong, China).

2.2. Preparation of SA-PEG Oil Repellent

To prepare a sodium alginate solution at a concentration of 3% (w/v), 1.5 g SA was added to 50 mL distilled water and mixed until completely dissolved. Added 0.13 g of p-TSA to the SA solution and stirred well. The stirred solution was transferred to a three-neck flask, heated at 50 °C and 135 r/min for 10 min, and then 0.5 g of polyethylene glycol was added and the temperature was raised to 60 °C for 10 h. The reactants were cooled to room temperature to obtain the SA-PEG oil repellent (SP).

2.3. Preparation of Water Repellent

To prepare a PDMS-OH solution at a concentration of 10% (w/v), 0.7 g of PDMS-OH was stirred with 7 mL of 95% (v/v) ethanol solution and stirred at 400 r/min for 20 min at room temperature. Then, added 0.6 g VTMS and 0.5 g APTES drop by drop into the PDMS-OH solution and stirred continuously for 3 h at 250 r/min to obtain water repellent (VAPDMS).

2.4. Preparation of Compound Water- and Oil-Repellent

Slowly added the VAPDMS into SP, mixed it in the volume ratio of 90:10, 80:20, 70:30 and 60:40, and stirred it at 25 °C and 300 r/min for 3 h. After removing impurities, compound water- and oil-repellent (SP/VAPDMS) was obtained.

2.5. Fabrication of SP/VAPDMS Coated Paper

Molded pulp samples were trimmed to 20 mm × 20 mm and weighed (w_0). The paper samples were then dipped into SP/VAPDMS solution for 1 min and then dried thoroughly in an oven at 60 °C, followed by pretreatment at 23 °C and 50% relative humidity (RH) and weighed again (w_1). The resulting SP/VAPDMS paper was obtained and its dip-coating load was calculated by Equation (1).

$$Dip - coating\ load\ (\%) = \frac{w_1 - w_0}{w_0} \times 100\% \tag{1}$$

2.6. Scanning Electron Microscopy (SEM)

A scanning Electron Microscopy (SUPRA 55, Carl Zeiss AG, Oberkochen, Baden-Württemberg, Germany) was used for imaging base paper and coated paper samples. Before SEM analysis, gold layer (15 nm) was coated on each sample by sputtering.

2.7. Fourier Transform Infrared (FTIR) Spectra

An ATR-FTIR spectrometer (AVAT-AR360, Thermo Nicolet Co., Madison, WI, USA) was used for the ATR-FTIR recording. 36 scans were performed in the spectral range of 4000–400 cm^{-1} to derive the spectrum.

2.8. Thermogravimetric Analysis (TGA)

The thermal stability test was performed on base paper and coated paper using a thermogravimetric analyzer (DSC214, Netzsch Instruments, Germany). The paper samples were heated within the temperature range of 30–600 °C with a heating rate of 10 °C/min to obtain TG and DTG curves.

2.9. Oil Resistance

Oil resistance was tested on base and coated paper samples according to an international test method, the TAPPI standard T 559 pm-96. The kit rating solutions by mixing different ratios of castor oil, heptane and toluene to obtain 12 liquids with different surface tensions. A drop of the solution was dropped onto the paper sample and wiped off with a cotton ball after 15 s. The paper sample was then inspected for signs or spots left by the droplets. The larger the kit rating number, the more resistant the sample is to oil. The data were obtained as the average of the experiments conducted with three samples.

2.10. High-Temperature Oil Resistance

Tissue was placed on the back of the coated paper (10 cm × 10 cm), and 1–2 mL of 82 °C soybean oil was dropped onto the surface of the paper with a pipette. After 10 min, the excess oil was wiped off with a cotton ball. The samples were considered to block the hot oil if no oil penetrated the paper towel.

2.11. Water Absorption

The coated paper samples were dried at 70 °C for 1 h and weighed (w_0) before testing. The paper was then immersed in deionized water and weighed (w_1) for each sample at 0.5, 1, 2, 3, 5 and 24 h after removing excess water with a clean tissue. The water absorption rate can be obtained from Equation (2).

$$Water\ absorption\ (\%) = \frac{w_1 - w_0}{w_0} \times 100\% \tag{2}$$

2.12. Water Vapor Transmittance (WVTR)

The WVTR of the paper samples were measured in $g \cdot (m^2 \cdot 24\ h)^{-1}$ by the Permatran-W (model 3/34, Mocon Inc., Minneapolis, MN, USA) system at 23 °C and 50% RH. Trimmed to 2.54 × 2.54 cm^2, the paper samples were placed on an aluminum plate with a circular opening of 5 mm diameter in the middle. Prior to characterization, the samples were pretreated with nitrogen for 1 h under testing conditions.

2.13. Mechanical Testing of Tensile Strength

The molded pulp was cut to 50 mm × 15 mm and tested on a universal testing machine (model 1185, Instron Limited, Wycombe, UK) at a speed of 10 mm/min for tensile testing, following the ISO 527-2, 1993 (E) standard.

2.14. Contact Angles (CAs)

Water and oil contact angles were measured using an OSA200-01 system (Optical Surface Analyzer, Ningbo NB Scientific Instruments Co., Ningbo, China). A 5 µL droplet was dropped onto the paper and the contact angle was measured after 0 s, 30 s and 5 min. Three replicates of each paper substrate were taken and the average value was used as the result.

2.15. Repulpability

Paper samples (base and coated paper) cut to 2 cm × 2 cm were immersed in warm water (50 °C) for 60 min, and then divided into four portions after being pulpized using a blender. Wash the first three portions with deionized water, 2% (v/v) acetic acid solution and 85% (v/v) ethanol solution, respectively. The fourth part was washed with 2% acetic acid solution and then 85% ethanol solution. Finally, the above samples were washed with distilled water to eliminate the residual solvent. The four pulp parts were drained and then flattened with a hot iron and dried at 70° for 2 h to obtain the recycled molded pulps. Subsequently, the recycled molded pulps were analyzed by ATR-FTIR spectroscopy to compare the characteristic peaks for exploring whether the coating had been removed.

3. Results

3.1. Fourier Transform Infrared (FTIR) Spectra

FTIR spectra were used to verify the cross-linking reactions of SP oil repellent and modification of VAPDMS water repellent (Figure 1). As can be seen from Figure 1a, the SP paper has a characteristic peak of −OH at 3300–3700 cm^{-1} compared to the base paper and the peak here is broadened, probably due to the overlap of the unreacted carboxyl and hydroxyl peaks in the alginate molecule. The sharp peak around 1601 cm^{-1} is attributed to the C=O peak generated by the esterification reaction between polyethylene glycol and alginate. Since only some of the carboxyl groups reacted with the hydroxyl group, the coexistence of carboxyl groups and ester bonds in the reaction product led to the overlapping of the C=O peaks of the two, resulting in a relatively large change in the width of the peak. In addition, a more obvious C−O stretching vibration peak appeared at 1086 cm^{-1}, which is a more obvious characteristic peak of the ester group. The above analysis results indicate that the chemical reaction between SA and PEG has occurred, and the polyethylene glycol-based oil repellent has been successfully prepared in this experiment. From Figure 1b, it can be seen that VAPDMS paper has a characteristic peak of Si−O−CH$_3$ at 2845–2975 cm^{-1} compared to base paper, and the −CH$_3$ group is considered to be a group with low surface energy [37]. Moreover, the presence of Si−CH$_3$ at 799 cm^{-1} confirms the presence of PDMS. The modified PDMS introduced C=C and C−N, so that the stretching vibration peaks of vinyl C=C and C−N appeared near 1590 cm^{-1} and 1260 cm^{-1}. These four characteristic absorption peaks were not seen in the base paper, but the modified PDMS has these three characteristic absorption peaks, which can be judged that VTMOS and APTES have been successfully attached to PDMS, and the hydrolytic condensation reaction between PDMS and VTMOS and APTES has occurred, and the polydimethylsiloxane-based water repellent (VAPDMS) has been successfully prepared.

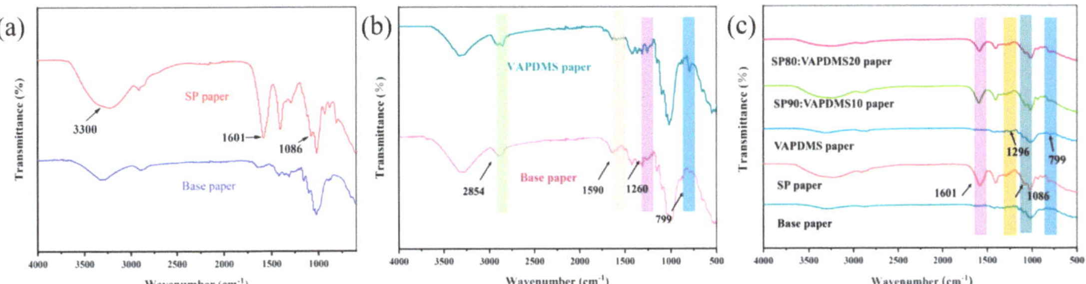

Figure 1. (**a**) FT−IR spectra of base paper and SP paper; (**b**) FT−IR spectra of base paper and VAPDMS paper; (c) FT−IR spectra of base paper, SP paper, VAPDMS paper, SP90:VAPDMS10 paper and SP80:VAPDMS20 paper.

Figure 1c demonstrates the FT−IR spectra of base paper, as well as paper samples coated with oil repellent (SP), water repellent (VAPDMS), and compound water- and oil-repellent (SA/VAPDMS). SA/VAPDMS paper had the characteristics of SP and VAPDMS papers peaks. Still, its infrared spectral characteristics were closer to those of SP due to its higher ratio. What's more, SP90:VAPDMS10 paper was more relative to SP than those of SP80:VAPDMS20. In addition, the intensity of the characteristic peaks of the O−H bond of SA/VAPDMS was significantly weaker, and the peak shape was wider compared with that of SP and VAPDMS, indicating the existence of a relatively strong hydrogen bonding between the water repellent and the oil repellent.

3.2. SEM

In order to understand the barrier mechanism of the coated paper at the microscopic level, SEM images of base paper, SP90:VAPDMS10 paper, SP80:VAPDMS20 paper and SP70:VAPDMS30 paper were recorded (Figure 2). It can be clearly observed from the SEM image of Figure 2a that the virgin paper fibers were mis-woven and had significant pores, which could allow the passage of water and oil. Figure 2b–d showed SEM images of the SP/VAPDMS paper with a film formed on the surface, which masked most of the pores. Furthermore, it can be observed that the higher percentage of SP in compound water- and oil-repellent, the better coverage and smoothness it would have.

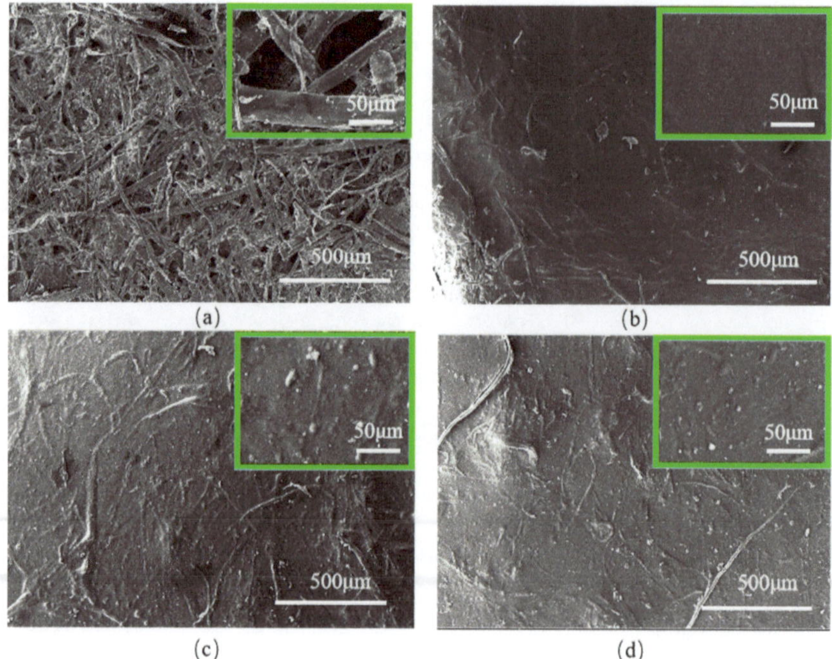

Figure 2. SEM images (×100) with zoomed-in pictures as insets (×1000) in the upper right corners of base paper (**a**), SP90:VAPDMS10 paper (**b**), SP80:VAPDMS20 paper (**c**) and SP70:VAPDMS30 paper (**d**).

3.3. Thermogravimetric Analysis (TGA)

The thermal stability of the coating layer was investigated by thermogravimetric analysis (TGA). TG and DTG curves of base paper and SP/VAPDMS paper were shown in Figure 3. Since the coating load (5.3%) was small, the overall trend of the TGA curves for the base and coated papers was not significantly different. It was analyzed that the mass loss observed below 120 °C corresponded to the evaporation of water from the paper. In the range of 250 °C to 375 °C, the significant mass loss was attributed to the decomposition of the paper and coating material. After 480 °C, the SP/VAPDMS paper lost less mass than the base paper, which could be due to the enhanced thermal stability of the compound water- and oil-repellant. In conclusion, as coated paper has no noticeable mass loss below 200 °C, it indicated that the coated paper was thermally stable for high temperature applications.

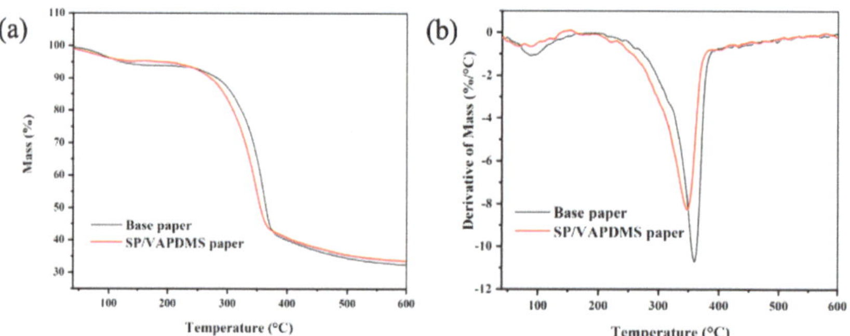

Figure 3. The (**a**) TG and (**b**) DTG curves of base paper and SP70:VAPDMS30 paper.

3.4. Oil Resistance

The oil resistance of SP/VAPDMS was investigated by dip-coating different loads of SP70:VAPDMS30 on the paper, and the results were illustrated in Figure 4a. The base had a kit rating of zero, indicating no oil resistance. The kit rating for 9.6% dip-coating load coated paper was 5/12. It met the requirements of greaseproof paper with a kit rating higher than or equal to 5 according to the TAPPI standard [38]. The kit rating for 20.9% dip-coating load coated paper was 11/12. It can be seen from the data that the dip-coating load improved the kit rating very significantly. Due to increased dip-coating load, the coating thickness will increase, which can better cover the paper fiber pores to prevent grease penetration. Therefore, the coating's integrity, uniformity, and compactness of the coating had an important role to play in the oil repellency of the paper [39,40].

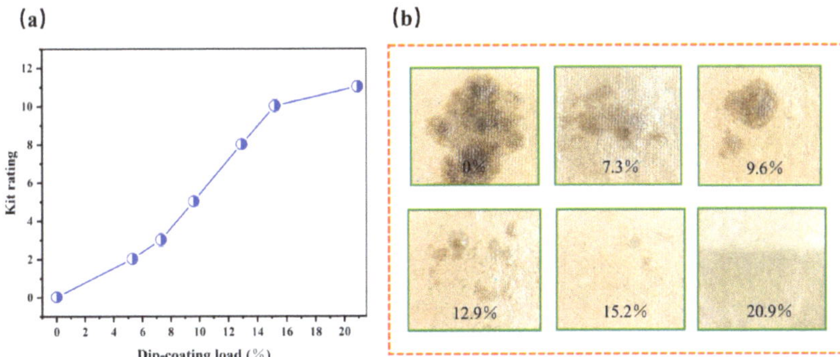

Figure 4. (**a**) The influence of dip-coating load on the oil repellency of SP/VAPDMS paper; (**b**) Thermal oil resistance testing of SP/VAPDMS paper. x% represents the load of SP70:VAPDMS30 on paper.

3.5. Thermal Oil Resistance

Food packaging will inevitably encounter high-temperature food in daily use, so the resistance of molded pulp to high-temperature grease significantly impacts its applicability. In this regard, we tested the paper against thermal oil at different dip-coating loads (Figure 4b). When the dip-coating load was less than 12.9%, the oil droplets penetrated the paper, and the dark spots on the surface were prominent. When the dip-coating load was 15.2%, the area of dark spot area was significantly reduced, and only a tiny amount of oil drops penetrated the paper. There were no black spots when the dip-coating load was increased to 20.9% and the paper achieved the best high-temperature grease resistance.

3.6. Water Absorption

The water repellency of SP/VAPDMS paper could be demonstrated by the water absorption rate (Figure 5a). The water absorption rate of the base paper was 93.1% at the beginning of 0.5 h and reached 136% after 24 h. The SP/VAPDMS paper showed remarkably low water absorption compared to the base paper, and the water absorption decreased as the proportion of water repellent increased. SP60:VAPDMS40 demonstrated a water absorption of 25.8% after 24 h, which was about 110% lower than the base paper. It was due to the hydrophobic group in the water repellent that allowed the paper to obtain low surface energy [41], thus considerably reducing hydrophilicity.

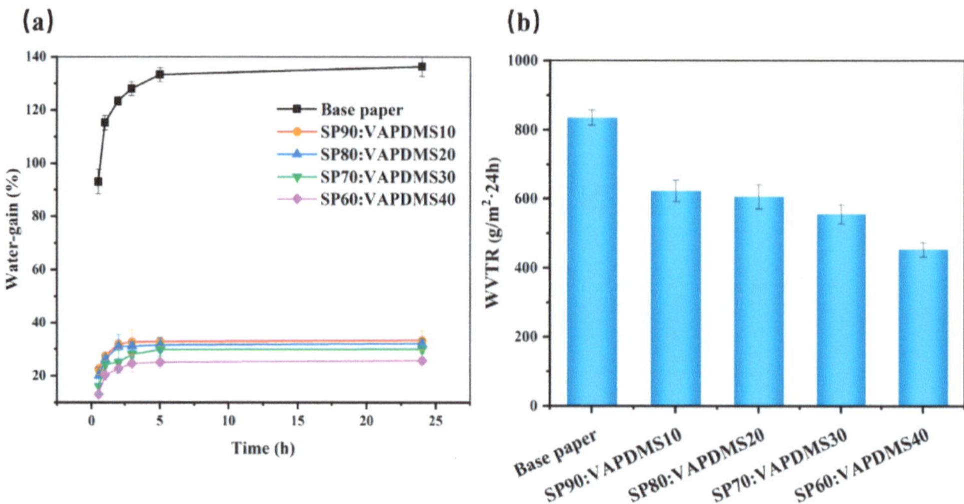

Figure 5. (**a**) Water absorption of base paper, SP90:VAPDMS10, SP80:VAPDMS20, SP70:VAPDMS30 and SP60:VAPDMS40 paper; (**b**) WVTR (g·(m^2·24 h)$^{-1}$) of base paper, SP90:VAPDMS10, SP80:VAPDMS20, SP70:VAPDMS30 and SP60:VAPDMS40 papers.

3.7. Water Vapor Transmittance (WVTR)

The WVTR of base and coated paper were also tested with a desiccant method (Figure 5b). The water vapor permeability of virgin paper was large, however, the permeability of SP/VAPDMS paper was greatly reduced compared to base paper. This was observed due to the hydrophobic nature of VAPDMS and the coverage of SA, which resulted in lower water vapor transmission and led to lower water vapor permeability. In addition, the increasing proportion of VAPDMS in the SP/VAPDMS contributed to a decrease in water vapor permeability. The lowest WVTR at 454 g·(m^2·24 h)$^{-1}$ was found in SP60:VAPDMS40 paper, which represented a 46% decline compared to 836 g·(m^2·24 h)$^{-1}$ for base paper. Overall, the content of the water repellent affected the water vapor barrier properties. The hydrophobic functional groups attached to the paper surface and the film formed by the oil repellent worked together to block the entry of water vapor.

3.8. Water/oil Contact Angles

To evaluate the wettability of base paper and coated paper to water and castor oil, an optical surface analyzer was used to test their water and oil contact angles at 0 s, 30 s and 5 min (Figure 6a,b). The base paper demonstrated moderate resistance to wetting at 0 s and 30 s, but as time elapsed, water completely penetrated the base paper after 5 min, with a WCA of 0°. SP paper was barely waterproof, the WCA at 0 s was found to be 21.3 ± 1.8°. VAPDM paper had the greatest wettability resistance, and the WCA was always stabilized at about 150 ± 3.1°. For different volume ratios of SP/VAPDMS (90:10, 80:20, 70:30, 60:40), the WCA increased with the proportion of water repellent. The SP90:VAPDMS10 paper showed a WCA of 56.4 ± 3.5° at 30 s, while the SP60:VAPDMS40 paper showed a WCA of 121.3 ± 1.6° at 30 s. The decreasing order of WCA was VAPDMS paper > SP/VAPDMS paper > Base paper > SP paper. In summary, the hydrophobicity of the coated paper depended mainly on the content of the waterproofing agent. The percentage of water repellents in the compound water-and oil-repellents increased, causing the hydrophobicity to increase. Still, the presence of hydrophilic oil repellents led to a lower WCA than VAPDMS. From the figures, it was evident that the WCA of SP70:VAPDMS30 and SP60:VAPDMS40 were stabilized and were more than 90° even after 5 min. For manufacturing budget reasons, the SP70:VAPDMS30 was chosen to validate its practical application. As shown in Figure 6c, water penetrated the base paper after 5 min, while the SP70:VAPDMS30 paper blocked the permeation of water, showing excellent water repellency.

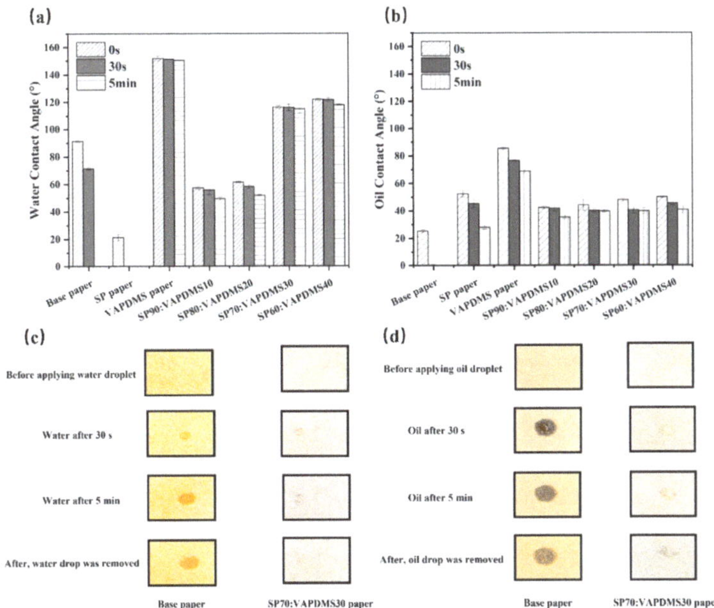

Figure 6. (**a**) Water contact angles (WCA) for base paper SP90:VAPDMS10 paper, SP80:VAPDMS20 paper, SP70:VAPDMS30 paper, and SP60:VAPDMS40 paper for 0 s, 30 s, and 5 min, respectively; (**b**) Oil contact angles (OCA) for base paper SP90:VAPDMS10 paper, SP80:VAPDMS20 paper, SP70:VAPDMS30 paper and SP60:VAPDMS40 paper for 0 s, 30 s, and 5 min, respectively; (**c**) Water droplet behavior on the base paper and SP70:VAPDMS30 paper, respectively; (**d**) Oil droplet behavior on the base paper and SP70:VAPDMS30 paper, respectively.

The oil contact angle (OCA) of base paper and coated paper, was shown in Figure 6b. An oil contact angle of 0° after 30 s was observed for the base paper, which suggested that it had no oil resistance. VAPDMS exhibited the highest oil contact angle, however, when the surface oil droplets were wiped away, the VAPDMS paper surface darkened over a large area, indicating that it was not impermeable to oil. For different volume ratios of SP/VAPDMS (90:10, 80:20, 70:30, 60:40), the oil contact angles were similar and stabilized at about 40° after 5 min. Compared with SP paper and VAPDMS paper, the OCA of SP/VAPDMS paper was lower than theirs, which could be due to the presence of the more reactive polar groups when SP and VAPDMS were mixed [42]. The SP70:VAPDMS30 was used to demonstrate its practical application, as shown in Figure 6d. The oil had wholly penetrated the base paper after 30 s, while the SP70:VAPDMS30 paper prevented the oil from penetrating. When the oil droplets were wiped off after 5 min, there were no areas of darkening on the surface, indicating that it had the potential to be used in food packaging with good oil repellency.

3.9. The Mechanical Properties

The assessment of the mechanical properties of the paper sample was shown in Figure 7. The tensile strength of the base paper was poor because the paper was a network structure consisting of natural fibers connected by hydrogen bonds, which can easily crack when subjected to mechanical stress. The coating on the surface of SP paper is due to the evaporation of water after drying to form a continuous dense film making the molecules of the substance closer to each other, which in turn increases the tensile strength of the food-contactable packaging. In addition, the coating not only forms a film on the surface of the paper, but also partially penetrates into the pores, strengthening the bond between the paper fibers and ensuring the quality of the packaging material. There is a difference in the tensile strength of base paper compared to VAPDMS paper, which is mainly related to the strength of hydrogen bonds between paper fibers. The higher stress in the VAPDMS paper is attributed to the fact that the addition of the water repellent makes the internal structure of the molded pulp change and strengthens the bonds with hydrogen [43], so the tensile strength of the VAPDMS paper is further improved.

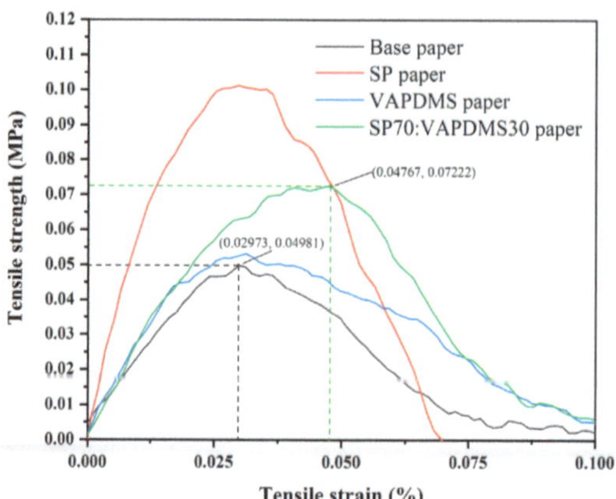

Figure 7. The stress and strain of base paper, SP paper, VAPDMS paper and SP70:VAPDMS30 paper.

Compared to base paper, SP70:VAPDMS30 paper exhibited significant tensile properties with a tensile strength of 0.07222 MPa, an improvement of 45%. Since the volume ratio of oil repellent in SP70:VAPDMS30 paper occupies much, the curve is similar to the stress-strain curve in SP paper, which fully indicates that the mixing of water and oil repellent does not reduce its tensile performance and can be applied to the packaging industry as well. In addition, the tensile curve of paper molded pulp showed multiple wave peaks, the possible reason for this is that paper molded pulp is a fiber network structure with a certain thickness, and the fracture is not completed in an instant, but the process of constantly breaking the internal hydrogen bonding force.

3.10. Repulpability

The most difficult part of the pulp recycling process is the separation of the pulp from the coating, especially when the coating is made of synthetic polymers. As a result, coated paper is generally disposed in landfills, which results a waste of resources and may harm the ecosystem. Since the coating materials are mainly manufactured from eco-friendly materials, it is possible to recycle pulp by separating the coating from the paper as needed. Considering these factors, we washed the dip-coated paper with several solvents and successfully removed the SP/VAPDMS coating from the pulp. From Figure 1c, it can be seen that VAPDMS paper introduced C=C and C−N, so that the stretching vibration peaks of vinyl C=C and C−N appeared near 1590 cm^{-1} and 1260 cm^{-1}. Comparing the infrared spectrogram of the water- and oil-repellant with that of the water repellant, it can be seen that the water- and oil-repellant appears C=C and C−N. The C=C of the water- and oil-repellant appears at 1601 cm^{-1}, and the wavenumber moves to a high wavenumber compared with the C=C of the water repellant, and the intensity of the absorption peak is weakened, indicating the existence of hydrogen bonding between the water repellant and oil repellant. Water- and oil-repellant at 1296 cm^{-1} carbon and nitrogen characteristic absorption peak compared with the water repellent in the carbon and nitrogen to move to a high wavenumber, and absorption intensity is reduced, indicating that the carbon and nitrogen bonds involved in the formation of intermolecular hydrogen bonding. The Si−CH$_3$ at 799 cm^{-1} of the water- and oil-repellant coincides with the characteristic peak in the water repellent, so it is not involved in the formation of intermolecular hydrogen bonds. Comparison of the ATR-FTIR spectra (Figure 8) of washed and unwashed recycled paper was used to detect the presence of residual coatings in the pulp. The spectra of the unwashed repulped paper were compared with those of the coated repulped paper washed with ethanol and acetic acid. After washing coated paper pulp with 85% ethanol aqueous (E), the peaks corresponding to ester groups (1601 cm^{-1} and 1086 cm^{-1}) and C−N, Si−CH$_3$ (1296 cm^{-1} and 799 cm^{-1}) disappeared, and the infrared spectrum were the same as that of base paper, which confirmed that the coating was basically washed off.

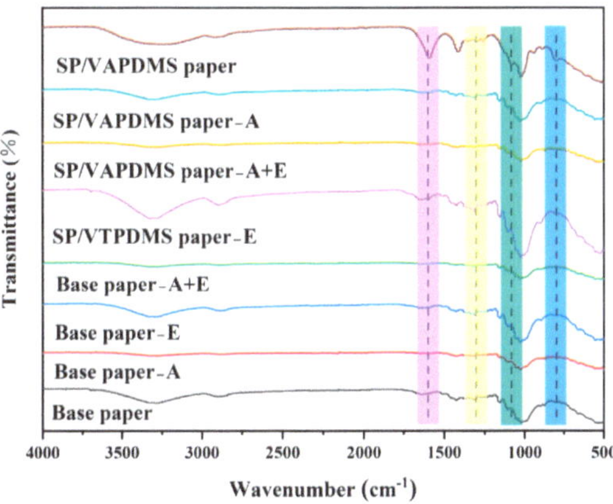

Figure 8. FTIR spectra of repulped papers made from fibers of base paper and SP70:VAPDMS30 paper, with 2% (*v*/*v*) acetic acid solution, 85% (*v*/*v*) ethanol solution and water.

4. Conclusions

In conclusion, we have developed a novel method to create an environmentally friendly water- and oil-resistance paper. The coated paper demonstrated excellent oil resistance (with a kit rating of 11/12) and outstanding water resistance (with a water contact angle of 121.9° and water absorption of 25.8%). The water vapor barrier performance of the coated paper was significantly improved, by 48.1% relative to the base paper. SEM analysis confirmed the disappearance of pores on the paper fiber surface after dip-coating. In addition, the coated paper exhibited good thermal stability and excellent mechanical properties. After washing and repulping tests, ATR-FTIR spectra of recycled paper demonstrated the recyclability of the pulp. This food-safe paper coating using environmentally friendly, biodegradable ingredients can be used to replace fluorinated coatings in the packaging industry, which will offer corresponding economic and environmental benefits.

Author Contributions: Conceptualization, X.W.; formal analysis, N.L. and W.H.; investigation, X.W. and W.H.; supervision, M.L.; validation, N.L. and M.L.; writing—original draft preparation, X.W. and W.H.; writing—review and editing, X.W. and M.L. All authors have read and agreed to the published version of the manuscript.

Funding: This work was financially supported by the Key Projects of Science and Technology Innovation in Fujian Province, China (No. 2021G2001) and External cooperation projects (cooperation) of Fujian Provincial Science and Technology, China (No. 2020I0043).

Institutional Review Board Statement: Not applicable.

Informed Consent Statement: Not applicable.

Data Availability Statement: All data generated or analyzed during this study are included in the present article.

Conflicts of Interest: The authors declare no conflict of interest.

References

1. Pålsson, H.; Hellström, D. Packaging logistics in supply chain practice—Current state, trade-offs and improvement potential. *Int. J. Logist. Res. Appl.* **2016**, *19*, 351–368. [CrossRef]
2. Abbas, M.; Buntinx, M.; Deferme, W.; Peeters, R. (Bio)polymer/ZnO nanocomposites for packaging applications: A review of gas barrier and mechanical properties. *Nanomaterials* **2019**, *9*, 1494. [CrossRef] [PubMed]
3. Beitzen-Heineke, E.F.; Balta-Ozkan, N.; Reefke, H. The prospects of zero-packaging grocery stores to improve the social and environmental impacts of the food supply chain. *J. Clean. Prod.* **2017**, *140*, 1528–1541. [CrossRef]
4. Lebreton, L.; Andrady, A. Future scenarios of global plastic waste generation and disposal. *Palgrave Commun.* **2019**, *5*, 6. [CrossRef]
5. Mohamed, S.A.A.; El-Sakhawy, M.; El-Sakhawy, M.A.-M. Polysaccharides, protein and lipid -based natural edible films in food packaging: A review. *Carbohydr. Polym.* **2020**, *238*, 116178. [CrossRef]
6. Gautam, A.M.; Caetano, N. Study, design and analysis of sustainable alternatives to plastic takeaway cutlery and crockery. *Energy Procedia* **2017**, *136*, 507–512. [CrossRef]
7. Yu, H.; Hou, J.; Namin, R.B.; Ni, Y.; Liu, S.; Yu, S.; Liu, Y.; Wu, Q.; Nie, S. Pre-cryocrushing of natural carbon precursors to prepare nitrogen, sulfur co-doped porous microcellular carbon as an efficient ORR catalyst. *Carbon* **2021**, *173*, 800–808. [CrossRef]
8. Mujtaba, M.; Lipponen, J.; Ojanen, M.; Puttonen, S.; Vaittinen, H. Trends and challenges in the development of bio-based barrier coating materials for paper/cardboard food packaging; a review. *Sci. Total Environ.* **2022**, *851*, 158328. [CrossRef]
9. Bastante, C.C.; Silva, N.H.C.S.; Cardoso, L.C.; Serrano, C.M.; Martínez de la Ossa, E.J.; Freire, C.S.R.; Vilela, C. Biobased films of nanocellulose and mango leaf extract for active food packaging: Supercritical impregnation versus solvent casting. *Food Hydrocoll.* **2021**, *117*, 106709. [CrossRef]
10. Wever, R.; Twede, D. The history of molded fiber packaging: A 20th century pulp story. In Proceedings of the 23rd IAPRI Symposium on Packaging, Windsor, UK, 3–5 September 2007.
11. Zhang, Y.; Duan, C.; Bokka, S.K.; He, Z.; Ni, Y. Molded fiber and pulp products as green and sustainable alternatives to plastics: A mini review. *J. Bioresour. Bioprod.* **2022**, *7*, 14–25. [CrossRef]
12. Lang, C.V.; Jung, J.; Wang, T.; Zhao, Y. Investigation of mechanisms and approaches for improving hydrophobicity of molded pulp biocomposites produced from apple pomace. *Food Bioprod. Process.* **2022**, *133*, 1–15. [CrossRef]
13. Wang, S.; Jing, Y. Effects of formation and penetration properties of biodegradable montmorillonite/chitosan nanocomposite film on the barrier of package paper. *Appl. Clay Sci.* **2017**, *138*, 74–80. [CrossRef]
14. Rastogi, V.K.; Samyn, P. Bio-based coatings for paper applications. *Coatings* **2015**, *5*, 887–930. [CrossRef]
15. Kansal, D.; Hamdani, S.S.; Ping, R.; Sirinakbumrung, N.; Rabnawaz, M. Food-safe chitosan–zein dual-layer coating for water- and oil-repellent paper substrates. *ACS Sustain. Chem. Eng.* **2020**, *8*, 6887–6897. [CrossRef]
16. Nair, A.; Kansal, D.; Khan, A.; Rabnawaz, M. Oil- and water-resistant paper substrate using blends of chitosan-graft-polydimethylsiloxane and poly(vinyl alcohol). *J. Appl. Polym. Sci.* **2021**, *138*, 50494. [CrossRef]
17. Liu, D.; Duan, Y.; Wang, S.; Gong, M.; Dai, H. Improvement of oil and water barrier properties of food packaging paper by coating with microcrystalline wax emulsion. *Polymers* **2022**, *14*, 1786. [CrossRef]
18. Du, Y.; Zang, Y.-H.; Du, J. Effects of starch on latex migration and on paper coating properties. *Ind. Eng. Chem. Res.* **2011**, *50*, 9781–9786. [CrossRef]
19. Schaider, L.A.; Balan, S.A.; Blum, A.; Andrews, D.Q.; Strynar, M.J.; Dickinson, M.E.; Lunderberg, D.M.; Lang, J.R.; Peaslee, G.F. Fluorinated compounds in U.S. fast food packaging. *Environ. Sci. Technol. Lett.* **2017**, *4*, 105–111. [CrossRef]
20. Kotthoff, M.; Müller, J.; Jürling, H.; Schlummer, M.; Fiedler, D. Perfluoroalkyl and polyfluoroalkyl substances in consumer products. *Environ. Sci. Pollut. Res. Int.* **2015**, *22*, 14546–14559. [CrossRef]
21. Stahl, T.; Mattern, D.; Brunn, H. Toxicology of perfluorinated compounds. *Environ. Sci. Eur.* **2011**, *23*, 38–90. [CrossRef]
22. Liu, H.; Gao, S.W.; Cai, J.S.; He, C.L.; Mao, J.J.; Zhu, T.X.; Chen, Z.; Huang, J.Y.; Meng, K.; Zhang, K.Q.; et al. Recent progress in fabrication and applications of superhydrophobic coating on cellulose-based substrates. *Materials* **2016**, *9*, 124. [CrossRef]
23. Zhao, C.; Li, J.; He, B.; Zhao, L. Fabrication of hydrophobic biocomposite by combining cellulosic fibers with polyhydroxyalkanoate. *Cellulose* **2017**, *24*, 2265–2274. [CrossRef]
24. Li, Z.; Rabnawaz, M.; Sarwar, M.G.; Khan, B.; Krishna Nair, A.; Sirinakbumrung, N.; Kamdem, D.P. A closed-loop and sustainable approach for the fabrication of plastic-free oil- and water-resistant paper products. *Green Chem.* **2019**, *21*, 5691–5700. [CrossRef]
25. Nakayama, N.; Hayashi, T. Preparation and characterization of poly(l-lactic acid)/TiO₂ nanoparticle nanocomposite films with high transparency and efficient photodegradability. *Polym. Degrad. Stab.* **2007**, *92*, 1255–1264. [CrossRef]
26. Buzarovska, A.; Grozdanov, A. Biodegradable poly(L-lactic acid)/TiO$_2$ nanocomposites: Thermal properties and degradation. *J. Appl. Polym. Sci.* **2012**, *123*, 2187–2193. [CrossRef]
27. Khairuddin, K.; Nurhayati, N.; Shodik, I.; Pham, T. Water vapour and grease resistance properties of paper coating based starch-bentonite clay. *J. Phys. Conf. Ser.* **2019**, *1153*, 012090. [CrossRef]
28. Kansal, D.; Hamdani, S.S.; Ping, R.; Rabnawaz, M. Starch and zein biopolymers as a sustainable replacement for PFAS, silicone oil, and plastic-coated paper. *Ind. Eng. Chem. Res.* **2020**, *59*, 12075–12084. [CrossRef]
29. Kansal, D.; Rabnawaz, M. Fabrication of oil-and water-resistant paper without creating microplastics on disposal. *J. Appl. Polym. Sci.* **2020**, *138*, 49692. [CrossRef]
30. Kopacic, S.; Walzl, A.; Zankel, A.; Leitner, E.; Bauer, W. Alginate and chitosan as a functional barrier for paper-based packaging materials. *Coatings* **2018**, *8*, 235. [CrossRef]

31. Tyagi, P.; Hubbe, M.A.; Lucia, L.; Pal, L. High performance nanocellulose-based composite coatings for oil and grease resistance. *Cellulose* **2018**, *25*, 3377–3391. [CrossRef]
32. Aulin, C.; Gällstedt, M.; Lindström, T. Oxygen and oil barrier properties of microfibrillated cellulose films and coatings. *Cellulose* **2010**, *17*, 559–574. [CrossRef]
33. Huntrakul, K.; Yoksan, R.; Sane, A.; Harnkarnsujarit, N. Effects of pea protein on properties of cassava starch edible films produced by blown-film extrusion for oil packaging. *Food Package Shelf Life* **2020**, *24*, 100480. [CrossRef]
34. Li, Z.; Rabnawaz, M. Oil- and water-resistant coatings for porous cellulosic substrates. *ACS Appl. Polym. Mater.* **2019**, *1*, 103–111. [CrossRef]
35. Li, Z.; Rabnawaz, M.; Khan, B. Response surface methodology design for biobased and sustainable coatings for water- and oil-resistant paper. *ACS Appl. Polym. Mater.* **2020**, *2*, 1378–1387. [CrossRef]
36. Xie, J.; Xu, J.; Cheng, Z.; Chen, J.; Zhang, Z.; Chen, T.; Yang, R.; Sheng, J. Facile synthesis of fluorine-free cellulosic paper with excellent oil and grease resistance. *Cellulose* **2020**, *27*, 7009–7022. [CrossRef]
37. Han, J.Y.; Hou, Y.J.; Zuo, Q.; Guo, J.; Wang, Y.Q. Hydrophobic modification of ZSM-5 molecular sieve and its catalytic performance for phenol oxidation with H_2O_2. *Saf. Environ. Eng.* **2017**, *24*, 91–96. (In Chinese) [CrossRef]
38. Sheng, J.; Li, J.; Zhao, L. Fabrication of grease resistant paper with non-fluorinated chemicals for food packaging. *Cellulose* **2019**, *26*, 6291–6302. [CrossRef]
39. Vrabič-Brodnjak, U.; Tihole, K. Chitosan solution containing zein and essential oil as bio based coating on packaging paper. *Coatings* **2020**, *10*, 497. [CrossRef]
40. Long, Z.; Wu, M.; Peng, H.; Dai, L.; Zhang, D.; Wang, J. Preparation and oil-resistant mechanism of chitosan/cationic starch oil-proof paper. *BioResources* **2015**, *10*, 7907–7920. [CrossRef]
41. Wu, G.; Liu, D.; Chen, J.; Liu, G.; Kong, Z. Preparation and properties of super hydrophobic films from siloxane-modified two-component waterborne polyurethane and hydrophobic nano SiO_2. *Prog. Org. Coat.* **2019**, *127*, 80–87. [CrossRef]
42. Malakhova, Y.N.; Buzin, A.I.; Chvalun, S.N. Linear and cyclolinear polysiloxanes in the bulk and thin films on liquid and solid substrate surfaces. *J. Surf. Investig.* **2018**, *12*, 339–349. [CrossRef]
43. Yin, F.; Tang, C.; Li, X.; Wang, X.B. Effect of moisture on mechanical properties and thermal stability of meta-aramid fiber used in insulating paper. *Polymers* **2017**, *9*, 537. [CrossRef] [PubMed]

 *coatings*

Article

Paper Coatings Based on Polyvinyl Alcohol and Cellulose Nanocrystals Using Various Coating Techniques and Determination of Their Barrier Properties

Alicja Tarnowiecka-Kuca [1,*], Roos Peeters [2], Bram Bamps [2], Magdalena Stobińska [1,*], Paulina Kamola [1], Artur Wierzchowski [1], Artur Bartkowiak [1] and Małgorzata Mizielińska [1]

[1] Center of Bioimmobilisation and Innovative Packaging Materials, West Pomeranian University of Technology, Klemensa Janickiego 35, 71-270 Szczecin, Poland; pdzieciol@zut.edu.pl (P.K.); wa50253@zut.edu.pl (A.W.); artur-bartkowiak@zut.edu.pl (A.B.); malgorzata.mizielinska@zut.edu.pl (M.M.)

[2] Materials and Packaging Research & Services, Hasselt University, Wetenschapspark 27, 3590 Diepenbeek, Belgium; roos.peeters@uhasselt.be (R.P.); bram.bamps@uhasselt.be (B.B.)

* Correspondence: alicja.tarnowiecka-kuca@zut.edu.pl (A.T.-K.); magda.stobinska@zut.edu.pl (M.S.)

Abstract: The goal of this work was to improve the barrier properties of selected papers against water, grease and oil or gases (water vapor and oxygen) by covering them with biodegradable commercial coating carriers based on cellulose nanocrystals (CNCs) and polyvinyl alcohol (PVOH). The aim was also to obtain cellulose recyclable packaging materials with improved barrier characteristics. The properties of paper coatings based on CNCs and PVOH were characterized. Various paper coating techniques (flexographic printing, rotogravure printing and blade printing) were evaluated with respect to the final properties of the surface-modified paper with different starting grammages ($40 \, \text{g/m}^2$, $70 \, \text{g/m}^2$, $100 \, \text{g/m}^2$). Functional properties, such as the barrier against oxygen, water vapor, water and grease; mechanical properties; and seal characterization of coated paper were examined. The results of this study demonstrated that the covering of the paper may improve the water, grease and oil barrier and that the best results were obtained for Gerstar $70 \, \text{g/m}^2$ coated with J12 coatings using the flexographic technique.

Keywords: coated paper; coating methods; PVOH; CNCs; barrier properties

Citation: Tarnowiecka-Kuca, A.; Peeters, R.; Bamps, B.; Stobińska, M.; Kamola, P.; Wierzchowski, A.; Bartkowiak, A.; Mizielińska, M. Paper Coatings Based on Polyvinyl Alcohol and Cellulose Nanocrystals Using Various Coating Techniques and Determination of Their Barrier Properties. *Coatings* **2023**, *13*, 1975. https://doi.org/10.3390/coatings13111975

Academic Editors: Jun Mei, Jing Xie and Biao Zhang

Received: 15 October 2023
Revised: 13 November 2023
Accepted: 14 November 2023
Published: 20 November 2023

1. Introduction

Food packaging is one of the most important processes in the food supply chain. The main function of the packaging materials is to protect and preserve the quality and safety of food products. It is also important for extending their shelf life [1]. Currently, the food packaging sector is faced with the necessity of the proper selection of the most appropriate eco-packaging materials taking into account the recent changes in both regulations with respect to circular economy and end-user sustainability expectations. Packaging materials that are nontoxic, recyclable and biodegradable and have high barrier properties against moisture, gas, and oil are in high demand. Paper as a sustainable packaging material arouses great interest, but due to inadequate barriers against water vapor and oxygen, its application in the processed food packaging sector is very limited [2]. EU policy requires a gradual increase in the amount of recycled materials used in packaging [3]. It should be noted that packaging made of cellulose in the form of paper or cardboard is one of the oldest types of packaging and has been successfully recycled for a long time [2,3].

Unmodified paper is characterized by its fiber-based and as a result porous structure. It can easily absorb moisture due to its hygroscopic structure, providing a low barrier against water, water vapor, oxygen or grease. These characteristics disqualify it as a material for use in a wide range of applications. To overcome this limitation, paper is usually coated with other materials, including plastics or aluminum, that provide improvements in the barrier properties at the expense of its eco-friendly and biodegradable nature [4].

One of the final processes that affects paper properties is calendering. During the calendering process, a paper sheet is pressed between two or more cylinders at a certain pressure and temperature [5]. The main purpose of calendering is to modify the surface structure [6]. In the case of printing on either cardboard or paper, the main calendering effect is the tightening of the holes, which reduces the surface roughness to achieve a good print quality.

In this case, various treatments are used to obtain a cellulose material with enhanced measurable barrier properties. To be applicable as food packaging, these materials are covered with one or more coatings offering separate barrier properties against gases and moisture and good sealing and mechanical properties. The barrier properties of the packaging materials define the level of resistance of the materials to various external and internal factors. In modified packaging materials, all substances that have contact with food products should be approved for contact with food [7]. All technological processes, with precautions taken for writing paper as well as packaging paper in general, depending on the area of application, are designed to prevent the penetration of certain chemical or physical agents.

Naturally renewable biopolymers including proteins, polysaccharides, lipids or combinations of those components have already been evaluated as barrier coatings for various paper materials. In most cases, they maintain the favorable recyclable and environmentally friendly characteristics of the paper packaging materials [8]. Biopolymers such as polysaccharides and proteins are regarded as the most promising substances due to their advantages of providing an appropriate barrier against gases and having a positive effect on mechanical properties [9]. Cellulose nanocrystals are bio-based nanoparticles of high crystallinity obtained from a cellulosic material (e.g., wood, cotton). The amorphous part of cellulose is removed by acid hydrolysis, and pure nanocrystals are obtained after the sonication of the batch. CNCs are biodegradable materials with excellent mechanical properties. In some studies, they were evaluated as coating additives in the investigation of the surface and barrier properties of paper [10].

Poly(vinyl alcohol) (PVOH) is a synthetic water-soluble polymer. It is mainly known for its high barrier capacity against oxygen, good thermal resistance, great adhesive properties and biocompatibility. Other characteristics it displays are good water solubility and biodegradability, excellent film-forming capacity and transparency [11]. Polyvinyl alcohol coatings have been implemented in a variety of applications. It is an ideal coating material to use in applications such as paper-based packaging and barrier materials due to its high content of hydroxyl groups. Additionally, the flexibility, high mechanical strength and long-term thermal stability of PVOH are the most important beneficial properties for the aforementioned applications [12].

The goal of this work was to improve the barrier properties of selected types of paper against water, grease and oil or gases (water vapor and oxygen) by covering them with biodegradable commercial coating carriers based on CNCs and PVOH.

2. Materials and Methods

2.1. Materials

Three commercial types of paper (reference papers) were used in this research:

- Kraft paper (grammage 40 g/m^2), (Nordic Paper, Greåker, Norway);
- White paper (grammage 70 g/m^2), (Gerstar HDS, Ahlstrom, France);
- Kraft paper (grammage 100 g/m^2), (Mondi, Poperinge, Belgium).

Two commercial coating carriers were used to cover the papers:

- C03—aqueous dispersion of cellulose nanocrystals (CNCs) (Eurikas, Vilvoorde, Belgium);
- J12—aqueous dispersion of polymers based on polyvinyl alcohol (FOBARO GROUP, Poznań, Poland).

The chemical composition of both coating carriers (C03 and J12) was not disclosed by the producer.

2.2. Coating Carrier Preparations

The aqueous solution/suspension of CNCs (C03) and the aqueous dispersion of PVOH were diluted with distilled water at a ratio of 4:1. Additionally, the C03 coating carrier was diluted with distilled water at a ratio of 1:1.

2.3. Coating Carrier Analysis

The coating carriers were analyzed by determination of solid content (moisture analyzer, Radwag, Radom, Poland), pH (Elmetron, Zabrze, Poland), appearance and viscosity. The viscosity was analyzed using two methods. As the first step of the experiments, a Ford viscosity cup No. 4 was used. Next, the viscosity was analyzed using an AR-G2 Rheometer (TA Instruments, New Castle, DE, USA) at 25 °C.

2.4. Paper Covering

A4 sheets of selected papers were covered with the coatings using three different types of coaters: Unicoater 409 (Erichsen, Hemer, Germany) with 4, 6, 12 and 24 µm diameter bars; SUMET (Messtechnik, Freiburg, Germany) with 25° and 40° blades using the drawn down speed 5 m/min; the hand flex-coating K-lox manual applicator (RK Print Coat Instrument, Royston, UK) with 400/5 (transfer volume 4.3 m^3/m^2) and 100/185 (application volume 39.1 m^3/m^2) engraved rollers. The coating was performed at 25 °C. All coated materials were dried for 4 min at a temperature of 80 °C.

2.5. Uncoated and Coated Paper Property Analysis

In this study, 3 types of uncoated (control samples) papers and paper covered with the C03 and the J12 coatings using different methods were thoroughly characterized. The following tests were performed: thickness (Sylvac, Yverdon-les-Bains, Switzerland), paper grammage and grammage of the obtained coatings according to PN-EN ISO 536:2020-08 [13]. Water absorption (Cobb60) analysis was carried out according to ISO 535:2023 [14]. This analysis allowed the determination of the water barrier properties. Grease barrier properties were measured using the KIT test according to ISO 16532-2-2007—oil and grease resistance [15].

The coating samples with a grammage lower than 5% of the total paper grammage were selected for further analysis. The oxygen transmission rate (OTR) of these samples was measured according to the ASTM F1927-20 [16] method (temperature 23 °C, 0% RH). The water vapor transmission rate (WVTR) was measured according to ASTM F1249-20 standard [17] (temperature 23 °C, 85% RH).

To analyze the structure of the uncoated paper and to verify the presence and integrity of the obtained coatings, a microscopic analysis of the selected papers was performed. The paper/coated paper samples were then placed on pin stubs, and a thin layer of gold in a sputter coater was used to cover them at room temperature (Quorum Technologies Q150R S, Laughton, UK). Then, the samples were analyzed with a scanning electron microscope (SEM). The microscopic analysis of the samples was carried out through the use of a Vega 3 LMU microscope (Tescan, Brno, Czech Republic). The analysis took place at 24 °C with the use of a tungsten filament with an accelerating voltage of 10 kV to create SEM images of the uncoated and covered papers. The specimens were then examined from above.

The seal strength of coated surfaces was determined in accordance with the ASTM F88/F88M-21 standard [18] using the RDM HSE-3 sealer with flat bars. Following the sealing of 30 mm wide samples, the central 15 mm width was subsequently excised for the assessment of seal strength, 4 h after sealing. The temperature of the bars was varied, while seal time and pressure were maintained at 0.3 s and 2.0 N/mm^2, respectively. An average and standard deviation were calculated based on three individual measurements.

3. Results

The results of this study demonstrated that C03 and J12 coating carriers were characterized by a high viscosity (Table 1). Unfortunately, it is impossible to cover paper with too-viscous carriers (Ford cup 4.0 > 60 s) using flexographic coating techniques. In order to decrease the viscosity of both the C03 and J12 aqueous dispersions, the analyzed dispersions were diluted with distilled water at a ratio of 4:1. As was shown in Table 1, a viscosity lower than 30 s was observed for diluted J12 carrier, while no significant decrease in viscosity was noticed for C03 diluted carrier. Therefore, in the case of C03, a 1:1 dilution was applied. Unfortunately, the high dilution of C03 did not decrease its viscosity but resulted in an undesirable decrease in the dry mass of the aqueous coating dispersion.

Table 1. Characteristics of the commercial coating carriers.

Specifications	C03	J12
Solid content (%)	25.6	20.0
After dilution	–	18
pH	2.5–6.5	5.5
Appearance	Liquid, transparent	Liquid, yellow−transparent
Viscosity (mPa·s at 23 °C)	800	500
After dilution	800	180
Viscosity (s at 23 °C)	>60	>60
After dilution	>60	<30

Three types of unmodified paper were characterized as control samples in relation to the modified papers (Table 2). It was important to analyze the paper before covering it. The results of this study demonstrated that the grease barrier of the uncoated paper was very low (Table 2). The highest water absorption was noticed for the uncovered Gerstar, while the lowest water absorption was observed for the Kraft paper (grammage 40 g/m^2) samples. In summary, none of the analyzed papers had a barrier against both grease/oil and water.

Table 2. Characteristics of the uncoated papers.

No.	Name	Grammage (g/m^2)	Thickness (mm)	KIT Test *	Cobb60 (g/m^2)
1	Kraft	40 ± 1.140	0.064 ± 0.002	1	27.4 ± 1.853
2	Gerstar	70 ± 0.713	0.065 ± 0.003	1	36.3 ± 0.192
3	Kraft	100 ± 1.141	0.143 ± 0.003	1	32.4 ± 3.412

* 1 = low grease and oil barrier. 12 = high grease and oil barrier.

It was assumed in this study that an expected, upper mass limit/grammage of the surface modification/coating should be lower than 5% wt. of the mass/grammage of the paper (before covering). Based on this assumption, the possible coating grammage for the three types of paper should be ≤2.0 g/m^2 for Kraft 40 g/m^2, ≤3.5 g/m^2 for Gerstar 70 g/m^2, and ≤5.0 g/m^2 for Kraft 100 g/m^2.

The results of the experiments that were performed using the UNICOATER 409 demonstrated that the grammage values of the J12 and C03 layers applied on the surface of the Kraft 40 g/m^2 paper were 4.2 g/m^2 (J12) and 8.7 g/m^2 (C03) (Table 3). This means that the coatings' grammages/weights exceeded 5% of the paper grammage even when the smallest application bar thickness (4 μm) was used. Similar results were noticed for the Gerstar 70 g/m^2 paper that was covered with the C03 layer (Table 4). The obtained findings resulted in the rejection of the covered Kraft 40 g/m^2 paper and Gerstar 70 g/m^2 paper coated with the C03 layer for further analysis (Tables 3 and 4). The results of the experiments led to the observation that none of the obtained J12 coatings applied on the surface of the Gerstar 70 g/m^2 paper was higher than 3.5 g/m^2. This means that none of them exceeded 5% of the paper grammage. Additionally, it was observed that covering the Gerstar 70 g/m^2

paper with J12 coatings improved its barrier properties. In the analysis of the grease and oil barrier changes, it was noticed that these barriers increased from 1 to 5 or from 1 to 12 when compared to uncoated Gerstar 70 g/m^2 paper. As was emphasized in Table 3, the use of a 4 μm thickness bar allowed a J12 layer of 1.2 g/m^2 grammage to be obtained. The thinnest coating improved the grease and oil barrier properties, but a higher improvement was observed for the layers with higher grammages. Comparing the J12 layers that were obtained using 6 μm, 12 μm and 24 μm thickness bars, it was demonstrated that the grease and oil barrier increased from 1 to 12, confirming that the grammage of the coating (in the range of 1.9–3.5 g/m^2) had no influence on the barrier improvement. Comparing the water barrier properties of J12 coatings, it was observed that the increase in the water barrier was the highest for the coating with the grammage of 3.5 g/m^2. The lowest improvement was observed for the layer with the grammage of 1.2 g/m^2. The findings proved that the grammage of the J12 coating on the surface of Gerstar 70 g/m^2 paper had an influence on its water barrier properties. It is worth mentioning that the grease and oil barrier properties and water barrier properties observed for the J12 coating with the grammage of 1.9 g/m^2 and the J12 coating with the grammage of 3.3 g/m^2 were the same. This means that from the economical point of view, it would be better to use a 6 μm thickness bar than a 12 μm thickness bar to cover Gerstar 70 g/m^2 paper with the J12 coating.

Table 3. Characterization of the papers coated with J12 layers using a laboratory coater (UNICOATER 409 Erichsen).

No.	Sample	Wet Coating Thickness (μm)	Paper Grammage (g/m^2)	Thickness (mm)	Coating Grammage (g/m^2)	KIT Test *	Cobb60 (g/m^2)
1	Kraft 40 g/m^2	4	44 ± 1.886	0.073 ± 0.003	4.2 ± 1.886	–	–
2	Gerstar 70 g/m^2	4	71 ± 0.825	0.069 ± 0.003	1.2 ± 0.825	5	22.9 ± 0.409
		6	71 ± 0.236	0.070 ± 0.001	1.9 ± 0.825	12	20.2 ± 1.677
		12	73 ± 0.589	0.070 ± 0.002	3.3 ± 0.589	12	20.2 ± 0.147
		24	73 ± 0.118	0.074 ± 0.003	3.5 ± 0.118	12	19.0 ± 0.071
3	Kraft 100 g/m^2	4	105 ± 2.121	0.169 ± 0.004	5.5 ± 2.121	5	39.1 ± 1.387
		6	106 ± 1.414	0.170 ± 1.414	6.3 ± 1.414	6	37.9 ± 3.928
		12	105 ± 0.471	0.172 ± 0.004	5.5 ± 0.471	5	4.0 ± 0.318
		24	106 ± 1.061	0.174 ± 0.009	6.1 ± 1.061	6	4.0 ± 0.078

* 1 = low grease and oil barrier. 12 = high grease and oil barrier.

Table 4. Characterization of the papers coated with C03 layers using a laboratory coater (UNICOATER 409 Erichsen).

No.	Sample	Wet Coating Thickness (μm)	Paper Grammage (g/m^2)	Thickness (mm)	Coating Grammage (g/m^2)	KIT Test *	Cobb$_{60}$ (g/m^2)
1	Kraft 40 g/m^2	4	48 ± 1.875	0.071 ± 0.002	8.7 ± 1.875	–	–
2	Gerstar 70 g/m^2	4	78 ± 1.875	0.077 ± 0.002	8.2 ± 1.711	–	–
3	Kraft 100 g/m^2	4	106 ± 2.828	0.169 ± 0.006	6.0 ± 2.282	1	37.4 ± 5.091
		6	105 ± 2.475	0.170 ± 0.006	5.8 ± 2.475	5	39.3 ± 3.323
		12	106 ± 1.367	0.172 ± 0.008	6.0 ± 1.367	5	35.7 ± 1.131
		24	107 ± 2.663	0.74 ± 0.010	–	–	–

* 1 = low grease and oil barrier. 12 = high grease and oil barrier.

The results of the work indicated that in the case of Kraft 100 g/m^2 paper, covering the paper with the J12 and C03 coatings using 4 μm and 6 μm thickness bars decreased the water barrier properties of the paper and increased the grease and oil barrier properties but not significantly (Tables 3 and 4). Similar results were noticed for the C03 layer that was applied on the surface of the Kraft 100 g/m^2 paper using a 12 µm thickness bar. As shown in Table 3, the grease and oil barrier increased from 1 to 5 and 6 when 12 µm and 24 µm thickness bars were used to coat the analyzed paper with the J12 layers, meaning that the coatings improved the barrier properties of the paper. Moreover, these coatings caused a decrease in water absorption (Table 3) from 32.4 g/m^2 to 4 g/m^2, meaning that the improvement of the water barrier properties observed for these two coatings was significant. However, it has to be underlined that both of the J12 coatings (with grammages of 5.5 g/m^2 and 6.1 g/m^2) exceeded 5% of the paper grammage.

Surface modification using flexographic printing is an effective and economically feasible technique only when the applied coatings have low viscosities (Ford cup < 30 s) and optimum high dry mass contents (>40% d.m.). The results of this study confirmed that the viscosity of the C03 coating carrier was higher than 60 s when measured using a Ford cup. It was described above that the dilution of the carrier did not decrease its viscosity but resulted in an undesirable decrease in the dry mass of the aqueous coating dispersion. This is why the C03 coating carrier was rejected and omitted from the further tests using K-lox. Table 5 summarizes the results observed for the J12 coatings applied on the surfaces of three analyzed papers. The results of the work indicated that the grammages of the Gerstar 70 1 g/m^2 paper and of the Kraft 100 1 g/m^2 paper were not changed when an anilox 4.3 cm^3/m^2 coater was used to cover the paper samples. As shown in Table 5, the J12 coating grammages applied on the surfaces of all papers for a theoretical applicator volume of 39.1 cm^3/m^2 were noticed to be lower than 1 g/m^2. It is worth mentioning that the grammages of the coatings obtained using the flexographic technique were lower than the grammages of J12 layers obtained using the UNICOATER 409.

Table 5. Characterization of the papers coated with J12 layers using a hand K-lox manual flex-coating applicator.

No.	Sample	Application Volume (cm^3/m^2)	Paper Grammage (g/m^2)	Thickness (mm)	Coating Grammage (g/m^2)	KIT Test *	Cobb$_{60}$ (g/m^2)
1	Kraft 40 g/m^2		40 ± 0.118	0.067 ± 0.003	0.4 ± 0.118	6	13.9 ± 0.212
2	Gerstar 70 g/m^2	4.3	70 ± 0.354	0.067 ± 0.003	–	–	–
3	Kraft 100 g/m^2		100 ± 0.354	0.158 ± 0.003	–	–	–
1.1	Kraft 40 g/m^2		40 ± 0.354	0.067 ± 0.004	0.4 ± 0.354	6	13.9 ± 0.212
2.1	Gerstar 70 g/m^2	39.1	70 ± 0.118	0.068 ± 0.003	0.2 ± 0.118	12	14.4 ± 0.141
3.1	Kraft 100 g/m^2		101 ± 0.707	0.147 ± 0.002	0.8 ± 0.707	1	45.3 ± 1.125

* 1 = low grease and oil barrier. 12 = high grease and oil barrier.

In an analysis of the water absorption of the J12 layers obtained using the flexographic method, it was determined that coatings applied on the surface of the Kraft 40 g/m^2 paper and on the surface of the Gerstar 70 g/m^2 paper improved the water barrier properties of the paper samples. However, this coating increased the water absorption of the Kraft 100 g/m^2 paper. Even though the grammage of the J12 coatings on the surface of the Kraft 40 g/m^2 paper and on the surface of the Gerstar 70 g/m^2 paper was low (Table 5), these coatings significantly improved the grease and oil barrier properties of these papers. It was shown that the grease and oil barrier increased from 1 to 6 for the Kraft 40 g/m^2 samples, while a higher increase (from 1 to 12) in grease and oil barrier properties was noticed for the Gerstar 70 g/m^2 paper. In summary, the highest barrier property improvements were observed for the Gerstar 70 g/m^2 paper covered with J12 coating. Comparing the J12 coatings applied on the surface of Gerstar 70 g/m^2 paper, it should be underlined that the layers with the grammages of 1.9 g/m^2 and 3.3 g/m^2 (obtained using the UNICOATER 409) and with the grammage of 0.2 g/m^2 (obtained using the K-lox) influenced the decrease in the grease and oil permeability and the water absorption (Tables 3 and 5), confirming that the coatings' barrier property improvement did not depend on their grammages significantly.

As shown in Table 6, the grammages of the J12 and C03 coatings applied on the surfaces of Kraft 40 g/m^2, Gerstar 70 g/m^2 and Kraft 100 g/m^2 papers using the blade technique did not exceed 5% of the papers' grammages. In a comparison of the three covering techniques that were used in this study, it was confirmed that the lowest coating grammages (lower than 1 g/m^2) were obtained when the flexographic technique was used (Tables 4–6).

Table 6. Characterization of the papers coated with J12 and C03 layers using a SUMET blade coater.

No.	Sample	Coating	Blade (°)	Paper Grammage (g/m^2)	Thickness (mm)	Coating Grammage (g/m^2)	KIT Test *	Cobb$_{60}$ (g/m^2)
1	Kraft 40 g/m^2		40	41 ± 0.354	0.072 ± 0.003	1.0 ± 0.354	4	29.7 ± 2.192
2	Gerstar 70 g/m^2	J12		71 ± 0.753	0.066 ± 0.002	1.4 ± 0.753	7	22.5 ± 2.893
3	Kraft 100 g/m^2		25	103 ± 1.633	0.150 ± 0.005	3.5 ± 1.633	1	23.6 ± 1.485
1.1	Kraft 40 g/m^2			41 ± 0.937	0.064 ± 0.004	1.3 ± 0.354	6	31.0 ± 1.872
3.1	Kraft 100 g/m^2	C03	40	103 ± 1.267	0.160 ± 0.009	3.5 ± 1.633	1	40.2 ± 0.849

* 1 = low grease and oil barrier. 12 = high grease and oil barrier.

The results of this study indicated that the water absorption of the Kraft 100 g/m^2 paper increased when it was covered with the C03 layer. However, the water slightly decreased after it was covered with the J12 coating. It was also observed that neither of these two coatings influenced the grease and oil barrier improvement (Table 6). As shown in Table 6, the grease and oil barrier of the Kraft 40 g/m^2 paper increased from 1 to 4 (J12 layer) and to 6 (C03 layer). Despite the grease and oil barrier improvement, both of these coatings were found to increase the water absorption of these paper samples. Opposite results were obtained for the Gerstar 70 g/m^2 paper covered with the J12 layer, which caused a decrease in water absorption when compared to uncovered paper. Additionally, the J12 coating applied on the surface of the Gerstar 70 g/m^2 paper was observed to improve the paper's grease and oil barrier properties.

To perform SEM analysis, Kraft 40 g/m^2 paper and Gerstar 70 g/m^2 paper coated with J12 layers were selected. The Kraft 100 g/m^2 paper covered with a C03 layer was selected as well. The paper samples were coated using a Sumet blade coater. It should be underlined that the SEM analysis was carried out to demonstrate if the coatings were present on the surface of the papers. The goal was also to show if the coatings were smooth and if they covered the whole surface of the papers. The results of the work demonstrated that the whole surface of the Gerstar 70 g/m^2 paper was covered with the J12 coatings. Although the grammage of the J12 coating was low (1.4 g/m^2, Table 6), the cellulose fibers were not visible, meaning that the coating carrier was well distributed on the surface of the paper (Figure 1b). It is worth mentioning that covering the Gerstar 70 g/m^2 paper with the J12 coatings led to the improvement of the water, grease and oil barrier. It can be suggested that the smooth and well-distributed coating had a significant influence on the water absorption and grease and oil permeability decrease. An SEM image (Figure 1b) shows that small breaks and scratches were visible on the coating's surface. This may lead to the conclusion that if the breaks and scratches had not been visible on the J12 surface, the barrier property improvement would have been higher. Different results were obtained for the Kraft 40 g/m^2 paper covered with a J12 layer. SEM analysis confirmed that the whole surface of the paper was not covered with the J12 coating. Partially covered paper samples and uncoated cellulose fibers are well visible in Figure 1a. In summary, the SEM analysis confirmed that the Kraft 40 g/m^2 paper surface was not completely covered with the J12 coating, which could have resulted in increased absorption of water by the paper (Tables 2 and 6).

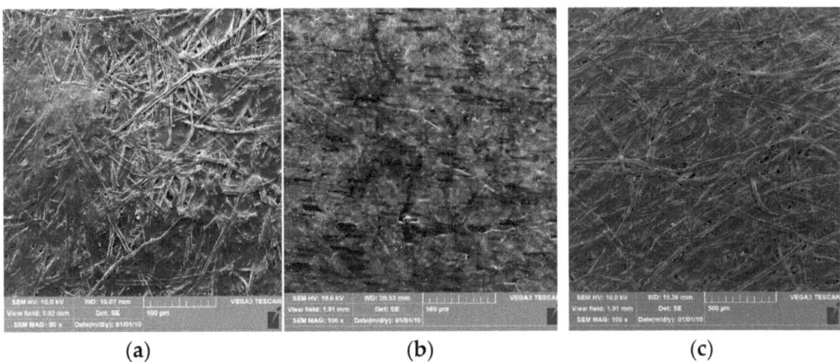

(a)   (b)   (c)

Figure 1. SEM images of papers covered with coatings using a Sumet blade coater: (**a**) J12 applied on Kraft 40 g/m^2, (**b**) J12 applied on Gerstar 70 g/m^2, (**c**) C03 applied on Kraft 100 g/m^2.

Similar results were observed for the Kraft 100 g/m^2 paper covered with a C03 layer. SEM analysis indicated that uncovered cellulose fibers were well visible, as shown in Figure 1c. The high viscosity of the C03 coating carrier could lead to difficulties during surface covering which could have resulted in the increase in the absorption of water by the analyzed paper.

The barrier properties against water vapor and oxygen were analyzed only for selected samples. The results of the WVTR and OTR analysis showed (Table 7) that the water vapor and oxygen permeability of the Kraft 40 and Kraft 100 papers covered with the J12 and C03 coatings decreased when compared to uncoated paper samples. However, it was confirmed that neither covered papers nor uncoated samples have a gas barrier against both water vapor and oxygen.

Table 7. WVTR and OTR results for papers covered with J12 and C03 coatings.

Coated Paper	WVTR (g/m^2·day)	OTR (cm^3/m^2·day)
Uncoated Kraft 40 g/m^2 J12	>5000.0	>20,000.0
Uncoated Kraft 100 g/m^2	>5000.0	>20,000.0
Kraft 40 g/m^2 coated with J12	>1000.0	>10,000.0
Kraft 100 g/m^2 coated with C03	>1000.0	>10,000.0

In Figure 2, the sealing results are presented for both paper coatings. One notable observation is the shift of the sealing curve towards higher temperatures in the case of C03. Apart from the distinct coating, which likely implies differences in glass transition and melting temperatures, the choice of paper also plays a significant role. For C03, a paper with a substantially higher basis weight of 100 g/m^2 was utilized, whereas a paper with a weight of 40 g/m^2 was employed for J12. Due to the brief sealing time of 0.3 s and the insulating properties of paper, it is rational that heavier papers necessitate more energy to attain a sufficient temperature at the interface where sealing occurs. For both C03 and J12, following the initiation of sealing, there exists a temperature window within which a maximum plateau value appears to be attained. In the case of C03, this plateau appears to persist, although there is, of course, a subsequent decrease in strength after temperatures at which the coating material would be pushed away or degrade. However, this decline is not discernible in the graph due to the limitation of not further elevating the temperature, as temperatures exceeding 250 °C are not employed in industrial packaging processes. In the case of J12, a decline is already evident at 180 °C, indicating that this material is optimally sealed at a lower temperature.

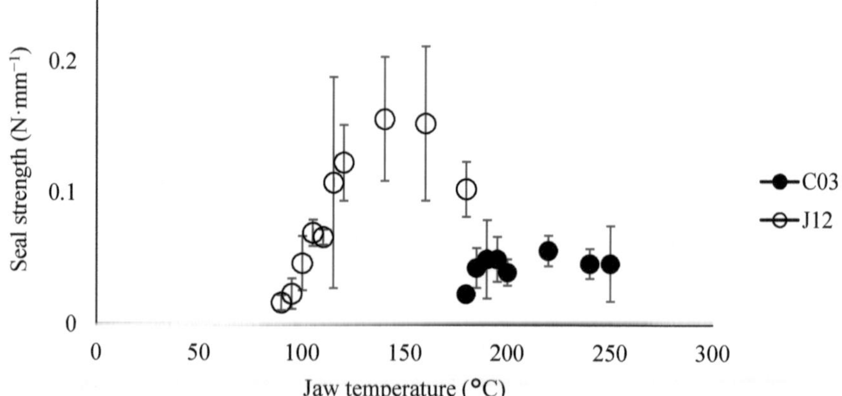

Figure 2. The influence of jaw temperature on seal strength of coated papers.

Additionally, it is crucial to highlight the seal failure mechanisms as they are linked to the numerical values for strengths. The C03-coated paper exhibited a peeling mechanism at all temperatures, whereas the J12-coated paper demonstrated fiber tearing starting from temperatures of 115 °C. In prior studies, fiber tearing has been acknowledged as a favorable outcome after sealing [19,20]. While peeling primarily involves the failure of the coating, in the case of fiber tearing, it is the paper itself that becomes the weak factor during the peel test.

4. Discussion

Water-soluble polymer and biobased filler coatings on paper packaging materials are very promising solutions for the future improvement of food packaging. They have potential environmental advantages over conventional synthetic paper coatings. Additionally, low cost, availability and renewability expand the use of nanocellulose as a coating carrier. It is a nontoxic, biobased substance that can be chosen as a coating component among other green alternatives [21–25]. Furthermore, CNCs have a high modulus of elasticity and a high tensile strength, demonstrating excellent mechanical properties [18]. The C03 carrier used in this study is a commercial coating carrier that has good oxygen barrier properties according to its producer. The results of the experiments that were performed using the UNICOATER 409 demonstrated that the layers obtained by covering papers with C03 layers decreased the water barrier properties of the paper and increased the grease and oil barrier properties, but not significantly. In the analysis of the use of the flexographic technique, it has to be highlighted that the C03 coating carrier was rejected and omitted from the tests using the K-lox due to its high viscosity. It was observed in this study that the C03 coatings that were applied on the surfaces of Kraft 40 g/m^2, Gerstar 70 g/m^2 and Kraft 100 g/m^2 papers using the Sumet blade technique did not exceed 5% of the papers' grammages. The results also indicated that the water absorption of the Kraft 100 g/m^2 paper increased when it was coated with the C03 layer, but the coating influenced the grease and oil barrier improvement. The results were confirmed by SEM analysis, which indicated the presence of uncovered cellulose fibers. The SEM analysis performed by Mazega et al. [26] confirmed that not all cellulose nanofibrils (CNFs) remained on the surface of the paper as a coating. Many of them penetrated the sheet transversally, established hydrogen bonds with inner fibers and increased the amount of bound water. The high viscosity of a CNC-based coating carrier could lead to difficulties during surface coating processes which could cause an increase in the absorption of water by the paper. Additionally, the C03 coating was confirmed to improve the barrier against gases (water vapor and oxygen). However, the improvement was not significant enough. These results were in agreement with those of Mazega et al. [26]. The results of this study indicated that the C03 coating applied on the surface of the paper had no barrier against water vapor and oxygen. Nair et al. [27] suggested that films made purely of mechanically fibrillated CNFs may have very high air and oxygen barrier properties. However, Tayeb et al. [28] mentioned that they can lose their barrier properties once placed in humid conditions. Herrera et al. [4] improved the oxygen and water vapor barrier properties of coatings applied on the surface of paper with the addition of sorbitol as a plasticizer and cross-linking. It is worth mentioning that due to the properties of CNCs, it may be proposed to use them in multilayer systems or in mixed systems, e.g., with shellac, which reduces water vapor permeability [22]. An increase in water barrier properties can also be noticed when CNCs are used together with Ag organic and beeswax particles. Coating paper with such a mixture leads to a modification of the paper surface towards an increase in its hydrophobicity [23]. Therefore, it may be concluded and even suggested that CNC coating is still a promising material with high potential for modification of cellulosic packaging materials [24].

Many scientific reports indicate that PVOH coatings have good properties against oxygen, water vapor, grease and oil. The commercial J12 coating used in this study is a mixture of polyvinyl alcohols. According to the producer protocol, the J12 coating carrier should provide both a close surface and an oxygen barrier on packaging materials. It can be applied by a variety of methods, including gravure roller (forward, reverse and offset, flex), return roller, air knife or Mayer bar. Polyvinyl alcohol (PVA) is a suitable coating carrier with good film-forming and biodegradable properties [29]. Additionally, one of the most important factors affecting the barrier properties of coated paper is the coating grammage. In this study, it was assumed that from the economical point of view, the coating grammage should not exceed 5% of the paper grammage, which is also important from an environmental point of view, meaning these materials may be recyclable.

The results of this study demonstrated that that the J12 coatings applied (using the UNICOATER 409) on the surface of the Gerstar 70 g/m^2 did not exceed 5% of the paper grammage. Additionally, this coating improved the paper's barrier properties against water, grease and oil. The coating grammage of the Gerstar 70 g/m^2 paper covered with the J12 coating using the K-lox was noticed to be lower than 1 g/m^2, which was important from the economical point of view, and it meant that the grammages of the coatings obtained using the flexographic technique were lower than the grammages of J12 layers obtained using the UNICOATER 409. Additionally, J12 coatings applied on the surface of the Kraft 40 g/m^2 paper and on the surface of the Gerstar 70 g/m^2 paper improved the water, grease and oil barrier properties of these papers. The Gerstar 70 g/m^2 paper covered with a J12 layer using a Sumet blade coater was also confirmed to exhibit decreased water absorption and decreased grease and oil permeability. It was also determined that covering the selected papers with the coatings based on PVOH influenced the WVTR and OTR of the papers. However, the barrier of the coatings against water vapor and oxygen was not high enough. Many authors confirmed that PVOH coatings may improve the barrier properties of paper; however, modification of a PVOH carrier can provide additional advantages. Christophliemk et al. [30] confirmed that an ethylene-modified PVOH grade was found to provide lower oxygen transmission rates at high relative humidity, as compared to a standard PVOH grade. Shuman et al. [31] have shown that calendering of a paper substrate at high roll temperatures improved the homogeneity of a PVOH coating that was applied on the paper surface. The authors suggested that the improved homogeneity was reflected in better barrier properties of the coated paper, such as lower OTR and WVTR and better grease and oil resistance.

In summary, the results of this study confirmed that the J12 coating carrier was a better carrier than the C03 coating carrier due to its lower viscosity and the possibility of the development of effective coatings (against water, grease and oil) with a low grammage. This is why it can be concluded that a PVOH coating carrier is a promising material with a high potential for modification of cellulosic packaging materials for food product application.

5. Conclusions

The results of this study demonstrated that C03 and J12 coating carriers were characterized by a high viscosity. The dilution of the carriers with distilled water at a ratio of 4:1 led to a decrease in the viscosity, but only for the J12 carrier. Kraft 40 g/m^2, Kraft 100 g/m^2 and Gerstar 70 g/m^2 papers were found to have a low grease and oil barrier and a high water absorption. The results of the experiments led to the observation that J12 coatings applied on the surface of the Gerstar 70 g/m^2 did not exceed 5% of the paper grammage. Additionally, this paper covered with J12 layers exhibited a significant improvement in its barrier properties against water, grease and oil. Moreover, the grammage of the coating (in the range of 1.9–3.5 g/m^2) had no influence on the barrier improvement. On the other hand, coating the Kraft 100 g/m^2 paper with J12 and C03 coatings decreased the water barrier properties of the paper and increased the grease and oil barrier properties but not significantly. Similar results were noticed for the C03 layer that was applied on the surface of the Kraft 100 g/m^2 paper using a 12 μm thickness bar. The paper samples covered with a J12 layer but with a higher grammage exhibited improved barrier properties. However, both of the J12 coatings exceeded 5% of the paper grammage.

The coating grammage of Gerstar 70 g/m^2 paper covered with the J12 coating using the K-lox with a theoretical applicator volume of 39.1 cm^3/m^2 was noticed to be lower than 1 g/m^2. This means that the grammages of the coatings obtained using the flexographic technique were lower than the grammages of J12 layers obtained using the UNICOATER 409. Additionally, J12 coatings applied on the surface of the Kraft 40 g/m^2 paper and on the surface of the Gerstar 70 g/m^2 paper improved the water, grease and oil barrier properties of these papers. It has to be mentioned that the highest barrier property improvements were observed for the Gerstar 70 g/m^2 paper coated with J12 and that they did not depend on the grammage significantly. The Gerstar 70 g/m^2 paper covered with a J12 layer using a Sumet blade coater was also found to exhibit decreased water absorption and decreased grease and oil permeability when compared to uncovered paper. SEM analysis of the three paper samples demonstrated that only Gerstar 70 g/m^2 paper was well coated with J12 layers and that the coating carrier was well distributed on the whole surface of the paper samples even if the grammage of the J12 coating was low (1.4 g/m^2). It was also demonstrated that covering the selected papers with the coatings based on CNCs and PVOH influenced the WVTR and OTR of the Kraft 40 and Kraft 100 papers. However, it was confirmed that neither covered papers nor uncoated samples have a barrier against water vapor and oxygen. Sealing test results for papers coated with J12 and C03 layers indicated that in the case of C03, the sealing curve shifted towards higher temperatures. In the case of J12, a decline was already observed at 180 °C, indicating that this material was optimally sealed at a lower temperature. Additionally, paper covered with the C03 layer exhibited a peeling mechanism at all temperatures, whereas the J12-coated paper demonstrated fiber tearing starting from temperatures of 115 °C.

Author Contributions: Conceptualization, R.P., A.B., A.T.-K., M.S. and P.K.; methodology, A.T.-K., P.K. and B.B.; validation, M.S. and P.K.; analysis, A.T.-K., M.S., P.K., B.B. and A.W.; data curation, M.M.; writing—original draft preparation, A.T.-K.; writing—review and editing, M.M. and M.S.; supervision, R.P. and A.B. All authors have read and agreed to the published version of the manuscript.

Funding: This research was funded in whole by the National Centre for Research and Development (NCBR), Poland within Project Repac2 no. CORNET/31/11/REPAC2/2022 (CORNET-31).

Data Availability Statement: The data presented in this study are available on request from the corresponding author.

Conflicts of Interest: The authors declare no conflict of interest.

References

1. Robertson, G.L. *Food Packaging and Shelf Lifee: A Practical Guide*; CRC Press: Boca Raton, FL, USA, 2009; pp. 1–16.
2. Saroha, V.; Khan, H.; Raghuvanshi, S.; Dutt, D. Preparation and characterization of PVOH/kaolin and PVOH/talc coating dispersion by one-step process. *J. Coat. Technol. Res.* **2022**, *19*, 1171–1186. [CrossRef]
3. European Commission. *Towards a Circular Economy: A Zero Waste Programme for Europe*; European Commission: Brussels, Belgium, 2014.
4. Herrera, M.A.; Mathew, A.P.; Oksman, K. Barrier and mechanical properties of plasticized and cross-linked nanocellulose coatings for paper packaging applications. *Cellulose* **2017**, *24*, 3969–3980. [CrossRef]
5. Granberg, A.; Nylund, T.; Rigdahl, M. Calendering of a moistened woodfree uncoated paper. *Nord. Pulp Pap. Res. J.* **1996**, *11*, 132–136. [CrossRef]
6. Sönmez, S.; Özen, Ö. Barrier Properties of Paper and Cardboard. *Acad. Res. Archit. Eng. Plan. Des. Eng. Plan. Des.* **2018**, 171–183.
7. *Regulation (EU) 2019/1381 of the European Parliament and of the Council of 20 June 2019 on the Transparency and Sustainability of the EU Risk Assessment in the Food Chain and Amending Regulations (EC) No 178/2002, (EC) No 1829/2003, (EC) No 1831/2003, (EC) No 2065/2003, (EC) No 1935/2004, (EC) No 1331/2008, (EC) No 1107/2009, (EU) 2015/2283 and Directive 2001/18/EC*; The European Parliament and the Council of the European Union: Luxembourg, 2019.
8. Khwaldia, K.; Arab-Tehrany, E.; Desobr, S. Biopolymer Coatings on Paper Packaging Materials. *Compr. Rev. Food Sci. Food Saf.* **2010**, *9*, 85–91. [CrossRef] [PubMed]
9. He, Y.; Li, H.; Fei, X.; Peng, L. Carboxymethyl cellulose/cellulose nanocrystals immobilized silver nanoparticles as an effective coating to improve barrier and antibacterial properties of paper for food packaging applications. *Carbohydr. Polym.* **2021**, *252*, 117–156. [CrossRef] [PubMed]
10. Shen, Z.; Rajabi-Abhari, A.; Oh, K.; Yang, G.; Youn, H.J.; Lee, H.L. Improving the Barrier Properties of Packaging Paper by Polyvinyl Alcohol Based Polymer Coating—Effect of the Base Paper and Nanoclay. *Polymers* **2021**, *13*, 1334. [CrossRef]
11. Idris, A.; Muntean, A.; Mesic, B.; Lestelius, M.; Javed, A. Oxygen Barrier Performance of Poly(vinyl alcohol) Coating Films with Different Induced Crystallinity and Model Predictions. *Coatings* **2021**, *11*, 1253. [CrossRef]
12. Niinivaara, E.; Desmaisons, J.; Dufresne, A.; Bras, J.; Cranston, E.D. Thick Polyvinyl Alcohol Films Reinforced with Cellulose Nanocrystals for Coating Applications. *ACS Appl. Nano Mater.* **2021**, *4*, 8015–8025. [CrossRef]
13. *PN-EN ISO 536:2020-08*; Paper and Cardboard—Determination of Grammage. Polish Committee for Standardization: Warszawa, Poland, 2020.
14. *ISO 535:2023*; Determination of Water Absorptiveness. Cobb Method. ISO: Genève, Switzerland, 2023.
15. *ISO 16532-2:2007*; Paper and Board Determination of Grease Resistance. ISO: Genève, Switzerland, 2007.
16. *ASTM F 1927-20*; Standard Test Method for Determination of Oxygen Gas Transmission Rate, Permeability and Permeance at Controlled Relative Humidity Through Barrier Materials Using a Coulometric Detector. American Society for Testing Materials: Philadelphia, PA, USA, 2020.
17. *ASTM F 1249*; Standard Test Method for Water Vapor Transmission Rate Through Plastic Film and Sheeting Using a Modulated Infrared Sensor. American Society for Testing Materials: Philadelphia, PA, USA, 2020.
18. *ASTM F88/F88M-21*; Standard Test Method for Seal Strength of Flexible Barrier Materials. American Society for Testing Materials: Philadelphia, PA, USA, 2021.
19. Lahtinen, K.; Kotkamo, S.; Koskinen, T.; Auvinen, S.; Kuusipalo, J. Charachterization for water vapor barrier and heat sealability properties of heat-treated paperboard/polylactide structure. *Packag. Technol. Sci.* **2009**, *22*, 451–460. [CrossRef]
20. Tuominen, M.; Ek, M.; Saloranta, P.; Toivakka, M.; Kuusipalo, J. The effect of flame treatment on surface properties and heat sealability of low-density polyethylene coating. *Packag. Technol. Sci.* **2013**, *26*, 201–214. [CrossRef]
21. Spagnuolo, L.; D'Orsi, R.; Operamolla, A. Nanocellulose for Paper and Textile Coating: The Importance of Surface Chemistry. *Chem. Eur. Chem. Plus Chem.* **2022**, *87*, 4–26. [CrossRef] [PubMed]
22. Yang, S.Y.; Jeong, K.M.; Won, J.M.; Kyu, L.Y. Application of CNC as a Coating Additive. *J. Korea Tech. Assoc. Pulp Pap. Ind.* **2019**, *51*, 114–120.
23. Hult, E.L.; Iotti, M.; Lenes, M. Efficient Approach to High Barrier Packaging Using Microfibrillar Cellulose and Shellac. *Cellulose* **2010**, *17*, 575–586. [CrossRef]
24. Liu, K.; Liang, H.; Nasrallah, J.; Chen, L.; Huang, L.; Ni, Y. Preparation of the CNC/Ag/Beeswax Composites for Enhancing Antibacterial and Water Resistance Properties of Paper. *Carbohydr. Polym.* **2016**, *142*, 183–188. [CrossRef]
25. Gicquel, E.; Martin, C.; Yanes, J.G.; Bras, J. Cellulose nanocrystals as new bio-based coating layer for improving fiber-based mechanical and barrier properties. *J. Mater. Sci.* **2017**, *52*, 3048–3061. [CrossRef]
26. Mazega, A.; Tarrés, Q.; Aguado, R.; Pèlach, M.À.; Mutjé, P.; Ferreira, P.J.T.; Delgado-Aguilar, M. Improving the Barrier Properties of Paper to Moisture, Air, and Grease with Nanocellulose-Based Coating Suspensions. *Nanomaterials* **2022**, *12*, 3675. [CrossRef]
27. Nair, S.S.; Zhu, J.; Deng, Y.; Ragauskas, A.J. High performance green barriers based on nanocellulose. *Sustain. Chem. Process.* **2014**, *2*, 23. [CrossRef]
28. Tayeb, H.A.; Tajvidi, M.; Bousfield, D. Enhancing the Oxygen Barrier Properties of Nanocellulose at High Humidity: Numerical and Experimental Assessment. *Sustain. Chem.* **2020**, *1*, 198–208. [CrossRef]
29. Huang, S.; Wang, X.; Zhang, Y.; Meng, Y.; Hua, F.; Xia, X. Water and Oil-Grease Barrier Properties of PVA/CNF/MBP/AKD Composite Coating on Paper. *Sci. Rep.* **2023**, *13*, 12292. [CrossRef]

30. Christophliemk, H.; Johansson, C.; Ullsten, H.; Järnström, L. Oxygen and water vapor transmission rates of starch-poly(vinyl alcohol) barrier coatings for flexible packaging paper. *Prog. Org. Coat.* **2017**, *113*, 218. [CrossRef]
31. Schuman, T.; Wikström, M.; Rigdahl, M. Coating of surface-modified papers with poly(vinyl alcohol). *Surf. Coat. Technol.* **2004**, *183*, 96. [CrossRef]

Article

Development of a Transparent Thermal Reflective Thin Film Coating for Accurate Separation of Food-Grade Plastics in Recycling Process via AI-Based Thermal Image Processing

Ali Salimian [1,*] and Uchechukwu Onwukwe [2]

1   Department of Computer Science, School of Engineering, London Southbank University, 103 Borough Road, London SE1 0AA, UK
2   Experimental Techniques Centre, Brunel University London, Kingston Lane, Uxbridge UB8 3PH, UK
*   Correspondence: salimiaa@lsbu.ac.uk

Abstract: This paper presents the development of a specific thin film coating designed to address the challenge of accurately separating food-grade plastics in the recycling process. The coating, created using a plasma sputtering process, is transparent to the visible spectrum of light while effectively reflecting infrared emissions above 1500 nm. Composed of a safe metal oxide formulation with a proprietary composition, the coating is applied to packaging labels. By employing thermal imaging and a computer vision AI model, the coated labels enable precise differentiation of plastics associated with food packaging in the initial stage of plastic recycling. The proposed system achieved a remarkable 100% accuracy in separating food-grade plastics from other types of plastics. This innovative approach holds great potential for enhancing the efficiency and effectiveness of plastic recycling processes, ensuring the recovery of food-grade plastics for future use.

Keywords: thermal imaging; infrared; sputtering; computer vision; plastic recycling

Citation: Salimian, A.; Onwukwe, U. Development of a Transparent Thermal Reflective Thin Film Coating for Accurate Separation of Food-Grade Plastics in Recycling Process via AI-Based Thermal Image Processing. *Coatings* **2023**, *13*, 1332. https://doi.org/10.3390/coatings13081332

Academic Editors: Jun Mei, Jing Xie and Biao Zhang

Received: 17 June 2023
Revised: 12 July 2023
Accepted: 14 July 2023
Published: 28 July 2023

1. Introduction

Plastic parts and products have become an integral part of our daily lives, coexisting with humans on a regular basis, and remain essential in various aspects of our everyday routines. However, once a product is utilized, it gives rise to environmental issues, particularly plastic waste pollution, which is a significant concern in today's world. Recent research indicates that, on average, humans could ingest 0.1–5 g of plastic per week, with the highest value equivalent to consuming a credit card weekly [1]. As of 2021, there were 24.4 trillion pieces of microplastics in the world's oceans, which is equivalent to $8.2 \times 10^4 \sim 57.8 \times 10^4$ tons [2]. Moreover, the global plastic market is projected to grow at an annual rate of 3.4% from 2021 to 2028 [3]. Plastic offers advantages, such as low weight, durability, and cost-effectiveness, which have contributed to the rapid growth of the plastic industry [4,5]. Astonishingly, humans produce approximately 500 million tons of plastic products each year, with 40% being single-use items [6]. Despite this substantial production and single-use rate, less than 16% of plastic is recycled. Insufficient recycling facilities and confusion regarding what can and cannot be recycled result in the majority of plastics ending up in landfill [7]. The process of classifying plastic waste, whether through manual or automated means, represents a significant challenge in waste management. Manual methods require more time, effort, and labour, making them less profitable compared with automated approaches.

Automated methods for plastic classification utilize sensors, radiation, or analysing the chemical and physical properties of different materials [8]. A common technique employed in automated classification processes is spectroscopy, which has been the subject of numerous studies aiming to enhance classification efficiency [9,10]. The most common spectroscopic methods investigated for this purpose are Raman spectroscopy, laser-induced breakdown spectroscopy (LIBS), infrared spectroscopy, and X-ray spectroscopy. However, the presence of additives, such as flame retardants, for instance, cannot be detected accurately using LIBS and Raman spectroscopy [11]. Infrared spectroscopy, particularly near-infrared spectroscopy (NIR) within the wavelength range of 0.8 to 2.5 μm, has gained significant attention globally for plastic identification and classification, and has already been implemented in some recycling facilities, exhibiting high performance in identifying polymer classes. However, it may not be suitable for classifying black plastics due to their high absorption within this wavelength range [12]. Fourier-transform infrared (FTIR) spectroscopy in the medium-wave infrared (MWIR) range can effectively classify polymers by type, including black plastics. However, it is highly sensitive to plastic shape and surface characteristics [13].

In contrast, X-ray spectroscopy can classify plastics based on traces, making it suitable for the identification of black plastics, but less effective in distinguishing between polymers of the same family due to their identical chemical composition [14].

In the field of plastic classification, researchers have explored non-invasive techniques, such as near-infrared (NIR) and mid-infrared (MWIR) spectroscopy, for material characterization. However, these spectroscopic methods are influenced by factors like surface topology, particle size, and orientation, which can affect the spectra. Various statistical methods have been employed to identify and classify polymer spectra. These efforts have aimed to differentiate between different types of plastics, but certain challenges remain. For example, the NIR domain has limitations in identifying resin collections and distinguishing between high-density polyethene (HDPE) and low-density polyethene (LDPE).

Recent studies have utilized techniques like attenuated total reflectance Fourier-transform infrared spectroscopy (ATR-FTIR) and mid-infrared hyperspectral imaging (MIR-HSI) for plastic classification. Researchers have developed reference spectral libraries and used ATR-FTIR measurements for polymer identification. Hyperspectral imaging has been employed for the classification of marine microplastics, and deep learning architectures have been utilized for the automatic counting and classification of microplastics. These studies indicate the potential for improving plastic classification using machine learning and advanced spectroscopic techniques.

Some researchers have initiated implementing machine learning and AI for the plastic classification process. Lorenzo-Navarro et al. developed a deep learning network architecture for the automatic counting and classification of microplastics between 1 and 5 mm in size, classifying them into fragments, pellets, and lines [15]. Their results suggest that deep learning architectures have the potential to improve further and can be applied to different spectral ranges. Jacquin et al. employed a mid-infrared hyperspectral imager (MIR-HSI) camera and machine learning algorithms to propose a cautious classification procedure for waste electrical and electronic equipment (WEEE) plastics, ensuring high purity of the classified samples [16]. However, the MWIR wavelength range did not work effectively for certain plastics of interest, limiting the classification to only four polymer classes. Wu et al. compared the performance of three algorithms—spectral angle mapper (SAM), partial least-squares discriminant analysis (PLS-DA), and linear discriminant analysis combined with principal component analysis (PCA-LDA)—for classifying NIR spectra from WEEE plastic samples [17]. While achieving good results, the NIR experiment was primarily suitable for light-coloured plastics, whereas most WEEE plastics are dark. Additionally, several studies have utilized hyperspectral imaging (HIS) in the short-wave infrared range (SWIR: 1000–2500 nm) for different imaging classifications [18–21]

Carrera et al. carried out an extensive implementation of machine learning models for polymer identification and classification using infrared spectra (NIR or MWIR) measured using different spectrometers and wavelengths. Their study focused on specific plastic types, but did not cover black plastics or differentiation between HDPE and LDPE [3].

Considering all the exciting approaches discussed above, there is an important issue in plastic recycling that needs to be addressed, especially when microwave-heating of foods is involved [22]. The mixing of plastics containing toxic materials with food-grade plastics poses a significant concern in terms of public health and safety. When different types of plastics, especially those intended for non-food applications, are mistakenly mixed with food-grade plastics, it can lead to contamination of the food packaging or storage containers [23]. Toxic substances present in certain plastics, such as phthalates, bisphenol A (BPA), or heavy metals, can leach into the food, posing potential health risks upon consumption. Therefore, strict separation and proper identification of plastics during the recycling process are crucial to prevent the mixing of toxic materials with food-grade plastics and ensure the integrity of food packaging and storage systems.

In this study, we propose a complementary first-stage separation technique for food-grade plastic packaging from any other plastics at the beginning of the recycling process, where food-grade plastics can be separated from any non-food plastics. We have developed a unique label coated with a thin film transparent to the visible spectrum of light, yet possessing infrared-reflecting properties. This thin film coating technology coupled with a computer vision-based system can identify food-grade plastic packaging at the first stage of the recycling process. This conceptual methodology will be used to classify images based on their thermal images and can identify and classify packaging plastics during the first stage of the recycling process via the label. A thermal camera will identify the special food-grade label from a conventional label, irrespective of the printed pattern on the label.

2. Experimental

In this study, we explore the application of a unique metal oxide coating on polyvinyl chloride (PCV) label sheets, deposited using a V6000 confocal sputter system (Manufactured by Scientific Vacuum Systems LTD., Reading, UK) equipped with an RF plasma source, to facilitate the classification of plastics during recycling. The coating, which remains transparent to visible light, exhibits a distinct property of reflecting mid-infrared (IR) radiation. As such, it will not interfere with the desired pattern or designs of the labels placed on food products. By utilizing these coated labels affixed to plastic packaging, we present a novel approach to plastic classification based on infrared thermal imaging.

The experimental setup involves the deposition of the metal oxide coating onto plastic packaging labels using the sputter deposition system. The nature of the metal oxide and its specific structure and deposition regime cannot be disclosed in full detail for commercial purposes. However, it is primarily based on zinc oxide with a particular doping and specific control of its deposition process to achieve maximum mid-infrared reflection. Infrared reflection by metal oxides occurs due to their unique electronic and structural properties. Metal oxides exhibit different levels of infrared reflectivity depending on factors such as composition, crystal structure, and surface morphology. Via the selective doping of metal oxides, the concentration of charge carriers can be manipulated, which ultimately translates to bandgap manipulation [24]. Different metal oxides, such as aluminium oxide (Al_2O_3), titanium dioxide (TiO_2), or zinc oxide (ZnO), exhibit different levels of infrared reflectivity and have unique mechanisms governing their behaviour. The deposition conditions and doping can lead to eliminating energy levels that typically absorb photons in the mid-infrared range and, thus, enhancing the reflectivity of infrared photons. ZnO can also absorb infrared radiation through the mechanism of free carrier absorption. When ZnO is doped or contains defects, it can generate free electrons or holes that can absorb infrared photons whose energy matches their levels. This absorption can reduce the overall reflectivity of ZnO in the infrared range [24,25]. Therefore, precise deposition conditions must be selected to prevent free carrier absorption. In our experiments, while developing these specific coatings, the following conditions were precisely fine-tuned for the desired reflective properties:

The deposition has to be carried out at a very specific RF plasma power, which translates to the kinetic energy of the argon atoms attacking the target and ejecting the zinc and doping element to be deposited on the substrate;

Very low chamber pressures. The best results were observed when depositing at pressures below 1.5×10^{-3} mbar. Above this chamber pressure, the infrared reflectivity was significantly reduced.

Generally, depositing under low pressure has certain advantages, such as increasing the mean free path, decreasing gas scattering, increased energy transfer, and decreased gas density variation. Gas density variation can lead to uneven deposition and hinder uniformity [26,27]. Considering that our objective was to achieve maximum infrared reflection, smoother and more even surface coatings were desired, hence why the pressure limit affected the reflectivity performance.

The metal oxide coating, with its specific composition and thickness, imparted the desired optical properties necessary for mid-IR reflection while maintaining transparency to visible light. The coated labels were then attached to multiple plastic objects representing different types of plastics commonly encountered in recycling processes.

To validate the effectiveness of the proposed classification method, we acquired thermal images of plastic samples both with and without the metal oxide-coated labels. These thermal images served as inputs for a computer vision-based image classification algorithm, which enabled the automated identification and sorting of plastic materials. By leveraging the distinctive mid-IR reflection properties of the metal oxide coating, the thermal images provided valuable information for accurate plastic classification.

The main objective of this research was to demonstrate the feasibility and effectiveness of using our unique transparent metal oxide coating in conjunction with infrared thermal imaging for plastic classification during recycling. By streamlining the sorting process through automation, we aimed to enhance the efficiency and accuracy of food plastic recycling, ultimately contributing to the reduction in toxins contained in food-grade plastics during recycling.

All the labels (coated and uncoated) were identical. The pattern of the labels is presented in Figure 1. The labels were coated under vacuum (under 1.5×10^{-3} mbar) in an Argon atmosphere with a plasma power of 100 w applied to a 6-inch magnetron fitted with a target material with a unique metal ZnO/doping complex. At an industrial scale, such label preparation can be carried out on a role-to-role basis with a plasma coating system that can be adequately designed for this objective to produce the labels very cheaply.

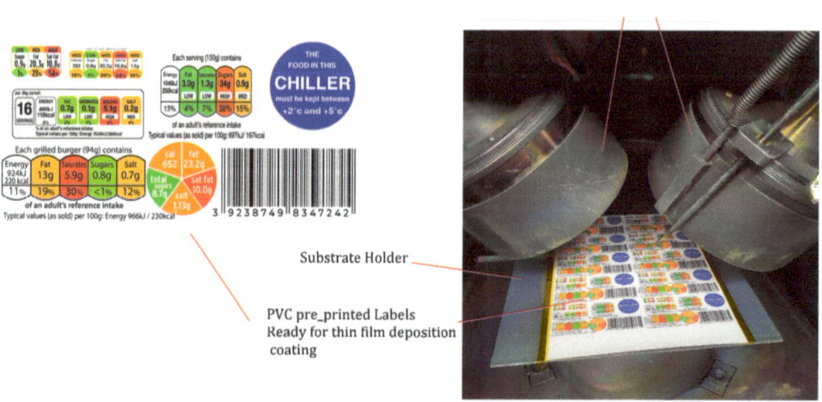

Figure 1. The printed pattern of the label. The labels were then placed at 45-degree angle to a 6-inch-diameter round magnetron loaded with the desired metal oxide target; the labels were then coated under a controlled argon atmosphere at a chamber pressure below 1.5×10^{-3} mbar.

3. Results

The Fourier-transform infrared (FTIR) transmittance spectra of the coated and un-coated labels are depicted in Figure 2. FTIR analysis provides valuable insights into the molecular composition and chemical characteristics of samples; however, in this study, our objective for using FTIR was to illustrate the comparative transmission of the coated labels and the uncoated labels in the IR region. The peaks present in Figure 2 are, as such, of no interest, as they are only related to the primary label material. From Figure 2, we can clearly see that our coating transmitted a significantly lower quantity of infrared radiation between 2000 and 16,500 nm, which was reflected and observed in the thermal images. The UV/Vis spectrum of a coated label in the visible to near infra-red region is presented in Figure 3.

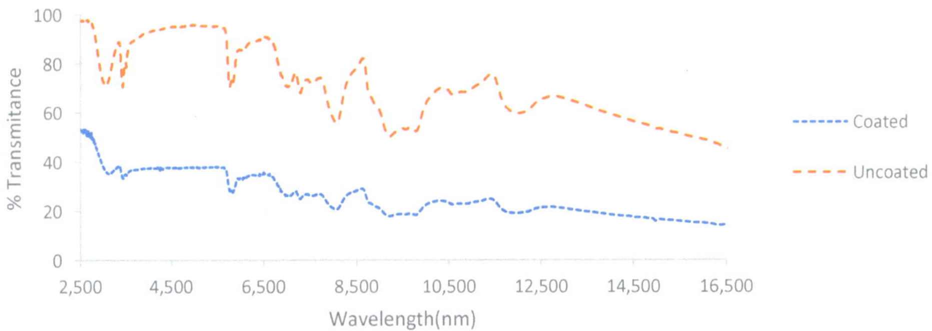

Figure 2. The Fourier-transform infrared (FTIR) transmittance spectra of the coated and uncoated labels. The peaks observed in this spectrum are associated with the material from the labels were made and, as such, were of no interest. We can see them present in both coated and uncoated samples. However, this image illustrates the effect of our metal oxide coating in terms of IR transmittance at different wavelengths being able to pass through the sample. We can see that transmittance in the coated sample was significantly restricted.

The labels coated with infrared reflecting film and uncoated labels were placed on various identical plastic objects (for coated and uncoated labels). The samples were then deformed randomly by hand to mimic their form of shape after having been disposed of, when on a recycling conveyer belt. A total of just under 450 images of various plastic objects with an IR-reflecting label and standard label were obtained using an FLIR E96 thermal visualization camera. The FLIR E96 offers an impressive thermal resolution of 640×480 pixels. This high-resolution thermal detector enables accurate temperature measurement and detailed thermal imaging. With a thermal sensitivity of less than 0.03 °C, the FLIR E96 can detect even subtle temperature variations. An example of objects labelled accordingly is illustrated in Figure 4. However, many more objects, as well as varying orientations and angles of view, were used to capture the dataset of images for this project.

The image data were split into training and validation sets, where 356 images were allocated for model training and 89 images were kept away to evaluate the performance of the computer vision models intended for the classification process. Two models were evaluated. The first model was built using the sequential API of the TensorFlow software library possessing a simple structure.

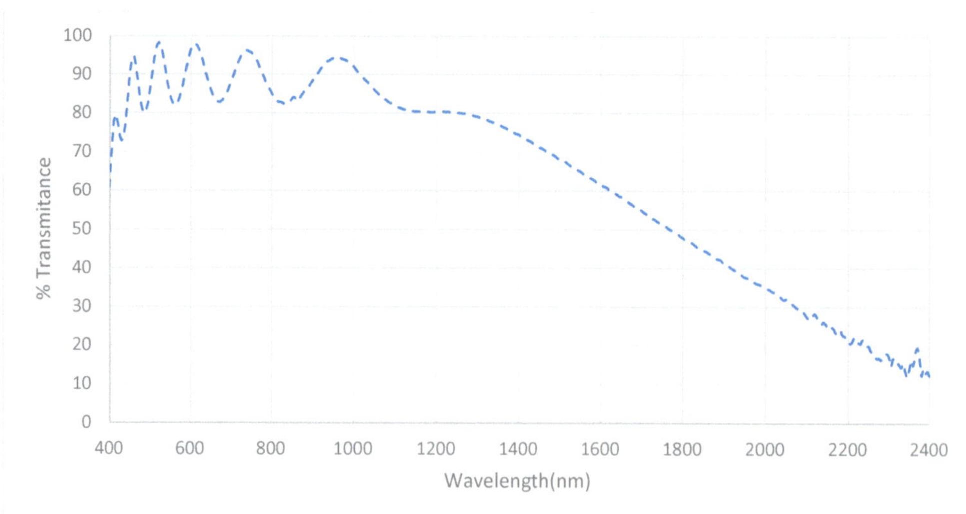

Figure 3. The UV-Vis spectroscopy plot of the coated label. The spectrometer was normalized with an uncoated label as a blank. The data clearly illustrate that the label was more than 80% transparent to the visible part of the spectrum.

Figure 4. An example of a set of plastic objects with standard labels and labels coated with the IR-reflecting oxide films. The objects on the right are fitted with uncoated labels, while the objects to the left are fitted with a coated label.

3.1. Basic Convolutional Neural Network Model

The model was a sequential deep learning model designed for image classification tasks. The model architecture consisted of several layers, including convolutional layers, max pooling layers, a flatten layer, dense layers, a dropout layer, and a final dense layer with a sigmoid activation function. The convolutional layers extracted relevant features from the input images, while the max pooling layers downsampled the feature maps to reduce their dimensions. The flatten layer converted the 2D feature maps into a 1D vector. The dense layers performed high-level feature extraction, and the final dense layer with sigmoid activation produced the classification output. The model was compiled with the binary cross-entropy loss function, the Adam optimizer (Adam optimization is a stochastic gradient descent method which is implemented in various deep learning applications, such as computer vision), and accuracy as the evaluation metric. This model can be trained to classify images into two classes based on the specified architecture and optimization setup; the model was trained for image recognition on the test data categorising infrared-reflecting samples from the conventional uncoated labels. The structure of this model is illustrated in Figure 5.

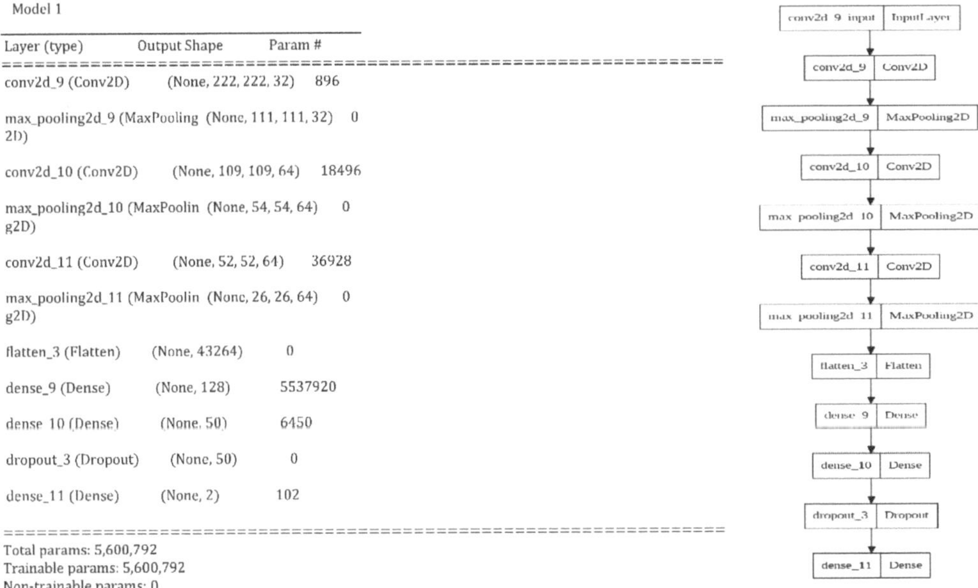

Model 1

Layer (type)	Output Shape	Param #
conv2d_9 (Conv2D)	(None, 222, 222, 32)	896
max_pooling2d_9 (MaxPooling 2D)	(None, 111, 111, 32)	0
conv2d_10 (Conv2D)	(None, 109, 109, 64)	18496
max_pooling2d_10 (MaxPoolin g2D)	(None, 54, 54, 64)	0
conv2d_11 (Conv2D)	(None, 52, 52, 64)	36928
max_pooling2d_11 (MaxPoolin g2D)	(None, 26, 26, 64)	0
flatten_3 (Flatten)	(None, 43264)	0
dense_9 (Dense)	(None, 128)	5537920
dense_10 (Dense)	(None, 50)	6450
dropout_3 (Dropout)	(None, 50)	0
dense_11 (Dense)	(None, 2)	102

Total params: 5,600,792
Trainable params: 5,600,792
Non-trainable params: 0

Figure 5. The model architecture consists of several layers, including convolutional layers, max pooling layers, a flatten layer, dense layers, a dropout layer, and a final dense layer with a sigmoid activation function.

The model performance metrics plots are illustrated in Figure 6 and, by tracking the validation loss plot, while good training accuracy was being achieved, the model was overfitting on the training data. Owing to the thermal signatures significantly varying on the samples depending on how they were crushed or deformed, and the limited features present in a thermal image compared to a standard image of objects, such a simple model would find it challenging to perform with 100% accuracy.

In the context of machine learning models, training loss, validation loss, accuracy, and validation accuracy are important metrics used to evaluate and monitor the performance of the model during the training process. Training loss refers to the measure of error or mismatch between the predicted output of the model and the actual target output on the training dataset. It quantifies how well the model is learning the patterns and relationships within the training data. The goal during training is to minimize the training loss, indicating that the model is improving its ability to make accurate predictions.

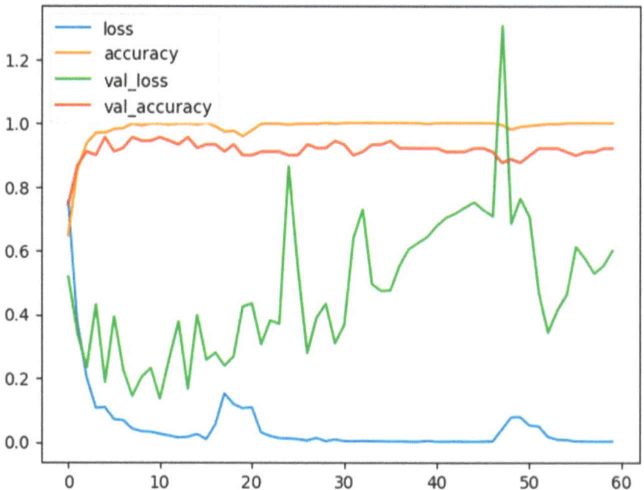

Figure 6. The performance metrics of the model during the training process over 60 epochs of training. The rise and variations observed in the validation loss indicate that the model struggled with correctly identifying unseen thermal images.

Validation loss, on the other hand, measures the error on a separate validation dataset that is not used during the training phase. It provides an estimate of how well the model generalizes to unseen data. Validation loss helps to assess if the model is overfitting or underfitting. Ideally, the validation loss should be like the training loss, indicating that the model is performing well on unseen data.

Accuracy is a metric that quantifies the overall correctness of the model's predictions. It calculates the percentage of correctly classified instances out of the total number of instances. In the context of training, the training accuracy is calculated using the training dataset and reflects how well the model predicts the correct labels for the training data. Validation accuracy measures the performance of the model on the validation dataset (in this case, thermal images of the plastic samples that the model has not encountered before). It indicates how accurately the model predicts the labels of the validation data. A high validation accuracy indicates that the model is performing well and generalizing effectively to unseen data.

During the training process, the objective is to minimize both training loss and validation loss, while maximizing accuracy and validation accuracy. Monitoring these metrics helps in assessing the progress of the model, identifying potential issues, such as overfitting, and making informed decisions for model optimization and improvement.

The validation data (set of thermal images of the plastics labelled accordingly, which were kept hidden from the model) were separated into three batches and the trained model was used to classify their images into IR-labelled or conventionally labelled. The three confusion matrices, as well as the associated F1 score, in Figure 7 illustrate the performance of the model on thermal images that it had never encountered before. The F1 score is a widely used metric for evaluating the performance of classification models. It combines precision and recall into a single value, providing a balanced measure of a model's accuracy. The F1 score considers both the model's ability to correctly identify positive instances (precision) and its ability to capture all positive instances (recall). It is calculated as the harmonic mean of precision and recall, ranging from 0 to 1, where a higher value indicates better performance. The F1 score is particularly useful when the dataset is imbalanced, where the number of instances in different classes varies significantly. By considering both precision and recall, the F1 score provides a comprehensive assessment of the model's ability to make accurate and comprehensive predictions.

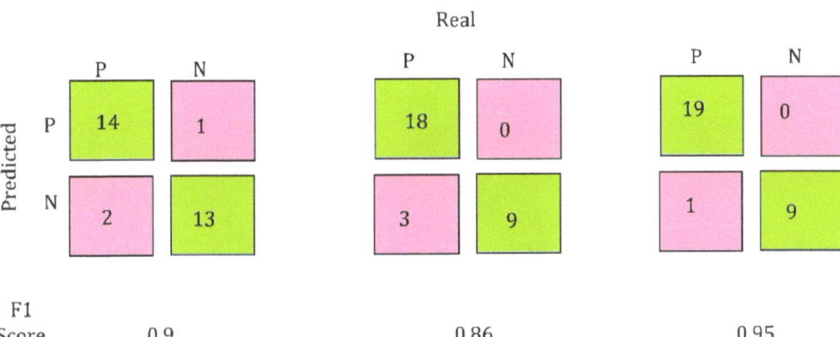

Figure 7. The predictive performance of the model on the three batches of unseen data.

A confusion matrix is a tabular representation that summarizes the performance of a classification model. It provides a visual representation of how well the model predicted the actual classes of the thermal images. The matrix consists of rows and columns, with each row representing the instances in a predicted class and each column representing the instances in an actual class.

The confusion matrix shows four key metrics:

True Positives (TP—Top green): The number of instances correctly predicted as positive by the model;

True Negatives (TN—Bottom green): The number of instances correctly predicted as negative by the model;

False Positives (FP—Top pink): The number of instances incorrectly predicted as positive by the model;

False Negatives (FN—Bottom pink): The number of instances incorrectly predicted as negative by the model.

From the confusion matrices presented in Figure 7, we can see that, while the model was able to detect the infrared reflections in the majority of the images, a certain number of misclassifications was made. We can see that the model seemed to achieve an accuracy of about 90%. Overall, out of 89 images, a total of 7 wrong classifications were made, which is not acceptable. Four of the misclassified images are illustrated in Figure 8. While visually, we can notice the reflection of the IR in the images, the model had wrongly classified them to be of the conventional uncoated label type.

As such, we can conclude that the model needs to be powerful enough to distinguish the shadows and brightness patterns of an infrared image. Therefore, we decided to implement transfer learning and benefit from a more sophisticated model.

3.2. Transfer Learning Model

ResNet-50 v2 is a convolutional neural network (CNN) architecture that is a variant of the original ResNet-50 model. ResNet stands for "Residual Network," and it was introduced by He et al. to address the degradation problem encountered in very deep neural networks [28]. ResNet-50 v2 improves upon the original ResNet-50 by introducing changes to the architecture to enhance performance and training efficiency. The ResNet-50 v2 architecture consists of 50 layers, including convolutional layers, pooling layers, fully connected layers, and shortcut connections. The core building block of the network is the residual block, which contains two or three convolutional layers, depending on the variant. The residual block introduces shortcut connections that allow the network to learn residual mappings, instead of attempting to learn the full transformation. This helps alleviate the vanishing gradient problem (the vanishing gradient problem occurs when gradients become extremely small during deep neural network training, leading to slow learning or ineffective training of earlier layers) and enables the training of very deep networks. During the recycling process, the plastic samples can be crushed or deformed in shape and, as such, the IR signature captured by the thermal camera can look significantly different, as observed in Figure 8; as such, a powerful model like ResNet will be better adapted for identifying various thermal signatures for the classification objective.

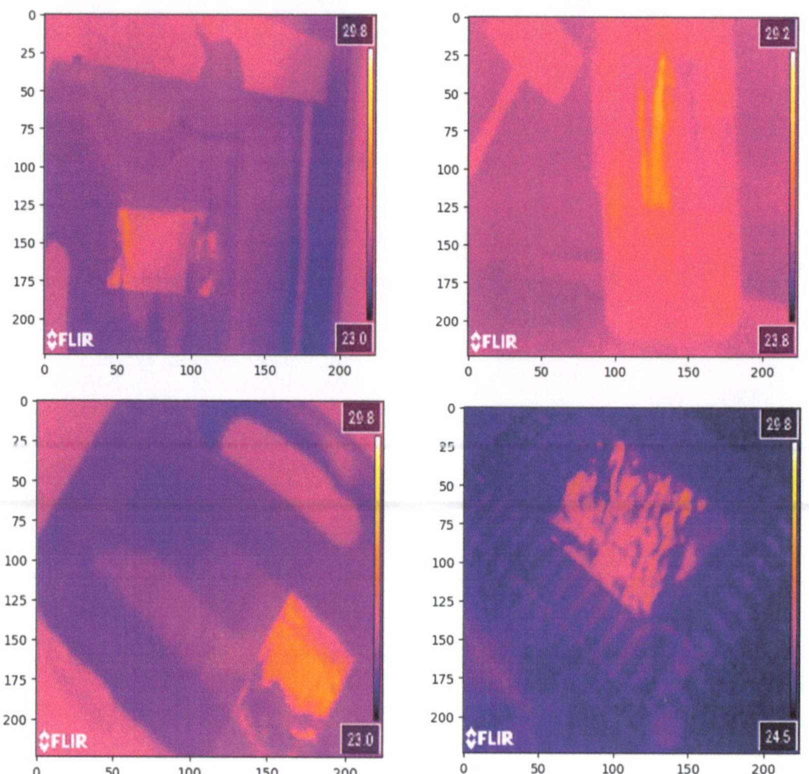

Figure 8. Four random examples of images that were misclassified by the deep learning model, indicating that the model will need a more complex structure to be able to make predictions with 100% accuracy.

In ResNet-50 v2, the residual blocks are organized into different stages. The first stage performs initial convolution and pooling operations, while subsequent stages consist of multiple residual blocks stacked together. The number of residual blocks per stage may vary depending on the architecture variant.

Another key feature of ResNet-50 v2 is the use of bottleneck blocks. These blocks are designed to reduce computational complexity by employing 1×1 convolutions to reduce the number of input channels, followed by 3×3 convolutions, and finally 1×1 convolutions to restore the number of channels. This bottleneck design allows the model to achieve better performance with fewer parameters. ResNet-50 v2 has demonstrated excellent generalization capabilities, allowing it to learn hierarchical features that are transferable across different datasets and tasks. This makes it suitable for transfer learning and fine-tuning on specific domain datasets. We modified the output layer of ResNet-50 v2 to match it to our set of binary data and kept all of the original weights frozen, only training the model for weights associated with the modified output layer. The model's structure is presented in Figure 9.

The transfer learning model was then trained with the same training data over 60 epochs, and the learning curve metrics of the model are presented in Figure 10. The model was then tested using the same validation data, and it was able to classify the infrared thermal images with 100% accuracy. The confusion matrices demonstrating the accuracy of the model's performance are presented in Figure 11.

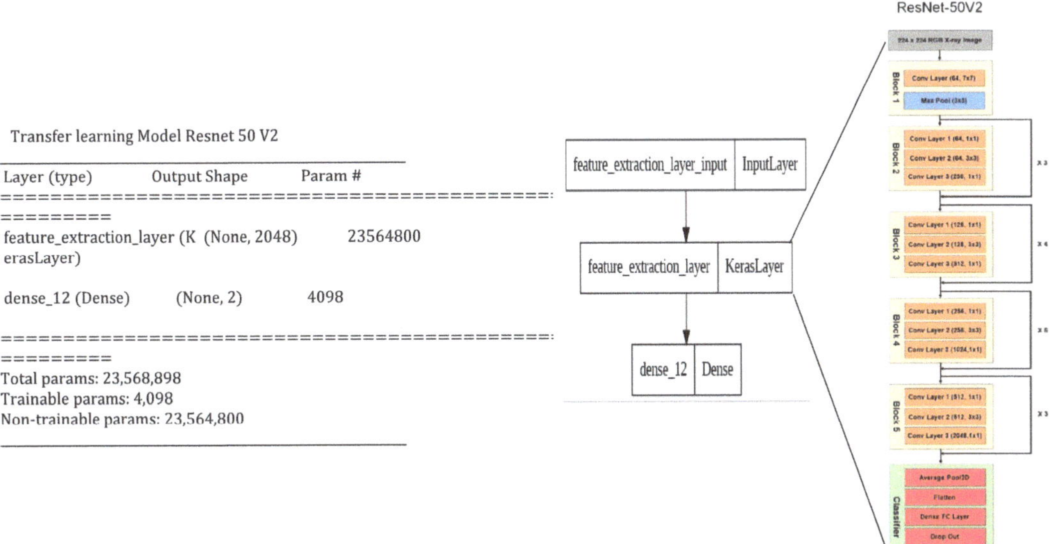

Figure 9. The structure of the transfer learning model based on the ResNet-50 v2 with a modified output layer.

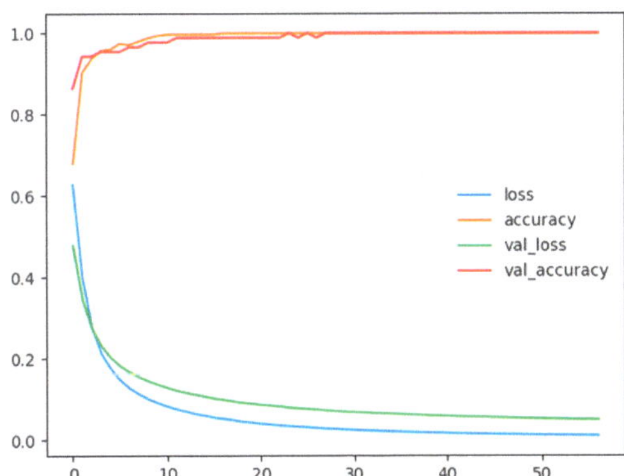

Figure 10. The performance metrics of the transfer learning model during the training process over 60 epochs of training. The model performed very well with both the training thermal images and the thermal images that it had not encountered before.

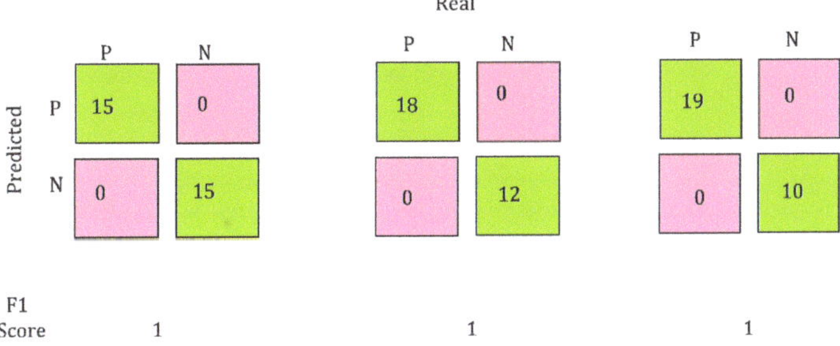

Figure 11. The predictive performances of the transfer learning model on the three batches of unseen data. The model classified all 89 thermal images correctly.

4. Discussion

By precisely manipulating the optoelectronic properties of a zinc oxide-based thin film via doping and precise deposition, its mid-infrared-reflecting properties can render the film useful for various applications. In our studies, the main challenge faced was identifying hot spots under the precise conditions of deposition, particularly the deposition chamber pressure and the RF plasma generating power.

Figure 12 illustrates the UV-Vis spectroscopy result associated with a portion of our samples during the fine-tuning process of achieving a thin film with the desired infrared-reflecting properties. Increasing plasma power would lead to a higher rate of deposition; however, we had to ensure that the power did not exceed levels that would damage the PVC labels. While 1 WCm^{-1} plasma power achieved faster coating on glass substrates, it would damage the labels. Hence, we had to identify a plasma power threshold that will yield optimal coating conditions without damaging the labels. The chamber pressure seemed interesting, as only under 1.5×10^{-3} mbar could we observe a drop in the IR absorption after the 1500 nm wavelength range.

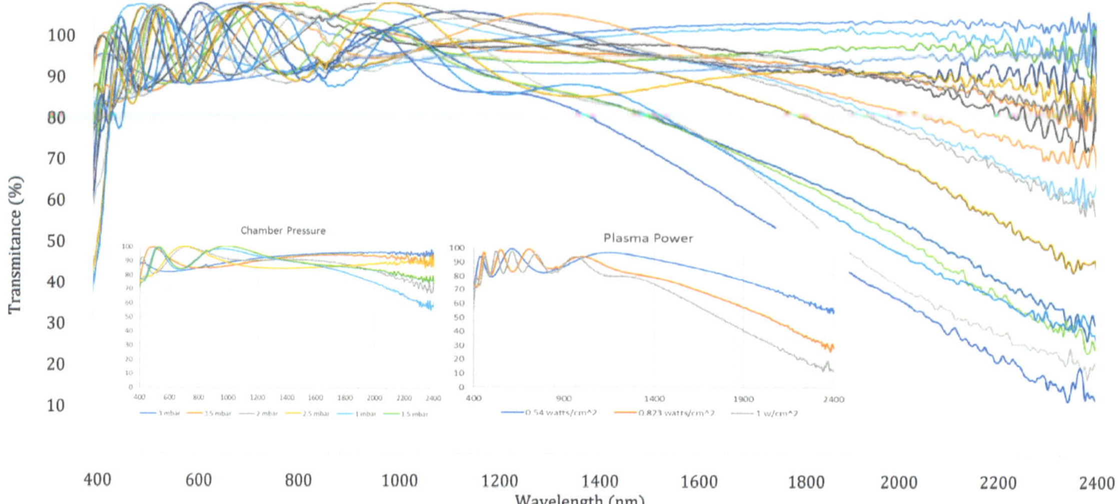

Figure 12. To reach the desired IR-reflecting properties of the packaging labels, numerous sputter-coating recipes were tested. The plasma power and chamber pressure had to be altered such that no damage to the labels could occur while ensuring minimum mid-IR transmittance.

However, when considering real-world applications, it is important to note that the labels coated with the film reflected infrared light, just as a mirror reflects visible light. This means that the angle of incidence from a thermal radiating source (in our experiments, a black-bodied object with a temperature of above 40 degrees centigrade) to the object carrying the label needs to be such that the thermal reflection falls within the receptive field of the thermal camera lens. Considering that such objects may be crushed and deformed, a single camera will fail to capture the thermal reflection; as such, a setup, as illustrated in Figure 13, will need to be considered where two thermal cameras need to be operating simultaneously. This would surely add additional one-off cost to the process, but will not be difficult to implement.

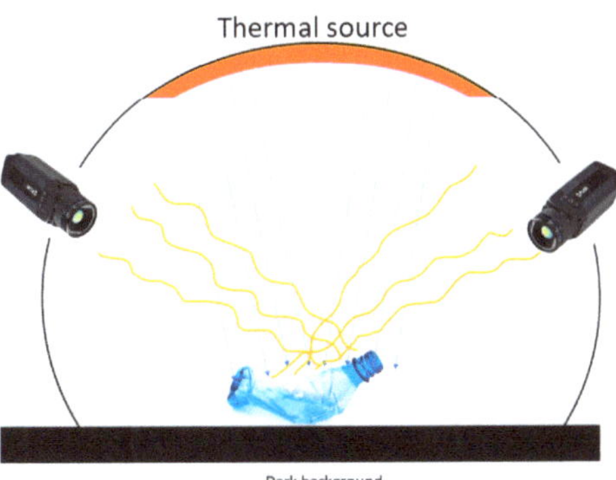

Figure 13. In real-life recycling classification applications, due to the deformities that plastic objects may endure, two thermal camera detectors will need to be implemented with a wide-angle thermal emitter facing the objects. The blue line represent the IR radiation incident on the surface and the yellow lines represent the IR reflection. This configuration will ensure that IR reflection from the object is detected, irrespective of deformities.

The thermal source can be a surface that bears a certain temperature, ideally above 40 degrees centigrade. The design of the thermal source can be carried out considering the reflectance of the IR radiation above the 1500 nm range and utilizing the Stefan–Boltzmann law and, in particular, Wien's law. Wien's law, also known as Wien's displacement law, is a fundamental principle in the study of black body radiation [29]. It establishes a relationship between the temperature of a black body and the wavelength at which the intensity of its radiation is at its maximum (peak wavelength). According to Wien's law, the product of the peak wavelength (λ_{max}) and the temperature (T) of a black body is a constant. Mathematically, it can be expressed as $\lambda_{max} = C/T$, where C is Wien's displacement constant (2.898×10^{-3} m·K) and T is the temperature of the black body (the thermal radiating source for our experiment) in Kelvins. For example, a black body with a body temperature of 40 degrees Celsius will emit thermal radiation with a peak wavelength of 9254 nm.

The Stefan–Boltzmann law describes the total power radiated by a black body and states that the power emitted per unit area is proportional to the fourth power of the absolute temperature [30]. Mathematically, it can be expressed as $P = \sigma \times A \times T^4$, where P is the power, σ is the Stefan–Boltzmann constant (approximately 5.67×10^{-8} W/(m^2·K^4)), A is the surface area, and T is the absolute temperature. This means, for example, a thermal source bearing one square meter of area and 9254 degrees Celsius will be emitting around 400 watts of power. As such, designing a detection module as depicted in Figure 12 will not be hard to achieve and will be within economical scope.

The successful application of the thin film coating process to create infrared reflecting labels coupled with an AI-based classification model for thermal image classification can open ideas in the field of plastic recycling. What is rather interesting is that, by using a ResNet50 model that has been trained on 1000 standard images, we achieved the best results. ResNet50 has not been trained on thermal images, yet it was able to successfully capture features from the thermal reflections to be able to accurately classify the labelled samples accordingly. By harnessing the properties of the unique metal oxide coating, our study leverages the power of AI-based deep learning models for image classification. On a final note, the labels produced in this study were, in fact, based on a plastic that is ultimately a pollutant itself. More recently, biodegradable materials for food packaging have been explored [31–33], and this labelling technology can potentially be coupled with these novel materials in terms of labelling them or, more interestingly, exploring the actual label material to be made of such biodegradable materials. Such an approach will make this labelling concept be further in line with the objective of minimizing environmental plastic pollution.

5. Conclusions

In this report, we conducted an explorative study that showcases the potential of infrared-reflecting labels created via the plasma sputtering of a thin film coating on the labels. Our research demonstrates that, when combined with AI-based deep learning models for image classification, these labels hold significant promise in effectively separating food-grade plastics during the initial stages of plastic recycling. Although this methodology is in its nascent stages, further advancements and refinement can yield remarkable results, particularly in the context of segregating food-grade plastics from plastics used for other applications, in particular, exploring the implementation of our coating on biodegradable labels.

While the potential of this methodology is already evident, further research and development hold tremendous promise for its integration into existing plastic-recycling workflows. This advancement would enable the efficient separation of food-grade plastics, which often require stricter processing and handling protocols, from plastics used in other applications.

The merging of thin film material science with artificial intelligence (AI) is generating significant excitement and opening new possibilities in research and applications. By combining the expertise of material scientists with the power of AI, we can accelerate the discovery and development of novel thin film materials with enhanced properties and functionalities. AI algorithms can efficiently analyse large datasets, identify patterns, and make predictions, enabling the rapid screening and optimization of thin film materials for specific applications.

Author Contributions: Conceptualization, A.S.; methodology, A.S.; software, A.S.; validation, A.S. and U.O.; formal analysis, A.S.; investigation, A.S.; resources, U.O.; data curation, A.S.; writing—original draft preparation, A.S.; visualization, A.S.; supervision, A.S.; project administration, A.S.; funding acquisition, A.S. All authors have read and agreed to the published version of the manuscript.

Funding: Internal Funding, London Southbank University, School of Engineering sourced from Research England's HEIF (Higher Education Innovation Fund) funding for use in Knowledge Exchange (KE) by HEIs.

Institutional Review Board Statement: Not applicable.

Informed Consent Statement: Not applicable.

Data Availability Statement: All data that support the findings of this study are included within the article.

Acknowledgments: The authors would like to thank Ahsan Sadiqi for his assistance in the project and special thank George Fern for allowing access to equipment's at their facilities utilized for this this project.

Conflicts of Interest: The authors declare no conflict of interest.

References

1. Senathirajah, K.; Attwood, S.; Bhagwat, G.; Carbery, M.; Wilson, S.; Palanisami, T. Estimation of the mass of microplastics ingested—A pivotal first step towards human health risk assessment. *J. Hazard. Mater.* **2020**, *404*, 124004. [CrossRef] [PubMed]
2. Isobe, A.; Azuma, T.; Cordova, M.R.; Cózar, A.; Galgani, F.; Hagita, R.; Kanhai, L.D.; Imai, K.; Iwasaki, S.; Kako, S.; et al. A multilevel dataset of microplastic abundance in the world's upper ocean and the Laurentian Great Lakes. *Microplast. Nanoplast.* **2021**, *1*, 16. [CrossRef]
3. Carrera, B.; Piñol, V.L.; Mata, J.B.; Kim, K. A machine learning based classification models for plastic recycling using different wavelength range spectrums. *J. Clean. Prod.* **2022**, *374*, 133883. [CrossRef]
4. Andrady, A.L.; Neal, M.A. Applications and societal benefits of plastics. *Philos. Trans. R. Soc. B Biol. Sci.* **2009**, *364*, 1977–1984. [CrossRef]

5. Hopewell, J.; Dvorak, R.; Kosior, E. Plastics recycling: Challenges and opportunities. *Philos. Trans. R. Soc. B Biol. Sci.* **2009**, *364*, 2115–2126. [CrossRef]
6. Peng, Y.; Wang, Y.; Ke, L.; Dai, L.; Wu, Q.; Cobb, K.; Zeng, Y.; Zou, R.; Liu, Y.; Ruan, R. A review on catalytic pyrolysis of plastic wastes to high-value products. *Energy Convers. Manag.* **2022**, *254*, 115243. [CrossRef]
7. Yadav, V.; Sherly, M.; Ranjan, P.; Tinoco, R.O.; Boldrin, A.; Damgaard, A.; Laurent, A. Framework for quantifying environmental losses of plastics from landfills. *Resour. Conserv. Recycl.* **2020**, *161*, 104914. [CrossRef]
8. Ruj, B.; Pandey, V.; Jash, P.; Srivastava, V. Sorting of plastic waste for effectiverecycling. *Int. J. Appl. Sci. Eng. Res.* **2015**, *4*, 564–571.
9. Gondal, M.A.; Siddiqui, M.N. Identification of different kinds of plastics using laser-induced breakdown spectroscopy for waste management. *J. Environ. Sci. Health Part A* **2007**, *42*, 1989–1997. [CrossRef]
10. Hollstein, F.; Wohllebe, M.; Arnaiz, S. Identification and sorting of plastics film waste by NIR-hyperspectral-imaging. In Proceedings of the 17th International Conference on Near Infrared Spectroscopy, Foz do Iguassu, Brazil, 18–23 October 2015.
11. Barbier, S.; Perrier, S.; Freyermuth, P.; Perrin, D.; Gallard, B.; Gilon, N. Plastic identification based on molecular and elemental information from laser induced breakdown spectra: A comparison of plasma conditions in view of efficient sorting. *Spectrochim. Acta Part B At. Spectrosc.* **2013**, *88*, 167–173. [CrossRef]
12. Zheng, Y.; Bai, J.; Xu, J.; Li, X.; Zhang, Y. A discrimination model in waste plastics sorting using NIR hyperspectral imaging system. *Waste Manag.* **2018**, *72*, 87–98. [CrossRef]
13. Mecozzi, M.; Pietroletti, M.; Monakhova, Y.B. FTIR spectroscopy supported by statistical techniques for the structural characterization of plastic debris in the marine environment: Application to monitoring studies. *Mar. Pollut. Bull.* **2016**, *106*, 155–161. [CrossRef] [PubMed]
14. Bezati, F.; Froelich, D.; Massardier, V.; Maris, E. Addition of X-ray fluorescent tracers into polymers, new technology for automatic sorting of plastics: Proposal for selecting some relevant tracers. *Resour. Conserv. Recycl.* **2011**, *55*, 1214–1221. [CrossRef]
15. Lorenzo-Navarro, J.; Castrillon-Santana, M.; Sanchez-Nielsen, E.; Zarco, B.; Herrera, A.; Martínez, I.; Gomez, M. Deep learning approach for automatic microplastics counting and classification. *Sci. Total Environ.* **2021**, *765*, 142728. [CrossRef] [PubMed]
16. Jacquin, L.; Imoussaten, A.; Trousset, F.; Perrin, D.; Montmain, J. Control of waste fragment sorting process based on MIR imaging coupled with cautious classification. *Resour. Conserv. Recycl.* **2021**, *168*, 105258. [CrossRef]
17. Wu, X.; Li, J.; Yao, L.; Xu, Z. Auto-sorting commonly recovered plastics from waste household appliances and electronics using near-infrared spectroscopy. *J. Clean. Prod.* **2020**, *246*, 118732. [CrossRef]
18. Serranti, S.; Palmieri, R.; Bonifazi, G.; Cozar, A. Characterization of microplastic litter from oceans by an innovative approach based on hyperspectral imaging. *Waste Manag.* **2018**, *76*, 117–125. [CrossRef]
19. Caballero, D.; Bevilacqua, M.; Amigo, J.M. Application of hyperspectral imaging and chemometrics for classifying plastics with brominated flame retardants. *J. Spectr. Imaging* **2019**, *8*, a1. [CrossRef]
20. Lorenzo-Navarro, J.; Serranti, S.; Bonifazi, G.; Capobianco, G. Performance Evaluation of Classical Classifiers and Deep Learning Approaches for Polymers Classification Based on Hyperspectral Images. In Proceedings of the 16th International Work-Conference on Artificial Neural Networks, Virtual Event, 16–18 June 2021; Springer: Berlin/Heidelberg, Germany, 2021; pp. 281–292.
21. Ni, C.; Li, Z.; Zhang, X.; Sun, X.; Huang, Y.; Zhao, L.; Zhu, T.; Wang, D. Online Sorting of the Film on Cotton Based on Deep Learning and Hyperspectral Imaging. *IEEE Access* **2020**, *8*, 93028–93038. [CrossRef]
22. Bastianoni, S.; Fantke, M.; Gamberi, C.V.; Winestrand, F.G. Migration of Chemical Compounds from Packaging Polymers during Microwave, Conventional Heat Treatment, and Storage. *Environ. Sci. Technol.* **2021**, *51*, 11328–11338.
23. Zgoła-Grześkowiak, M.; Szczęsny, B.; Czaja-Bulsa, D.M. Migration of Chemicals from Packaging Polymers into Food Simulants: A Review. *Food Addit. Contam. Part A* **2018**, *35*, 1297–1314.
24. Korotcenkov, G.; Sagadevan, S.; Podder, J. *Metal Oxides for Optoelectronics and Optics-Based Medical Applications*; Metal Oxide Series; Mohammad, F., Ed.; Elsevier: Amsterdam, The Netherlands, 2022.
25. Ashika, S.A.; Balamurugan, S.; Palanisami, N. Investigation on multifunctional binary oxides for near-infrared (NIR) reflective pigment applications. *Emergent Mater.* **2021**, *5*, 1183–1197.
26. Wojcieszak, D.; Mazur, M.; Pokora, P.; Wrona, A.; Bilewska, K.; Kijaszek, W.; Kotwica, T.; Posadowski, W.; Domaradzki, J. Properties of Metallic and Oxide Thin Films Based on Ti and Co Prepared by Magnetron Sputtering from Sintered Targets with Different Co-Content. *Materials* **2021**, *14*, 3797. [CrossRef] [PubMed]
27. Nafarizal, N. Precise Control of Metal Oxide Thin Films Deposition in Magnetron Sputtering Plasmas for High Performance Sensing Devices Fabrication. *Procedia Chem.* **2016**, *20*, 93–97. [CrossRef]
28. He, K.; Zhang, X.; Ren, S.; Sun, J. Deep Residual Learning for Image Recognition. In Proceedings of the IEEE Conference on Computer Vision and Pattern Recognition (CVPR), Las Vegas, NV, USA, 27–30 June 2016; pp. 770–778.
29. Planck, M. *The Theory of Heat Radiation*; Dover Publications: Mineola, NY, USA, 1914.
30. Stefan, J.; Boltzmann, L. On the Relationship between Heat Radiation and Temperature. *Ann. Phys.* **1889**, *273*, 497–525.
31. Flórez, M.; Guerra-Rodríguez, E.; Cazón, P.; Vázquez, M. Chitosan for food packaging: Recent advances in active and intelligent films. *Food Hydrocoll.* **2022**, *124 Pt B*, 107328. [CrossRef]

32. Mallamaci, G.; Brugnoli, B.; Mariano, A.; Scotto d'Abusco, A.; Piozzi, A.; Di Lisio, V.; Sturabotti, E.; Alfano, S.; Francolini, I. Surface modification of polyester films with polyfunctional amines: Effect on bacterial biofilm formation. *Surf. Interfaces* **2023**, *39*, 102924.
33. Bangar, S.P.; Whiteside, W.S.; Ashogbon, A.O.; Kumar, M. Recent advances in thermoplastic starches for food packaging: A review. *Food Packag. Shelf Life* **2021**, *30*, 100743.

Review

Progress in Fruit and Vegetable Preservation: Plant-Based Nanoemulsion Coatings and Their Evolving Trends

Teodora Cvanić, Olja Šovljanski , Senka Popović , Tamara Erceg, Jelena Vulić , Jasna Čanadanović-Brunet, Gordana Ćetković and Vanja Travičić *

Faculty of Technology Novi Sad, University of Novi Sad, Bulevar Cara Lazara 1, 21000 Novi Sad, Serbia;
teodora.cvanic@uns.ac.rs (T.C.); oljasovljanski@uns.ac.rs (O.Š.); madjarev@uns.ac.rs (S.P.);
tamara.erceg@uns.ac.rs (T.E.); jvulic@uns.ac.rs (J.V.); jasnab@uns.ac.rs (J.Č.-B.); gcetkovic@uns.ac.rs (G.Ć.)
* Correspondence: vanjaseregelj@tf.uns.ac.rs

Abstract: Innovative technologies in the food industry are focused on integrated approaches to improve the sustainability of the food system that cover the whole supply chain. Huge post-harvest losses of fruits and vegetables and the use of synthetic chemical preservatives for this purpose are a matter of grave concern for any country. High demands for safe and healthy food have contributed to maximizing efforts to investigate post-harvest technology. Since fruits and vegetables are extremely perishable foods, they require the best post-harvest methods to maintain their storage stability and increase shelf-life. A solution for this emerging problem was found in the application of nanoemulsion edible coatings, described as thin-layered edible coatings or films with the possibility to provide additional benefits such as antioxidant and antimicrobial properties. These coatings provide protection against moisture loss, respiration, gaseous exchange, microbial spoilage, etc., offering promising results to safeguard the physicochemical during the time of storage and transportation of fruits and vegetables. This review summarizes the newest studies of nanoemulsion coatings on fresh products, providing valuable information regarding preparation and application methods and applied polymers and bioactives. Moreover, it gives a detailed description of the influence of nanoemulsion coating application (shelf-life, weight loss, colour, etc.) on fresh fruits and vegetables during storage.

Keywords: post-harvest technology; nanoemulsion coatings; edible packaging; shelf life; fruits and vegetables; component analysis; preparation techniques; application methods; preservation effects

Citation: Cvanić, T.; Šovljanski, O.; Popović, S.; Erceg, T.; Vulić, J.; Čanadanović-Brunet, J.; Ćetković, G.; Travičić, V. Progress in Fruit and Vegetable Preservation: Plant-Based Nanoemulsion Coatings and Their Evolving Trends. *Coatings* **2023**, *13*, 1835. https://doi.org/10.3390/coatings13111835

Academic Editors: Jun Mei, Jing Xie and Biao Zhang

Received: 3 October 2023
Revised: 24 October 2023
Accepted: 25 October 2023
Published: 27 October 2023

1. Introduction

Current trends in people's lifestyles, fast urbanization, and improvements in technology lead to higher demands for safe and healthy food, with requirements to contain additional health benefits [1]. The global change in the consumption of fruits and vegetables has been followed by an increase in the knowledge and awareness of healthy compounds in these products, such as antioxidants and other micronutrients. Hence, it is inevitable that the expansion of fresh and minimally processed fruits and vegetables in the food industry will be followed by the consumers' high expectations regarding their freshness and quality. Unfortunately, changes in colour, browning, sweating, or other undesirable variations could turn down consumers from buying it [2,3]. In addition to the evident variations in organoleptic characteristics, ensuring the safety of food is of great importance. This concern becomes particularly pronounced when considering fruits and vegetables, as they are typically consumed directly after harvest and are frequently devoid of any sterilization treatments. The fundamental susceptibility of these commodities to microbial contamination and spoilage underscores the critical need to prioritize food safety measures in their production and distribution [4].

Fruits and vegetables are renowned for their sensory appeal, including taste, aroma, texture, and visual appeal [5,6]. Any deviation from their natural qualities due to contamination or spoilage not only diminishes their overall quality but can also pose significant health risks to consumers. Unlike many other food products that undergo various processing steps or cooking procedures before consumption, fruits and vegetables are often consumed raw, which heightens the importance of their initial safety [7–9]. Moreover, fruits and vegetables are frequently handled by multiple actors within the supply chain, from growers and harvesters to distributors and retailers, increasing the potential for contamination at various points along the journey from farm to fork [10,11]. Fruits and vegetables constitute a main part of a healthy diet, but very high amounts can be damaged by microbial spoilage, leading to food safety concerns and economic losses. Moreover, the Food and Agriculture Organization (FAO) estimates that more than 1.3 billion tonnes of food are wasted each year, with fruits and vegetables accounting for 40%–50% of the total waste produced. In the past decades, numerous conventional methods have been developed for food preservation, but there are only a few strategies for extending the shelf life of fruits and vegetables. One of the innovative and globally accepted methods for achieving this goal is the use of thin-layered edible coatings or films, which, throughout the extension of shelf life, provide good quality for selling, distribution, and storage. Moreover, the implementation of nanoemulsion technology empowers edible coatings and films due to their small size and high surface area, which can lead to encapsulation and target delivery of different compounds, providing additional benefits such as antioxidant and antimicrobial properties [12,13]. Furthermore, a significant feature of edible coatings is that their utilization is not limited to fruits and vegetables since there are other highly perishable types of food such as meat and poultry, dairy products, bakery products, and fish and seafood [14].

Since nanoemulsion-based edible coatings represent a promising strategy in food technology, many research and review articles on this topic have been published, such as [1,6,9,10,15]. These review articles offer insight into recent developments in edible films and coatings [10], advances in coating materials [6], potential applications [1], and the formulation of coatings for fruits and vegetables [9]. However, there are not many commercially available formulations of edible coatings, as well as just 16 patents regarding nanoemulsion edible coatings and films with either antifungal, antibacterial, or antioxidant properties for application on fruits and vegetables [12,16]. Therefore, there is a need for further advancements in composition, formulation, application methods, and practical applications of these coatings across diverse produce items. This review offers insight into the core concepts, techniques, and components of nanoemulsions. The special significance is reflected regarding the effects on shelf life and various physicochemical and microbial qualities of treated fruits and vegetables during the postharvest period. Moreover, this review article stands out as it covers applicability to real food and its effects in the real-time preservation of fruit and vegetable commodities.

2. Edible Coatings

Edible packaging, which includes edible films and edible coatings, is described as a thin layer for primary packaging made from edible components without any potential health risk for consumers [17,18]. These coatings represent a replacement for natural protective waxy coatings, which can be completely made from renewable materials like natural polymers [19]. Moreover, they can be produced as single- or multi-layer coatings based on desirable characteristics [20]. Edible coatings can be applied in liquid form directly to food, and they are considered an integral unit with food products, while edible films are carved into a solid sheet and commonly used as wraps, pouches, bags, capsules, and casings [21]. Besides application procedures, another difference between edible coatings and films is their thickness, where coatings are thinner than 0.025 mm, while films are usually thicker than 0.050 mm [22].

Awareness of important features of edible coatings, e.g., decreasing oxygen permeability, reducing respiration rates, etc., which represent a barrier to chemical and biological changes, can provide necessary compatibility between packaging and food products [23]. Furthermore, edible coatings can contain many active compounds, e.g., antimicrobials, antioxidants, nutrient dyes, and spices, resulting in improvements in exceed softening and surface browning, preventing microbial spoilage, and increasing nutritional and antioxidant properties [5,20]. Edible coatings and films demonstrated promising results in food preservation due to the controlled diffusion and gradual release of incorporated antimicrobial compounds to the surface of food. The choice of antimicrobial agents should be correlated with the type of food, chosen polymer, and target microorganism(s) [24].

When it is used for fresh fruits and vegetables, it is desirable to have low water vapour permeability in order to decrease desiccation, while oxygen permeability should be minimal to provide retard respiration, but anaerobic conditions must not be reached to avoid ethanol production and off-flavour formation [20]. Organoleptic and functional properties of edible coatings must be suitable for each type of food product, so their formulation depends on their expected purpose [23].

Considering all the above-mentioned characteristics of edible coatings, they can evidently play a crucial role in maintaining and improving food safety for several reasons; their potential roles, advantages, and limitations are summarized in Figure 1 [25–27].

Figure 1. Edible coating summary for application in food safety protocols.

Future research should focus on improving coating application methods, developing sustainable and cost-effective formulations, and expanding the range of food products that can benefit from these coatings [28]. Edible coatings offer a viable and sustainable solution for improving food safety by providing microbial barriers, reducing cross-contamination, controlling moisture levels, and extending the shelf life of various food products. While they come with certain limitations, their advantages in terms of naturalness, customization, improved quality, and waste reduction make them a valuable asset in the search for safer and longer-lasting food options [8,29].

3. Nanoemulsions

The use of nanotechnology has increased and gained an indispensable role in the areas of food and food packaging, along with promising results in many scientific fields in recent decades. Nanotechnology is engaged in the production, processing, and application of materials whose thickness is less than 1 nm [30]. The characteristics of nanomaterials (small size and high surface area) are the reasons for their widespread use in medicine and industrial applications [31]. Possible applications of nanotechnology are encapsulation and target delivery of particles (pharmaceutical, antimicrobial, and antioxidant), increasing the shelf life of food and food storage, detecting contamination, as well as upgrading the flavour and sensory properties. The aforementioned characteristics of nanotechnology explain why one of the most promising applications of nanoemulsions is the utilization of edible coatings and films [32].

These specific structures can change the nutritional content of food and have the ability to magnify the bioaccessibility of hydrophobic bioactive compounds [33]. Bioactive compounds show low stability and bioavailability, as well as degradation and deterioration because of oxidation [34]. Furthermore, many bioactive compounds break free easily from the matrix due to their small size, resulting in a reduction in efficiency and the shelf life of the food; therefore, it is important to design biodegradable packaging in a way that controls the release of bioactive compounds [35]. It can be concluded that it is difficult to design functional products, but nanoemulsions through the process of encapsulation of bioactive compounds can protect them from inadequate processing conditions (oxidation, light, pH, temperature variations, and other influences) [34].

The mentioned nano-scale structures contain two immiscible liquids whose size of droplets can be between 20 and 200 nm, depending on the composition and fabrication method. The formation of nanoemulsions begins with a dispersion of small droplets of one immiscible liquid in another immiscible liquid; therefore, the usual phases are oily and aquatic. Hereof, nanoemulsions can be the oil-in-water (O/W) and the water-in-oil (W/O) types [15,36]. The small size provides better stability to gravitational separation and droplet aggregation, as well as optical clarity and higher bioavailability of encapsulated bioactive chemicals while enhancing phytochemical properties [32,37]. The content of the oil phase includes essential oils (EOs), flavour oils, triglyceride oils, preservatives, flavours, colours, oil-soluble vitamins, nutraceuticals, and pharmaceuticals. Hence, oil phases demonstrate various physicochemical properties, where crystallization, melting behaviour, and thermal history are important parameters due to their influence on viscosity, melting point, polarity, refractive index, and interfacial tension of the nanoemulsion system. The water phase commonly consists of water, but it can possess other important components, such as co-solvents, buffers, salts, preservatives, carbohydrates, and proteins [31].

Even though nanoemulsions have a promising quality for edible packaging, the main issue is long-term stability because of thermodynamically unfavourable systems, in which the positive free energy wants to be released through flocculation, coalescence, Ostwald ripening, and gravitational separation [15]. To ensure the stability of nanoemulsions for commercial applications, it is necessary to control their composition and structure. Furthermore, the oil and water phases should be chosen carefully, along with the selection of the most suitable additives. The unavoidable particle of nanoemulsions is a surfactant that has a key role in maintaining the functional abilities of nanoemulsions, and it should be chosen carefully for each application and type of nanoemulsion. These added additives will protect nanoemulsions from the breakdown [15,31]. Specifically, a hydrophilic emulsifier is used for O/W nanoemulsions, while a lipophilic emulsifier is used for W/O nanoemulsions. Based on the location of the oil and water phases, nanoemulsions can have different structures and form simple or multiple nanoemulsions [31]. The most common surfactants are lecithin; sodium deoxycholate; tween 20, 40, 60, and 80; as well as span 20, 40, 60, and 80. Additionally used surfactants are casein, β-lactoglobulin, polyethylene glycol polysaccharides (e.g., gums and starch), sodium dodecyl sulfate, etc. [36].

3.1. The Main Features of Nanoemulsion-Based Coatings

Even though edible coatings have been used for centuries, their full potential was seen when the food processing sector adopted nanotechnology as a cutting-edge approach and opened fresh ideas for post-harvest storage methods. The use of nanoemulsion edible coatings is a biodegradable alternative approach to address problems during storage for agricultural commodities by creating barrier-enhancing protective nanoemulsions and nanoparticle coatings for use on fruits and vegetables that can help to reduce physiological disturbances on the organoleptic properties of fruits and vegetables (Figure 2). The nanoemulsion technology ensures the use of lower amounts of materials with improved efficiency due to an increase in surface area [38]. Furthermore, edible coatings should not hamper the minimum rate of respiration and impact the produce's colour and flavour. However, it should have high adhesive qualities and be thin enough to apply evenly to fruits. Moreover, for the commercial application of edible coatings, it is valuable that the main components are inexpensive and conveniently accessible [9].

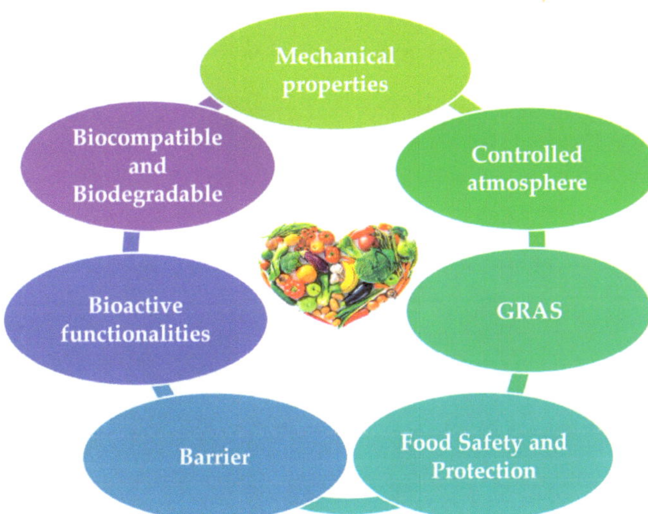

Figure 2. The main features of edible coatings and films.

Recent advancements in nanoemulsion-based edible coatings have focused on tailoring formulations to meet the specific needs of different fruits and vegetables. These innovations include the incorporation of natural antimicrobial agents, antioxidants, and flavour and texture enhancers to improve product quality, safety, and consumer appeal of different foods [25]. Formulation adjustments are made to ensure compatibility with the unique characteristics of each type of produce, such as fresh fruits and vegetables [28].

Nanoemulsion coatings can be applied using various methods, such as dipping, spraying, or brushing, depending on the produce type and desired outcome. These methods facilitate the uniform deposition of the coating, which serves as a protective barrier against moisture loss, microbial contamination, and oxidation [10]. The versatility of nanoemulsion coatings is showcased by their successful application on a wide array of fruits and vegetables, including but not limited to berries, citrus fruits, leafy greens, and root vegetables [29]. These coatings have been shown to extend shelf life, maintain freshness, and preserve nutritional quality across different produce categories. Recent advances have also focused on improving the effectiveness of edible coatings by optimizing factors such as coating thickness, drying methods, and post-harvest storage conditions. Additionally, considerations such as temperature, humidity, and the specific needs of different produce varieties have been considered to enhance the performance of these coatings [6].

3.2. Nanoemulsion-Based Edible Coatings' Polymers

3.2.1. Coatings Polymers

The flexibility of edible coatings' composition and performance creates new opportunities for the creation of innovative packaging techniques that satisfy the two important objectives of food preservation and quality improvement. Furthermore, edible coatings are a desirable option since they lessen the need for packaging made of plastic, which reduces waste and has a positive influence on the environment [27]. Edible packaging is typically made from edible polymers that can be divided into three groups: polysaccharides, proteins, and lipids [39], which are present in Figure 3. All compounds employed in coatings must be food-grade and safe for human consumption, as well as other additives. Furthermore, the choice of coating material is of greatest importance to ensure desirable characteristics for each type of fresh fruit and vegetable. However, one type of edible polymer can possess excellent properties in certain areas, but it can be inadequate in others. Hence, the utilization of a combination of edible polymers has most of the advantages of a single polymer, resulting in the improvement in technological functionality and biological characteristics of edible films and coatings [40–42].

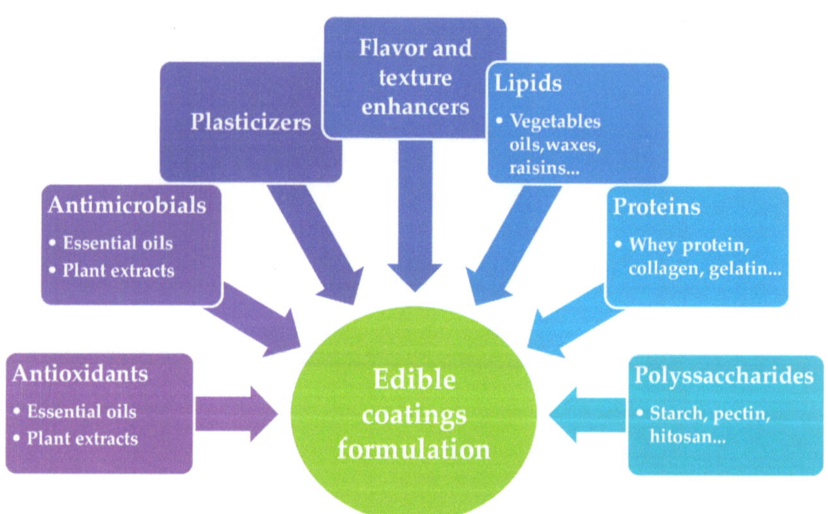

Figure 3. Main components in edible coating formulations.

Coating polymers should be reasonably priced and widely accessible. The coatings should be simple to apply, have good adhesive properties, dry rapidly, and have a consistent thickness. Furthermore, during long storage, the coating's functionality and structural stability must be preserved. The covering must be adaptable to certain morphological changes, e.g., mechanical damage or fruit shrinkage. Since the properties of coating polymers dictate their use and function (Table 1), it is important to know their main features [28].

Table 1. Examples of nanoemulsions polymer(s) for creating coatings for fruits and vegetables.

Nanoemulsion Polymer(s)	Active Compound	Functional Role	Fruits/Vegetable	Ref.
Polysaccharide-based				
Pectin	Ascorbic acid, citric acid, and sodium chlorite	Decreased microbial deterioration; little effect on the nutritional or sensory value.	Apple	[43]
Chitosan	Carvacrol	Prevented microbial growth for 13 days at 5 °C.	Cut carrot	[44]
Starch	Basil leaf extract	Extended the shelf life of eggplant up to 16 days; maintained firmness; reduced moisture loss; prevented the growth of total soluble solids, and prevented colour changes.	Eggplant	[45]
Alginate	Lemon EO * or orange EO	Inhibited bacterial and fungal growth for 15 days of storage.	Red raspberry	[46]
Xanthan	Cinnamic acid	Slowed down the process of browning; increased their shelf-life by up to 4 days and 8 days of storage.	Fresh-cut pear	[47]
Protein-based				
Collagen	/	Reduced the rate of gas transport, which made them useful instruments for extending shelf life.	Mangoes and apples	[48]
Whey protein	/	Maintained various fruit qualities, such as the total phenolic content, browning, and product weight loss.	Apples	[49]
Soy	Cinnamaldehyde	Proved to be an effective antioxidant and moisture barrier.	Banana	[50]
Zein	Carvacrol, thymol, and eugenol EO	Reduced growth of some major pathogens on the peel surfaces of melons.	Melon	[51]
Lipid-based				
Carnauba wax	Lemongrass oil	Prevented foodborne pathogens (*Salmonella typhimurium* and *Escherichia coli*) and prolonged shelf life.	Grape berries	[7]
Beeswax	/	Preserved different fruits' quality metrics and taste characteristics.	Mandarin	[52]
Soy lecithin	Lysophosphatidylethanolamine	Improved the shelf life of banana fruits; provided normal colour development and low ethylene production.	Banana fruits	[53]

* Essential oil.

3.2.2. Polysaccharide-Based Coatings

Polysaccharide-based coatings are characterized to be a poor moisture barrier due to their hydrophilic nature and have a selective permeability to O_2 and CO_2 and resistance to oils while expressing nontoxicity. They are a source of hardness, crispness, compactness, viscosity, adhesiveness, and gel-forming properties. Also, they can be easily found at low prices [54,55]. The presence of hydroxyl groups in the polysaccharides, which results in internal hydrogen bonding, is primarily responsible for their capacity to form films. Furthermore, polysaccharides are especially suitable for short-term food packaging applications. The most-used contributories for edible films and coatings are alginate, chitosan, cellulose, starch, pectin, alginate, and carrageenan [56]. As presented in Table 1, pectin-based nanoemulsion coatings show promise in reducing microbial deterioration in apples. While these coatings have little impact on the nutritional or sensory qualities of the fruits, they contribute to increased shelf life and reduced spoilage [43]. On the other side, chitosan-based nanoemulsion coatings are effective in preventing microbial growth in cut carrots, thus extending their shelf life during storage at lower temperatures [44]. One more polysaccharide is used for nanoemulsion coatings in starch in combination with basil leaves. A study conducted by Kumar et al. [45] showed benefits in extending the shelf life of eggplants, maintaining firmness, reducing moisture loss, and delaying undesirable changes in texture and colour. Alginate-based nanoemulsion coatings incorporating EOs of citrus exhibit antibacterial and antifungal properties that contribute to the preservation of red raspberries during storage. Finally, xanthan can also be considered one of the polysaccharides used for nanoemulsion systems since Sharma and Rao [47] published that coatings based on xanthan and cinnamic acid can slow down the browning process in fresh-cut pears.

3.2.3. Protein-Based Coatings

Proteins are macromolecules formed by amino acids in different molecular conformation structures that can form fibrous or globular forms. They are hydrophobic molecules incorporated with hydrophilic amino acid residues that influence permeability. The source of proteins for edible packaging can originate from either animals (casein, whey gelatin, etc.) or plants (soy protein, corn zein, etc.) [55]. Proteins-based films ensure mechanical stability, but plasticizers must be added to improve flexibility. They are excellent barriers, except for water, which can be overcome through the addition of hydrophobic components. Moreover, they provide additional nutritional value [54,57]. According to the overview in Table 1, some proteins have been used in nanoemulsion coating systems. For example, collagen-based nanoemulsion coatings are effective in reducing the rate of gas transport. This property makes them valuable for extending the shelf life of fruits like mangoes and apples [48]. In the work of Li et al. [50], soy coatings enriched with cinnamaldehyde were applied to bananas. The outcome of their research was confirmation of an effective moisture barrier and significant antioxidant features of the formulated coating. The utilization of whey protein has been widely presented and investigated. Feng et al. [49] observed significant preservation of apples as they protect total phenolic content, browning, and weight loss. Further, Boyaci et al. [51] formulated coatings assembled from zein protein and different EO, such as carvacrol, thymol, and eugenol EO. This combination of ingredients reduced the growth of some major pathogens on the peel surfaces of melons, providing effective protection against microbial spoilage.

3.2.4. Lipid-Based Coatings

Lipids are a heterogeneous group that can be divided into classes: simple lipids, derived lipids, and lipid components. In the food industry, simple lipids, including oils and fats, are mostly used. Lipids are used in edible packaging due to various antibacterial, antioxidant, mechanical, barrier, and functional properties, but they are not polymers. Therefore, they do not have a self-supporting structure and cannot function without other compounds. The homogenous distribution of the small size of lipid fragments can improve stability and decrease the water vapour permeability of edible films and coatings. The main advantage of lipid components is their possible compatibility with other polymers, resulting in an improved barrier and mechanical properties, but films and coatings loaded with lipid components can be sensitive to oxidation and retention of off-flavours and have a bitter aftertaste [55,58,59]. Table 1 shows a few examples of lipid-based coatings. Regarding the study of Kim et al. [7], the incorporation of lemongrass oil in carnauba wax showed promising results in the prevention of foodborne pathogens, such as *Salmonella typhimurium* and *Escherichia coli*, and improved the shelf life of grape berries. The application of beeswax as a coating solution on mandarins, due to the many health benefits of beeswax, could preserve many fruits' quality and taste characteristics. However, the use of beeswax is a bit limited due to potential consumer allergic reactions [52]. Ahmet and Palta [53] applied soy lecithin coatings on bananas. Their study revealed that the incorporation of lysophosphatidylethanolamine, as an active compound, led to improvements in the shelf life of tested fruits, as they succeeded in maintaining normal colour development and low ethylene production.

Overall, the nature of these materials can be mostly hydrophilic (polysaccharides), hydrophobic (lipids), or both (proteins), and the use of solvents is limited to water or ethanol to maintain edible properties as they are considered to be safe for human consumption [54]. The mechanism of linkage of biopolymers involves multiple intermolecular forces such as electrostatic, hydrophobic, ionic interactions, and covalent bonds. Therefore, the properties of edible films and coatings depend on the physical and chemical features of the chosen biopolymer or a combination of them. Additionally, plasticizers and other additives (Figure 3) can be added to the biopolymers to achieve desirable properties in the edible films and coatings. Plasticizers are small molecules that are added to films and coatings to reduce brittleness and flaking or enhance flexibility or toughness [60]. Functional additives are essential participants of films and coatings because they enhance, for example, nutritional, health, and sensory properties, as well as antimicrobial and antioxidant activity, which will be discussed in the next chapter. Other additives are used to solve technical problems that may occur during the process [40].

3.2.5. Composite Coatings

The utilization of a combination of edible polymers in the development of edible films and coatings offers several advantages over using a single polymer. This approach can lead to improvements in both technological functionality and biological characteristics, enhancing the overall effectiveness of these food-grade coatings [29]. For example, different polymers have varying abilities to create barriers against moisture, oxygen, and other gases. Combining polymers with complementary barrier properties can result in improved overall barrier performance. For instance, one polymer may excel at moisture resistance, while another is better at blocking oxygen. Together, they provide a more robust protective barrier for food products. Furthermore, edible films and coatings may need to exhibit specific mechanical properties, such as flexibility, tensile strength, or elasticity, depending on the application [61]. Combining polymers allows for the fine-tuning of these properties to meet the specific requirements of the food product. For example, a blend of polymers can result in a coating that is both flexible and strong. Some edible coatings are designed to release bioactive compounds, such as antimicrobials, antioxidants, or flavour compounds, over time [62,63]. By using a combination of polymers, it is possible to control the release kinetics more precisely, ensuring that the bioactive components are released when and where they are needed. One more important characteristic is adhesion to the food surface, and this parameter can be critical for the effectiveness of edible coatings. Combining polymers with different adhesive properties can enhance the coating's ability to stick to the food substrate, reducing the risk of delamination or cracking [25].

Certain combinations of polymers can exhibit synergistic effects, where the properties of the blend are greater than the sum of the individual components. This can result in coatings with unique characteristics that cannot be achieved with a single polymer. Combining polymers can lead to improved biological characteristics, such as increased antimicrobial activity or better compatibility with specific food products. Utilizing a combination of polymers can also lead to more sustainable coatings by reducing the need for single-use plastics or synthetic materials. Natural and biodegradable polymers can be blended to create coatings that are eco-friendly and have minimal environmental impact. Different food products have diverse requirements and challenges when it comes to coatings. Using a combination of polymers provides flexibility, allowing food scientists to tailor coatings to suit a wide range of applications [26,63].

3.3. Nanoemulsion-Based Coatings' Active Compounds

Searching sustainable and health-conscious solutions for the food industry, many researchers have made significant strides in identifying environmentally friendly active compounds with minimal health risks. This exploration is driven by the growing consumer demand for food products that are not only free from harmful pesticide residues but also aligned with a broader commitment to environmental well-being. In this context, plant-based extracts have emerged as a promising category of natural preservatives, offering a safer and more sustainable alternative to synthetic chemical pesticides and preservatives [64]. Particularly noteworthy is their role in preventing postharvest illnesses and preserving the quality of food products after harvesting, marking an essential alteration towards eco-friendly and health-conscious food preservation practices [65]. Raw vegetables, fruits, and herbs/spices contain active compounds that can express antimicrobial and antioxidant activity. Natural phytochemicals can be found in a variety of foods, including fruits and vegetables (such as guava, garlic, pepper, onion, and cabbage), seeds and leaves (such as olive leaves, nutmeg, parsley, caraway, fennel, and grape seeds), and herbs and spices (such as lemongrass, basil, oregano, ginger, rosemary, thyme, and cinnamon) [45,66]. As previously mentioned, nanoemulsion active compounds have emerged as essential ingredients in modern food industry approaches. Among these active compounds, EOs, derived from plant sources, have garnered significant attention due to their multifaceted applications and potential health benefits. Recognized for their aromatic and therapeutic properties, EOs serve as a valuable category of natural extracts with wide-ranging capabilities [22]. EOs are natural phytochemicals that have been widely used as a valuable source of bioactive components that have antioxidant and antibacterial properties, which pique the interest of the food industry as they have been recognized as safe (GRAS) for the environment and human health. As a result, there is a growing interest in using these oils for sustainable agriculture, and a lot of research has been performed that shows that plant EOs and extracts may often be used as medicines and food preservatives, as can be seen in Table 1, wherein many active ingredients are EOs. Studies on the use of EOs and their components against microbial diseases found in horticulture products have produced a collection of findings that can guide the development of practical solutions for food safety. Also, these EOs differ in terms of their biological activity, physical and chemical characteristics, and scent. Therefore, it is critical to choose the best option or combination for each individual application [67,68].

3.4. Nanoemulsion Coating Production Methods

Nanoemulsion coatings have gained significant prominence in various industries due to their ability to encapsulate and deliver active compounds, enhance product stability, and improve functionality. The production of nanoemulsion coatings involves several methods, each with its unique advantages and applications. These methods are fundamental to harnessing the potential of nanoemulsions for diverse industrial needs [1]. Methods for nanoemulsion production are separated into two groups: low-intensity and high-intensity methods (Figure 4). Low-intensity methods are based on the spontaneous formation of small droplets in surfactant–oil–water mixtures when change occurs in the system composition or environmental conditions, and methods within this group are the phase-inversion temperature, emulsion-inversion point method, and spontaneous and membrane emulsification [69]. On the other hand, high-intensity methods have found their use in industrial applications. These methods use mechanical devices, such as valve homogenization, micro fluidization, and sonication, to disrupt and merge the oil and water phases, creating intense turbulence, shear, and cavitation flow profiles [35,70]. The implementation of certain fabrication methods depends on the nature of the components and the required characteristics of nanoemulsions, but most approaches are high-energy ones. However, because nanoemulsions might experience destabilizing processes, a good formulation is essential for ensuring their long-term stability. The ability to withstand physical change is a definition of stability [71]. Coalescence and Ostwald ripening are the two methods by which emulsions are destabilized. According to Rao and McClements [72], Ostwald ripening occurs when the size of oil droplets increases as oil molecules diffuse from small to big droplets through the continuous phase in relatively water-soluble oils (such as EOs), while coalescence occurs due to weak stearic repulsion when two oil droplets merge into a single, bigger droplet.

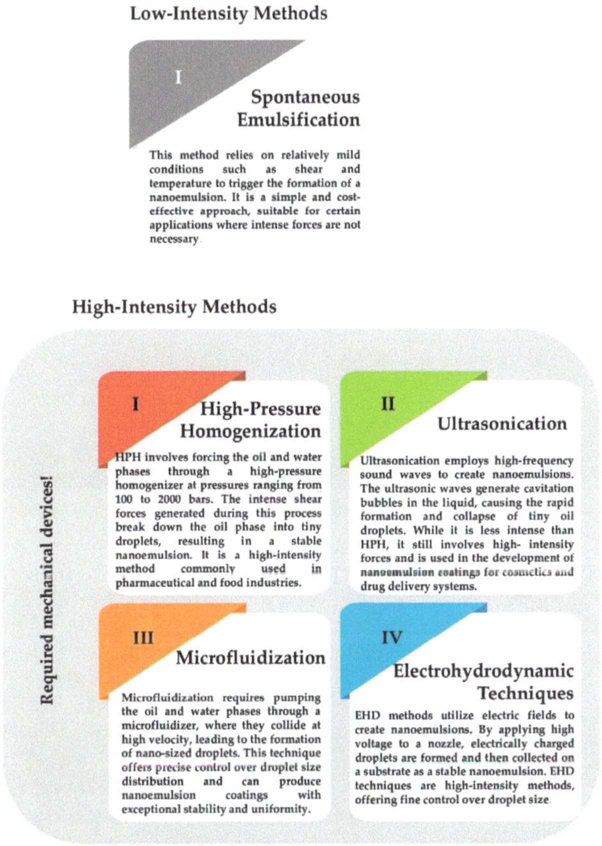

Low-Intensity Methods

High-Intensity Methods

Figure 4. Schematic diagram showing a comparison between the high- and low-energy production methods for nanoemulsion coatings.

The production of nanoemulsion films and coatings starts with dissolving polymers into an adequate solvent. Afterwards, in the solution of the chosen polymer or combination of polymers, active compounds are added (as part of nanoemulsions or unattended), along with plasticizers and additives, to make the final formulation of the coating. There are two process methods for the production of nanoemulsion coatings: the dry-process and wet-process methods. The first method relies on the thermoplastic characteristics of polymers and includes moulding or extrusion. This method does not use solvents that make it environmentally friendly, and thus the obtained films are unable to cover uneven surfaces. On the other hand, the most common method used on the laboratory scale is the casting method (Figure 5). In this wet-process method, a solution of coating is dispersed on the surface and then air dried for a certain amount of time in order to achieve evaporation of the solvent. The edible films are obtained when the solution is dispersed on a flat surface and afterwards peeled off. However, coatings are formed on the product surface if the solution is directly applied to food products via most common application methods (dipping or spraying) [73,74].

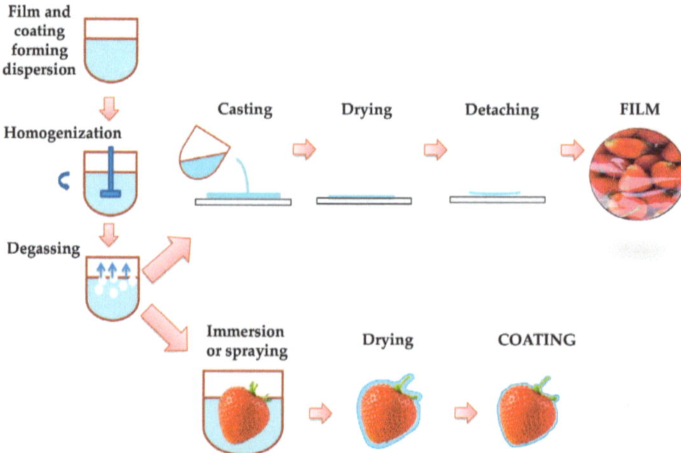

Figure 5. Scematic representation of the production of edible films (casting method) and coatings.

3.5. Nanoemulsion Coating Application Method

The application of nanoemulsion coatings and edible films in the food industry plays a pivotal role in enhancing the quality, safety, and shelf life of a wide range of products. While both coatings and films share the common goal of providing protection and functionality, they can be applied through various methods, each with its own advantages, disadvantages, and limitations. In this context, it is essential to understand these application methods in more detail [75]. The application of edible films is to be as sheets that surround the food, unlike edible coatings, where there are a few application methods, such as dipping, fluidized bed, panning and spraying. However, they may be created from the same raw materials, have the same goal, and serve the same function, with the only difference being how they cover the food [14]. Moreover, applications of edible coatings have demonstrated a considerable impact as a postharvest treatment on the physiological effect and physicochemical qualities. Their application is presented in Table 2, which summarizes the most recent main component of the edible coating's active component and polymers that have been used on fruits and vegetables, as well as producing techniques and application methods. Apparently, as can be seen in Table 2, the two most common methods for coating application are dipping and spraying.

One of the methods used for applying nanoemulsion coatings and edible films is the immersion or dipping method. This method is adequate for distributing an even coating from all sides on the target fruit/vegetable, even when it has a rough surface. Firstly, the fruit/vegetable is dipped in a coating solution for a specific amount of time to create a semipermeable covering on the surface. The typical dipping duration is between 30 s and 5 min. The following step can be submerging in a crosslinking solution for a certain amount of time to increase the coating's stability and toughness. Secondly, the fruit/vegetable is removed from coating solutions to allow the draining of surplus solution in favour of obtaining a uniform and controlled coating thickness [6,76]. Another common method for applying nanoemulsion coatings and edible films is spraying, which provides an equal distribution of the coating solution across the fruits/vegetables, resulting in an average thickness of coating. Furthermore, a coating solution is applied via a nozzle and can be applied as a layer-by-layer coating. Due to high pressure (about 60–80 psi), this method does not require a significant amount of coating solution, and the solution can be less viscous [75].

Table 2. Recent applications of edible coatings on fruits/vegetables based on different polymers with the addition of active compounds.

Target Fruits/Vegetable	Nanoemulsion Active Compound	Nanoemulsion Polymer	Producing Technique	Application Method	Ref.
Papaya	Ginger oil	Carnauba wax and hydroxypropyl methylcellulose	Ultra-Turrax homogenizer	Hand-coating procedure	[77]
	Oregano EO *	Sodium alginate	Ultrasonication	Dipping	[78]
	Eucalyptus staigeriana, Lippia sidoides, and *Pimenta pseudocaryophyllus* leaf extracts	Carboxymethylcellulose	/	Dipping	[79]
	Citral oil	/	Phase-inversion method	Dipping	[80]
Okra	Basil oil and aqueous extract of *Sapindus mukorossi*	Alginate	Ultrasonication	Dipping	[81]
Apple	Rosemary EO	Water chestnut starch	Ultra-Turrax homogenizer, followed by ultrasonication	Dipping	[82]
	Lemongrass oil	Carnauba–shellac wax	Ultra-Turrax homogenizer, then high-pressure homogenizer	Dipping	[83]
	Cinnamon EO *	Pectine from apple pomace	Ultra-Turrax homogenizer		[84]
	Lemon EO	Aloe vera gel and hydroxypropyl methylcellulose	Ultra-Turrax homogenizer	Spraying	[85]
Apricot fruits	Pomegranate peel extract	Chitosan	Ultrasonication	Dipping	[86]
Tomato	Sweet orange EO	Sodium alginate	Ultrasonication	Dipping	[87]
	Thymol	Quinoa-chitosan	Spontaneous emulsification, ultrasound, and a combination of both methods	Dipping	[88]
	Cardamom EO	Carboxymethyl cellulose	Ultrasonication	Dipping	[89]
Strawberry	Lime, fennel, and lavender	Yam starch	Ultra-Turrax homogenizer	Spraying	[90]
	Cymbopogon martinii and *Mentha spicata* EO	Arrowroot starch, cellulose nanocrystals, carnauba wax	Ultra-Turrax homogenizer	Dipping	[91]

Table 2. *Cont.*

Target Fruits/Vegetable	Nanoemulsion Active Compound	Nanoemulsion Polymer	Producing Technique	Application Method	Ref.
Fresh-cut red bell pepper	Tea tree oil	Chitosan	Magnetic stirrer	Dipping	[92]
Cherry	*Eryngium campestre* L. EO	Chitosan	Sonication with an ultrasonic bath, then Ultra-Turrax homogenizer	Dipping	[93]
Pineapple	Eugenol (clove EO) and Aloe vera gel	Chitosan	Spontaneous emulsification and ultrasonication	Spraying	[94]
	Citral and sesame oil	Sodium alginate	Ultrasonication	Dipping	[95]
Citrus fruits	Eugenol, carvacrol and cinnamaldehyde	Sodium alginate with citric acid, sucrose ester, Vitamin C, and potassium sorbate	Ultra-Turrax homogenizer, then passed through a microfludizer	Applied on the surface of the peel with a soft brush	[96]
Salacca fruits	Orange oil	Carnauba wax	Homogenizer	Spraying	[97]
Blackberries	Lactic and acetic acid	Chitosan	Ultrasonication	Spraying	[98]
Grape berries	Lemongrass oil	Carnauba	Ultra-Turrax homogenizer, then high-pressure homogenizer	Dipping	[7]
Garambullo fruits	Tomato oily extract	Gelatin	Ultra-Turrax homogenizer	Dipping	[99]
Peach	Aloe vera gel extract	Chitosan	/	Dipping	[100]
Fresh-cut kiwi fruits	*Opuntia ficus-indica Mucilage*	*Aloe arborescens* gel	Ultra-Turrax homogenizer	Dipping	[101]
	Ascorbic acid and vanillin	Alginate and carboxymethylcellulose	Ultra-Turrax homogenizer, followed by ultrasonication		[102]
Pear	Lemon peel EO	Guar gum and chitosan	/	Dipping	[103]
Jujube fruits	/	Aloe vera gel, carboxymethyl cellulose, and pectin	/	Dipping	[104]
Guava	*Cyclea barbata* leaves extract	Alginate	/	Dipping	[105]
Capsicum fruits	Mulberry leaf extract	Pectin	/	Dipping	[106]
Guava fruits	Ginger extract and garlic extract	Gum arabic and aloe vera gel coating	/	Dipping	[13]

Table 2. *Cont.*

Target Fruits/Vegetable	Nanoemulsion Active Compound	Nanoemulsion Polymer	Producing Technique	Application Method	Ref.
Melon	Citral oil	/	Phase-inversion method	Dipping	[80]
	Citral	Chitosan and carboxymethyl cellulose	Ultrasonication		[107]
Lettuce	Oregano oil	/	Ultrasonication	Dipping	[108]

* Essential oil.

The application process should be performed in hygienic surroundings to avoid contamination and guarantee the efficiency of the coating. Additionally, it must be applied evenly and successfully adhered to the produce's surface. Furthermore, it is important to be aware of the features of each application method. Therefore, in Figure 6, the main benefits and disadvantages of the two most common application methods are summarized [27]. The dipping method provides even application of coatings on fruits or vegetables' surfaces that are uneven or rough; therefore, it does not have limitations on the type of produce. However, the application is rather time-consuming due to the requirement to carefully perform immersion and drying steps. On the other hand, the spraying method allows uniform and consistent cover with explicitly precise distribution control but has difficulties when applied on severely uneven surfaces. However, it is efficient for large-scale applications and uses lower amounts of the coating solution, even though sometimes products can be over-sprayed, resulting in material wastes [109,110].

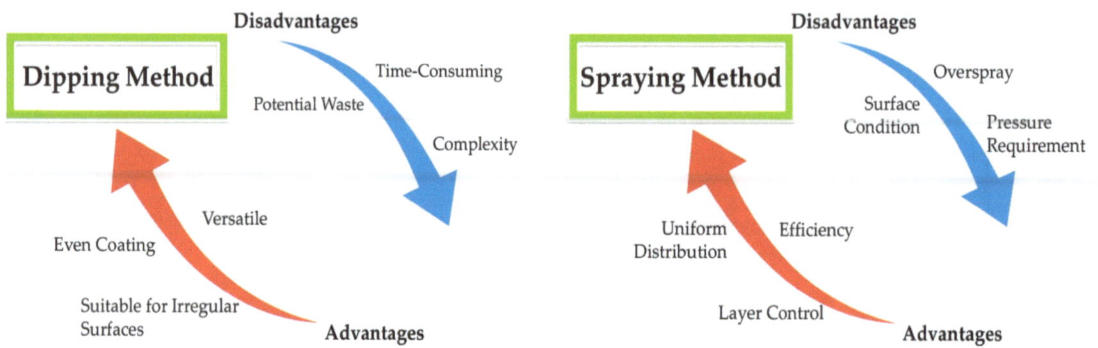

Figure 6. The advantages and disadvantages of dipping and spraying method.

As mentioned earlier, examples of the application of spraying and dipping methods are presented in Table 2. In the case of papayas, several nanoemulsion coating systems have been developed. In the study of [77], ginger oil is employed as the active compound to coat papayas. The coating formulation includes carnauba wax and hydroxypropyl methylcellulose as the nanoemulsion polymer. The producing technique involves the use of an Ultra-Turrax homogenizer, a high-powered device for emulsification. The application method chosen is a hand-coating procedure. This approach likely provides precise control over the coating process and ensures even coverage on the papaya surface. Ginger oil, known for its aromatic and potentially bioactive properties, could contribute to the preservation and flavour enhancement of the papayas. In the case of a recent study by Tabassum et al. [78], oregano EO is the selected active compound for coating papayas. The coating formulation incorporates sodium alginate-based edible coating, which is known for its ability to provide a protective and barrier layer. Ultrasonication is used as the producing technique, suggesting the creation of a stable and fine nanoemulsion. The chosen application method is dipping, which ensures comprehensive coverage and uniformity. Oregano EO is a natural ingredient with antimicrobial and antioxidant properties, making it a valuable choice for extending the shelf life and quality of papayas. Papayas with *Eucalyptus*, *Lippia sidoides*, and *Pimenta pseudocaryophyllus* leaf extracts developed by Zillo et al. [79] include a carboxymethylcellulose coating, which is likely selected for its film-forming and protective characteristics. Interestingly, the producing technique is not specified, suggesting that the method may vary or be proprietary. This study highlights the potential of utilizing a combination of natural extracts to create innovative nanoemulsion coatings for papayas, although specific details about the production process are not provided. In this study, citral oil is employed as the active compound, and the application method is dipping. Citral oil is known for its citrusy aroma and potential antimicrobial properties, which can contribute to both flavour enhancement and food safety in papayas. The producing technique involves a "phase inversion method", which is a technique that induces a change in the phase of the emulsion, often resulting in the formation of nano-sized droplets [80].

For instance, in the coating of pineapples, chitosan-based coatings incorporated with clove EO and aloe vera gel were formulated by Basumatary et al. [94]. Aloe vera gel is known for its gel-forming ability, while together with clove EO, provides many health benefits. The coatings were prepared via two methods, spontaneous emulsification and ultrasonication, which provided stable nanoemulsion coatings, which were applied via spraying. In the work of Prakash et al. [95], pineapples were dipped in a coating solution prepared from sodium alginate enriched with citral and sesame oil. Citral was added due to its antimicrobial activity, while sesame oil ensures the stability of nanoemulsion. Utilization of ultrasonication provided the nano distribution of droplets, resulting in even distribution of active components.

In Table 2, there are many examples of apple coatings; all reviewed research articles used different EOs, such as rosemary, lemongrass, cinnamon, and lemon EO. However, contrary to other scientists who chose the dipping method, Farina et al. [85] decided to spray their apple samples with aloe vera gel and hydroxypropyl methylcellulose coating, prepared using an Ultra-Turrax homogenizer, which provided an attractive natural shine to fruits. The same production method for coatings was involved in the work of Naqash et al. [84], who used pectin, as a main polymer, obtained from pomace, which is an environmentally and economically responsible practice. Additionally, the expressed antimicrobial activity of the formulated coating is due to the presence of cinnamon EO and its benefits. For the preparation of carnauba–shellac wax coating, firstly, an Ultra-Turrax homogenizer and then a high-pressure homogenizer were used, resulting in a stable nanoemulsion for even 5 months of storage [83]. In a recent study by Bashir et al. [82], for dipping apples, water chestnut starch coatings were prepared via ultrasonication, which provided an adequate solution for post-harvest protection against decay, weight loss, maintaining firmness and high concentrations of anthocyanins and ascorbic acid in corps.

Besides seeking improvements in obtaining optimal formulation of edible coatings (polymers, active compounds, additives), storage conditions, and producing methods, further research is necessary concerning the application method. In conclusion, a variety of parameters, including the type of produce, desired coating qualities, and manufacturing scale, determine whether to dip or spray edible films and nanoemulsion coatings. The best strategy must be chosen depending on the application needs because each method has a unique mix of benefits and drawbacks [79,81].

4. Studies of Nanoemulsion Coating Applications on Fruit/Vegetable Samples

Even though edible coatings have been used to preserve food since the 12th century in East Asia, nowadays, they have gained huge popularity due to the numerous benefits of edible packaging, such as the impact on colour, shelf life, firmness, and weight loss, as graphically presented in Figure 7 [111,112]. As presented in Table 1, the application of edible coatings on fruits/vegetables is in the expanding interest of scientists, as can be seen in a vast number of publications that give insights into different aspects of the effects of the implementation of different formulations of coatings.

4.1. pH

The pH is viewed as an important parameter for the evaluation of the quality of fruits and vegetables. Manzoor et al. [99] kept track of changes in the pH value of kiwi fruit slices and the influence of alginate and carboxymethylcellulose coatings incorporated with ascorbic acid and vanillin. They observed significant differences on the 7th day of storage between control and treated samples. Moreover, alginate coating containing 1% vanillin exhibited the highest preservation of pH values of kiwi slices compared to carboxymethylcellulose coatings. In the work of Prakash et al. [95], the pH of the pineapple samples was unaffected by the various coating treatments and stayed between 3.9 and 4.5 throughout storage. On the other hand, throughout the storage period, the pH values of all sweet cherries increased significantly regardless of their treatment. The pH differences were found to be less pronounced in coated cherries than in untreated cherries [93]. In the study of Anjum et al. [13], a significant change in pH was observed for guava fruits that increased up to day 9, then decreased from day 12 to day 15. The explanation for the rise in pH is due to the decrease in organic acids and potential sugar conversion.

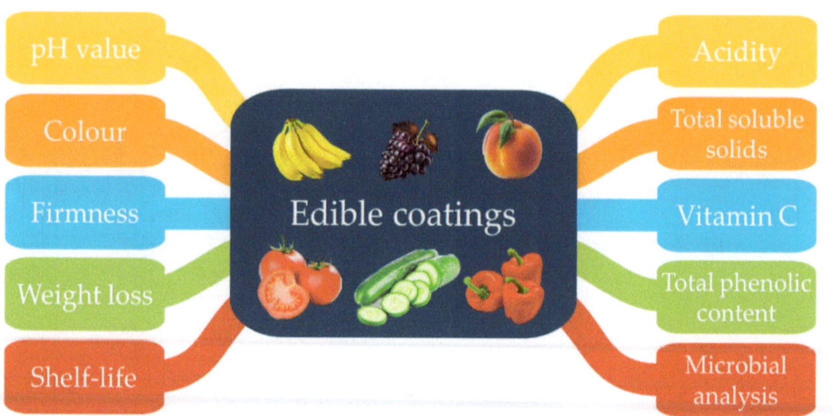

Figure 7. Listed impacts of edible coatings/films.

4.2. Total Phenolic Content

The hidden worry is that fresh fruit and vegetable storage causes nutritional loss because of water retention, oxidation, microbial breakdown, and enzymatic breakdown. Therefore, another important element in assessing the organoleptic and nutritional quality of fruits and vegetables is the levels of total phenolic content (TPC) [92]. In the study by Sathiyaseelan et al. [92], the TPC considerably changed between preservation methods and days of preservation. Nanoemulsion coating succeeded in maintaining the TPC in red bell pepper, at a high value of 92.37 mg/g of GAE, while the TPC value of control fruits was 64.59 mg/g of GAE. In the work of Oliveira Filho et al., 2022 [91], during storage, the total phenolic compound concentration declined in all samples. Strawberries coated with *Cymbopogon martinii* and *Mentha spicata* EOs showed the lowest TPC decrease among the treatments tested in this study (1.43–1.52 to 1.03–1.11 mg GAE/g of fruit and 1.57–1.68 to 1.07–1.13 mg GAE/g, respectively) as a comparison to treatments without using EOs, where TPC was 1.7–0.78 mg GAE/g of fruit. For example, the TPCs of cherries significantly dropped during the 21-day storage period at 4 °C, but the coating treatment had a notable inhibitory effect on this decreasing tendency. The highest TPC value (50.21 g/mL gallic acid) was measured in coated sweet cherries after 14 days of storage, while the lowest TPC value (10.36 g/mL gallic acid) was found in the control group after 21 days of storage. The outcomes also showed that the most efficient method for reducing TPC loss in the coated cherries was to encapsulate *Eryngium campestre* L. EO in chitosan coating, which kept the TPC value above 1.13 µg/mL gallic acid for the first seven days of storage [93].

4.3. Colour

The key factor encouraging a consumer's decision to consume fruits or vegetables is their appealing colour, flavour, and texture. As a result, keeping the colour of fruits and freshly cut items is essential for the food industry [92]. Changes in visual colour indicate fruit ripening. In the case of papaya, colour changes as the fruit ripens from green to yellow/orange due to the degradation of green chlorophyll and the production of yellow/orange carotenoids. Their study revealed that control fruits had lower hue values, which indicates papaya peel colour shift from green to yellow colour, than coated fruits, where the highest hue values were found for the carnauba-based coating with an

18% solid phase [77]. Contrarily, in the work of Sathiyaseelan et al. [92], significant colour changes in fresh-cut red bell peppers between the control and treatment samples were not detected. For example, in the study of Phothisuwan et al. [97], carnauba wax coatings with different orange oil concentrations were applied on fresh salacca fruits that initially had yellowish-white flesh that later turned yellowish-brown. The findings of this study disclosed that the salacca fruits had the maximum lightness and yellowness values when coated with 2.5% carnauba wax that contained 0.08% orange oil. Increased blackness in the salacca was associated with a higher content of orange oil (0.16%), which may be related to orange oil's influence on colour change. Additionally, coating with 0.16% orange oil resulted in the salacca flesh turning an unacceptably light shade of brown at the end of the test. The finding from this study revealed how important the concentration of EO in the formulation of the coating is.

4.4. Firmness

The texture profile is one of the most noticeable physiological qualities that is attributed to consumer acceptance. Firmness, one of many textural variables, more accurately depicts the texture of the tissue and, as a result, can better indicate fruit freshness. It is directly correlated with water loss rates as well as internal metabolic activities such as cellulose hydrolysis, membrane integrity loss, and depolymerization of starch and pectin [82,94]. Basumatary et al. [94] prepared chitosan-based edible coating with eugenol oil and aloe vera gel, and its characterization revealed that the firmness of the pineapple samples declined throughout the course of the 20-day storage period, but it was at its lowest within the first 4 days. After 12 days of storage, the coated pineapples maintained a firmness of around 43 N, which is roughly twice as firm as the control samples, whereas the hardness of the control samples decreased from about 60 N to 23 N. According to Sathiyaseelan et al. [92], fresh-cut red bell pepper maintained its hardness for 12 days when it was coated with a combination of calcium chloride, low molecular-weight chitosan, and tea tree oil. When compared to the control, the findings revealed that the combined treatment also preserved the springiness, gumminess, chewiness, and resilience of fresh-cut red bell pepper. However, in the work of Phothisuwan et al. [97], the control revealed the same trend as the salacca hardness preserved with carnauba wax with and without orange oil, but lower values were discovered. The study of Arabpoor et al. [93] showed that uncoated cherries lost stiffness more quickly than the coated ones, suggesting that chitosan nanoparticles with *Eryngium campestre* essential oil effectively delayed tissue softening. Compared to uncoated samples, fruit softening was likewise delayed in cherries treated with only chitosan nanoparticles. However, on day 14 of storage, the tissues of cherries coated with chitosan nanoparticles were significantly softer than those coated with chitosan nanoparticles with EO. The semi-permeable surface of coated fruits may have contributed to fruit firmness persistence by limiting metabolic gaseous exchange (oxygen and carbon dioxide) across the coating barrier, followed by a decrease in metabolic activity and oxidizing enzyme effectiveness [82].

4.5. Weight Loss

Weight loss rate is an important indicator for measuring the degree of preservation of fruits and vegetables. Therefore, Arabpoor et al. [93] studied the effects of chitosan coating loaded with *Eryngium campestre* L. EO on the preservation of cherries during a 21-day storage period. Their study revealed that coatings served as an efficient barrier to prevent water loss, extending the fruits' shelf life. Their results showed that the control fruits had the maximum weight loss during storage, while coated cherry fruits had a weight loss of only 16%. For example, Bashir et al. [82] observed minimum weight loss in apples of 22% coated with starch-based nanoemulsion coatings with rosemary EO, while weight loss in control samples was 35%. According to Miranda et al. [74], the weight loss of papaya fruits was minor when maintained in cold storage, regardless of the type of coating. Contrarily, fruits lost more weight when it was placed in a warmer environment

(22 °C or room temperature). They found out that carnauba-based coatings were more effective than hydroxypropyl methylcellulose due to their hydrophobic nature. In the work of Tabassum et al. [78], all the samples showed a notable increase in weight loss during storage, but the uncoated sample showed the greatest increase. In the study of Yang et al. [96], when stored until day 60, the weight loss rates of citrus fruits for the nanoemulsion treatment (3.14%) and control (4.12%) were significantly different, indicating that nanoemulsion coating had a better preservation effect on cold storage. According to Vilaplana et al. [98], compared to untreated and chemically fungicide-treated blackberries, fruits coated with chitosan made with lactic acid showed lower weight loss after 14 days of storage at 4 °C.

4.6. Shelf Life

Since fruits and vegetables are living tissues and extremely perishable foods, they require the best post-harvest methods to maintain their storage stability and increase their shelf life. After harvest, physiological processes limit the shelf life of fruits and vegetables [113]. The study on citrus fruits disclosed a statistically significant difference, in detail, during cold storage; rotten fruits were present in the control and nanoemulsion-treated samples from 20 to 30 days, respectively. When the fruits were preserved for 60 days, the decay rate of the nanoemulsion-treated fruits was 4.1%, compared to 7.4% for the control group [96]. With regard to the shelf life of fresh-cut or minimally processed fruits, it is important not to exceed 10^6 CFU/g of aerobic plate count, which has been reported as the limit value. In the study of Prakash et al. [95], pineapple shelf life was prolonged for 3 days, as uncoated samples reached the limit on the 9th day, while coated samples reached it on the 12th day. According to Anjum et al. [13], the shelf life of guava fruit extracts was a substantial effect of combined edible coatings and plant extracts. The average shelf life of control guava fruits was 6 days, while those of ginger extract + gum arabic, aloe vera gel + gum arabic, and garlic extract + gum arabic coatings were 9, 10, and 13.1 days, respectively. In the case of garlic extract + gum arabic coatings, the shelf life of guava was doubled.

4.7. Acidity

The organic acid content of fruits is directly related to titratable acidity (TA), which usually declines due to the conversion of organic acids into sugar during the reverse-glycolysis process of fruit ripening [86,94]. Also, acidity is a key element that significantly affects fruit flavour [102]. Based on the work of Basumatary et al. [94], the TA values of the control sample rose throughout the first 12 days of storage from roughly 0.58 to almost 0.98 but then sharply dropped to roughly 0.35 on day 20, indicating accelerated ripening and senescence. All coated samples had their TA values maintained for the first 12 days of storage, but the aloe vera gel coating had them maintained for the entire 20-day storage duration in this study. Gull et al. [83] reported that at the end of the storage period, control fruit samples showed the largest decline, with a value of 0.22%. But coating applications considerably prevented this decline. Coated samples of apricots were treated with chitosan-based coatings containing pomegranate peel extract in concentrations of 0.75% and 0.50%, which displayed TA values of 0.33 and 0.25%, respectively; fruits coated with 1% of pomegranate peel extract efficiently maintained high TA values (0.34%). This might be explained by the coatings' barrier properties, which limited the oxygen supply and slowed down the ripening process. Manzoor et al. [102] found that alginate coating more successfully maintained TA than carboxymethylcellulose coatings, but all coated kiwis showed higher TA than untreated samples during 7 days of storage. In the study of Yang et al. [96], the TA values decreased steadily during storage, and nanoemulsion treatment revealed higher TA content (0.58%) than control TA content (0.53%), which was expected given the use of organic acids in the respiratory process. After 10 days of storage, there was a significant difference between the control group, where the TA value fell abruptly from 62.30 to 35.52, while on the 13th day in the nanoemulsion treatment group, it decreased from 62.30 to 40.13. Based on the work of Sortino et al. [101], during 9 days of storage of kiwi fruit slices, it is notable that control samples showed a significant decrease in TA values, from 1.8 to 0.9 g/L, while TA values for samples coated by aloe gel with *Opuntia ficus-indica* Mucilage extract were between 1.8 and 1.4 g/L. According to Prakash et al. [95], the TA of the sliced pineapples, which varied between treatments during the storage period and ranged from 0.51 to 0.62, was not significantly affected by the various citral nanoemulsions used in the edible coatings.

4.8. Total Soluble Solids

It is suggested that the rise in soluble solids during ripening involves a mechanism of cell wall breakdown that provides a source of carbon for sugar production. Investigating the impact of coarse emulsion and nanoemulsion alginate-based coating containing oregano EO on the total soluble solids (TSS), scientists found out that the TSS of all the samples progressively grew in the first days of storage. All emulsion-coated samples displayed a rapid drop in the TSS after 12 days. Meanwhile, the nanoemulsion-coated samples displayed a rise in the TSS by the end of the storage period, preventing sugar degradation for a longer time and extending the shelf life of fresh-cut papaya as a result, while the uncoated sample had a sudden increase in the TSS until day 8 and then a quick reduction, indicating a limited shelf life [78]. Similarly, Sortino et al. [101] investigated the TSS in kiwi fruit slices and found out it did not substantially differ across treatments during storage; as for control and coated samples, the TSS increased from 12.9 Brix to 14.1 and 13.6 Brix, respectively. In the work of Yang et al. [96], the TSS content of untreated and nanoemulsion-treated fruits gradually increased throughout the first 30 to 40 days of storage and considerably decreased as storage time increased. On the 60th day, the TSS concentrations in the nanoemulsion treatment and control were 14.41% and 15.14%, respectively, indicating that the conversion rate of reducing sugars was significantly slowed down by the nanoemulsion covering. Regarding the TSS values of pineapple, coated samples were maintained at about 17 °Brix, while uncoated samples increased slightly from 15.96 to 18.93 °Brix during storage [95].

4.9. Vitamin C

Ascorbic acid or vitamin C (VC) is a natural antioxidant that is important in illness prevention; immune system stimulation; and the promotion of healthy skin, gums, tendons, and ligaments. However, due to its high degree of oxidation, it is unstable to a variety of conditions, including temperature, pH, oxygen, metal ions, and enzymes (such as ascorbate oxidase or peroxidase). Therefore, it frequently serves as a fruit quality indicator [102]. Yang et al. [96] analyzed the VC content of citrus fruits covered by alginate-based enriched with a cinnamaldehyde, eugenol, and carvacrol nanoemulsion. Their study revealed that VC contents of untreated and nanoemulsion-treated fruits, on the 60th day of storage, were 21.28 mg/100 g and 24.38 mg/100 g of VC, respectively. In the study of Prakash et al. [95], in pineapples, initial ascorbic acid content was determined after 12 days of storage. The initial values were between 18.60 and 19.38 mg/100 g, which, it was discovered, fell throughout the storage time. However, it disclosed that the VC content was notably lower (2.59 mg/100 g) in the uncoated samples at the end of the storage period, while in coated samples was between 10.6 and 12.57 mg/100 g. According to Anjum et al. [13], differences between coated and control samples in VC content of guava fruits were not observed, but there was an evident trend of decrease throughout storage time.

4.10. Microbial Analyses

Microbial analyses are important to evaluate fruit and vegetable safety. The cut surface of a fruit is exposed to environmental deterioration and more vulnerable to decomposition by numerous bacteria in the absence of a physical or chemical barrier in the form of a protective epidermis, which directly affects the health of consumers [78]. A study by Vilapana et al. [98] compared chemical antifungal products and chitosan coatings enriched with lactic and acetic acid in the post-harvest life of berries, which were impacted by the presence of fungi like *Mucor* sp. Their study revealed that blackberries treated with lactic acid coating had a greater disease reduction after 7 days at 4 °C than fruits treated with an ascorbic acid coating (40.29%), and both coatings considerably outperformed fruits sprayed with synthetic fungicide (34.32% disease reduction) in terms of soft mould control. Even though chitosan coatings with lactic and ascorbic acid were less efficient than synthetic fungicide treatment, Vilaplana et al. [98] pointed out that it still may be an alternative approach for the reduction of soft mould in blackberries. On the other hand, microbial analyses for total bacterial, yeast, and mould count on fresh-cut papaya were revealed to be greatly lower between the coated and untreated samples. On day 8 of storage, the bacterial count of the uncoated sample was 6.83 log CFU/g, which exceeded the critical limit. On the other hand, up until day 16 of storage, practically all of the samples containing alginate coating with oregano EO had a bacterial count that did not surpass 3 log CFU/g. The difference in the restriction of the yeast and mould growth for the course and nanoemulsion coating was in a 4-day prolongation for nanoemulsion coatings, which implicates the superiority of nanoemulsions compared to emulsions [78]. In the study by Robledo et al. [88], the effect of quinoa protein/chitosan coating with thymol EO against the growth of mould (*Botrytis cinerea*) on cherry tomatoes was evaluated. After 7 days at 5 °C, tomatoes coated with thymol nanoemulsion quinoa protein/chitosan displayed a significant reduction in fungal growth when compared to control samples. In control samples (uncoated and coated only with quinoa protein/chitosan), yeast and mould counts taken on days 0 and 2 revealed a population between 2.6 and 3.6 log CFU/g that grew to approximately 6.28–6.45 log CFU/g at day 7 of storage, while in coated samples reached 4.7 log CFU/g. In this regard, no inhibition of fungal proliferation was observed when quinoa protein/chitosan alone was used; however, on day 7, the thymol nanoemulsion of quinoa protein/chitosan displayed the inhibition of fungal development. In the study of Sortino et al. [101], microbiological analyses of aerobic mesophilic bacteria, pseudomonads, and yeasts were performed. It was found that at the end of storage, minimal concentrations of aerobic mesophilic bacteria in untreated and treated samples were 3.3 and 1 log CFU/g, respectively. The pseudomonads were discovered after 7 days, and their concentrations after 9 days were 4.5 and 1.5 log CFU/g in the control and coated samples, respectively. A similar pattern was seen with yeasts, with concentrations of 2 and 1.5 log CFU/g. It is evident that the amount of aerobic mesophilic bacteria, pseudomonads, and yeast on the kiwi fruit slices was dramatically reduced during storage time due to an aloe gel mucilage coating solution. In the study by Prakash et al. [95], a rise in the total plate count and yeast and mould count of all the samples was observed during the storage period. Regardless, alginate-based coating with citral nanoemulsion significantly reduced the microbial growth compared to control, where the most effective coating contained a higher concentration of citral nanoemulsion. The results showed that coating with 1% citral nanoemulsion reduced the growth of yeast and mould counts by 2.77 log CFU/g and the total plate count by 4.68 log CFU/g compared to the uncoated control during 12 days of storage. In the study of Abdalla et al. [114], after 12 days of storage, in the control sample, we found substantial contamination that prevented us from counting colonies. Along with the white and yellow colonies that looked like yeast, there were also colonies of black filamentous fungi that had a velvety aspect. The strawberry treated with the pectin coating was the second sample with the highest contamination (4.1 log CFU/g), which displayed three distinct forms of apparent contamination. Finally, there was no obvious contamination on the fruits coated with film chitosan and film pectin–chitosan with EO.

5. Is industrialization and Wider Application of Plant-Based Nanoemulsion Coatings for Fruits and Vegetables Possible?

Based on this review and all presented results in the scientific field, the industrialization and wider application of plant-based nanoemulsion coatings for fruits and vegetables are indeed possible and promising. However, several factors contribute to the feasibility of their industrial adoption.

5.1. Optimization of Nanoemulsion Coating Formulations

Current research and development efforts have led to important advancements in the formulation of plant-based nanoemulsion coatings. Researchers are continually refining the composition and properties of these coatings to meet industry-specific needs, as shown in Table 2. However, this includes optimizing the choice of plant-based polymers, active compounds, and stabilizers to ensure both effectiveness and economic feasibility. As the primary goal of these coating is to extend the shelf life of fruits and vegetables, reduce postharvest losses, and preserve product quality, the food industry is interested in this technology since minimizing waste and ensuring food safety is the priority for the current decade. This makes nanoemulsion coatings an attractive option, and the needs of modern consumers can be satisfied through this sustainable and eco-friendly food-related technology. By creating a protective barrier on the surface of the produce, these coatings can reduce water loss, inhibit oxygen penetration, and inhibit microbial growth. This can increase marketable timeframes, which are highly desirable outcomes for the food industry. Also, using natural antimicrobial and antioxidant agents makes this solution acceptable on many levels for modern society [1]. Many of these coatings incorporate natural antimicrobial agents, such as EOs, which can effectively inhibit the growth of spoilage and foodborne pathogenic microorganisms. It is very important to emphasize that many regulatory bodies have recognized the potential of plant-based nanoemulsion coatings, and some guidelines and safety assessments have been established in the recent past [24]. This step can lead to the adoption and standardization of this type of coating system in the food industry.

5.2. Application of Coatings

As innovation is necessary in the food industry, but also in coating application techniques and equipment, much research needs to be performed on easier and more cost-effective ways to apply nanoemulsion coatings on an industrial scale. This primarily refers to advancements in spraying and dipping methods [27]. Furthermore, as the food industry seeks to reduce its high environmental impact, the development of eco-friendly coatings using biodegradable materials is gaining special traction. This addresses concerns about the ecological footprint of food packaging and preservation, which can be improved with plant-based nanoemulsions. Despite these promising aspects, several challenges and considerations need to be addressed for the widespread industrialization of plant-based nanoemulsion coatings. One of the obstacles for producers, especially for small- and medium-sized enterprises, is the high cost of coating formulations, as well as specific application techniques. Therefore, achieving the cost-effectiveness of coatings while delivering the desired benefits is crucial for industrial adoption. One of the challenges during the industrialization of these types of coating systems can also be uniformity during application. Ensuring consistent coating coverage on a large scale, especially for irregularly shaped or textured fruits and vegetables, requires efficient application methods and equipment. Scaling up production from laboratory experiments to industrial levels can pose technical challenges that also need to be addressed [115]. However, social barriers can similarly affect the wide application of this product. Therefore, consumer education and understanding all the advantages of plant-based nanoemulsion coatings is necessary.

5.3. Research Needs

Summarizing these facts, it can be said that plant-based nanoemulsion coatings have the potential to improve food technology by addressing critical challenges related to the shelf life, food safety, and sustainability of targeted products. Their industrial and worldwide adoption depends on overcoming cost barriers, refining application methods, and navigating regulatory documents. As consumer preferences continue to evolve in the future, the food industry can make a solid base for this challenge using eco-friendly plant-based nanoemulsion coatings. Therefore, continued research, investment in technology, and regulatory support are going to play crucial roles in realizing the full potential of these coatings in the food industry. Profound studies of nanoemulsion coatings systems are crucial to understanding their properties to ensure the ability to manipulate and predict their behavior. Further research should focus on better understanding physicochemical processes occurring in fresh products, as well as the effects of nanoemulsion coatings, with the objective of securing mutual compatibility. Furthermore, it would be significant to broaden the spectrum of food products that can be coated and to enhance coating application techniques. Another important aspect is searching for new sources of bioactive compounds that would be more efficient and provide a satisfactory effect on the prolongation of shelf life. Further, performing more scale-up studies will provide beneficial information on the production of nanoemulsion coatings on an industrial scale, which could lead to wider implementation of edible coatings in the food industry.

6. Future Perspectives

The implementation of post-harvest techniques is necessary due to the high rate of food losses, especially for fresh-cut and minimally processed fruits and vegetables that require additional protection to maintain the highest quality and safety. The application of a nanoemulsion edible coating system on fruits and vegetables provides a substantial impact as a postharvest treatment and offers promising results to safeguard the physico-chemical during the time of storage and transportation, as well as to provide antibacterial and antifungal properties, retain nutrition, and guarantee the prolongation of shelf-life of products. These coatings act as effective barriers, shielding the produce from external factors such as oxygen, moisture, and microbial contaminants, ensuring the freshness and quality of fruits and vegetables for extended periods and contributing to both consumer satisfaction and a reduction in food waste. However, to make plant-based nanoemulsion coatings technologically feasible and economically available, advanced and thorough research is required regarding functionality, composition, sensory and mechanical properties, and the application aspect of coatings on real food. Future perspectives should include a comprehensive study of different sources of active compounds and their effects, a profound understanding of impacts regarding specific application to fruits and vegetables, and an adequate combination of components in coating formulations to fulfil the necessary qualifications for widespread application.

Author Contributions: Conceptualization, V.T. and O.Š.; methodology, V.T. and J.V.; validation, J.Č.-B., G.Ć., and S.P.; formal analysis, T.E.; investigation, T.E.; resources, T.C.; writing—original draft preparation, T.C.; writing—review and editing, V.T.; visualization, O.Š. and J.V.; supervision, V.T.; project administration, G.Ć.; funding acquisition, V.T. All authors have read and agreed to the published version of the manuscript.

Funding: This research study was funded by the Provincial Secretariat for Higher Education and Scientific Research of Autonomous Province Vojvodina (Serbia) within the project "New ecological phytopreparation for antifungal and antioxidant treatment of selected fruits and vegetables—EcoPhyt" (grant No. 142-451-3108/2022-01) and by the Ministry of Education, Science, and Technological Development of the Republic of Serbia (grant No. 451-03-47/2023-01/200134).

Institutional Review Board Statement: Not applicable.

Informed Consent Statement: Not applicable.

Data Availability Statement: Data are contained within the article. The data used to support the findings of this study can be made available by the corresponding author upon request.

Conflicts of Interest: The authors declare no conflict of interest.

References

1. Tripathi, A.D.; Sharma, R.; Agarwal, A.; Haleem, D.R. Nanoemulsions Based Edible Coatings with Potential Food Applications. *Int. J. Biobased Plast.* **2021**, *3*, 112–125. [CrossRef]
2. Sessa, M.; Ferrari, G.; Donsì, F. Novel Edible Coating Containing Essential Oil Nanoemulsions to Prolong the Shelf Life of Vegetable Products. *Chem. Eng. Trans.* **2015**, *43*, 55–60. [CrossRef]
3. Ramos, M.; Mellinas, C.; Solaberrieta, I.; Garrigós, M.C.; Jiménez, A. Emulsions Incorporated in Polysaccharide-Based Active Coatings for Fresh and Minimally Processed Vegetables. *Foods* **2021**, *10*, 665. [CrossRef] [PubMed]
4. Tomić, A.; Šovljanski, O.; Erceg, T. Insight on Incorporation of Essential Oils as Antimicrobial Substances in Biopolymer-Based Active Packaging. *Antibiotics* **2023**, *12*, 1473. [CrossRef]
5. Hassan, B.; Chatha, S.A.S.; Hussain, A.I.; Zia, K.M.; Akhtar, N. Recent Advances on Polysaccharides, Lipids and Protein Based Edible Films and Coatings: A Review. *Int. J. Biol. Macromol.* **2018**, *109*, 1095–1107. [CrossRef]
6. Jose, A.; Pareek, S.; Radhakrishnan, E.K. Advances in Edible Fruit Coating Materials. In *Advances in Agri-Food Biotechnology*; Sharma, T.R., Deshmukh, R., Sonah, H., Eds.; Springer: Singapore, 2020; pp. 391–408, ISBN 9789811528743.
7. Kim, I.-H.; Oh, Y.A.; Lee, H.; Song, K.B.; Min, S.C. Grape Berry Coatings of Lemongrass Oil-Incorporating Nanoemulsion. *LWT Food Sci. Technol.* **2014**, *58*, 1–10. [CrossRef]
8. Gupta, C.; Prakash, D. Chapter 10—Safety of Fresh Fruits and Vegetables. In *Food Safety and Human Health*; Singh, R.L., Mondal, S., Eds.; Academic Press: Cambridge, MA, USA, 2019; pp. 249–283, ISBN 978-0-12-816333-7.
9. Bajaj, K.; Adhikary, T.; Gill, P.P.S.; Kumar, A. Edible Coatings Enriched with Plant-Based Extracts Preserve Postharvest Quality of Fruits: A Review. *Prog. Organ. Coat.* **2023**, *182*, 107669. [CrossRef]
10. Liyanapathiranage, A.; Dassanayake, R.S.; Gamage, A.; Karri, R.R.; Manamperi, A.; Evon, P.; Jayakodi, Y.; Madhujith, T.; Merah, O. Recent Developments in Edible Films and Coatings for Fruits and Vegetables. *Coatings* **2023**, *13*, 1177. [CrossRef]
11. Shewfelt, R.L.; Prussia, S.E.; Sparks, S.A. Chapter 2—Challenges in Handling Fresh Fruits and Vegetables. In *Postharvest Handling*, 3rd ed.; Florkowski, W.J., Shewfelt, R.L., Brueckner, B., Prussia, S.E., Eds.; Academic Press: San Diego, CA, USA, 2014; pp. 11–30, ISBN 978-0-12-408137-6.
12. Travičić, V.; Cvanić, T.; Ćetković, G. Plant-Based Nano-Emulsions as Edible Coatings in the Extension of Fruits and Vegetables Shelf Life: A Patent Review. *Foods* **2023**, *12*, 2535. [CrossRef]
13. Anjum, M.A.; Akram, H.; Zaidi, M.; Ali, S. Effect of Gum Arabic and Aloe Vera Gel Based Edible Coatings in Combination with Plant Extracts on Postharvest Quality and Storability of 'Gola' Guava Fruits. *Sci. Hortic.* **2020**, *271*, 109506. [CrossRef]
14. Rios, D.A.d.S.; Nakamoto, M.M.; Braga, A.R.C.; da Silva, E.M.C. Food Coating Using Vegetable Sources: Importance and Industrial Potential, Gaps of Knowledge, Current Application, and Future Trends. *Appl. Food Res.* **2022**, *2*, 100073. [CrossRef]
15. Hasan, S.M.K.; Ferrentino, G.; Scampicchio, M. Nanoemulsion as Advanced Edible Coatings to Preserve the Quality of Fresh-Cut Fruits and Vegetables: A Review. *Int. J. Food Sci. Technol.* **2020**, *55*, 1–10. [CrossRef]
16. Vargas, M.; Pastor, C.; Albors, A.; Chiralt, A.; González-Martínez, C. Development of Edible Coatings for Fresh Fruits and Vegetables: Possibilities and Limitations. *Fresh Prod.* **2008**, *2*, 32–40.
17. Galus, S.; Kadzińska, J. Food Applications of Emulsion-Based Edible Films and Coatings. *Trends Food Sci. Technol.* **2015**, *45*, 273–283. [CrossRef]
18. Silva-Weiss, A.; Ihl, M.; Sobral, P.J.A.; Gómez-Guillén, M.C.; Bifani, V. Natural Additives in Bioactive Edible Films and Coatings: Functionality and Applications in Foods. *Food Eng. Rev.* **2013**, *5*, 200–216. [CrossRef]
19. Mahajan, B.V.C.; Tandon, R.; Kapoor, S.; Sidhu, M.K. Natural Coatings for Shelf-Life Enhancement and Quality Maintenance of Fresh Fruits and Vegetables—A Review. *J. Postharvest Technol.* **2018**, *6*, 12–26.
20. Otoni, C.G.; Avena-Bustillos, R.J.; Azeredo, H.M.C.; Lorevice, M.V.; Moura, M.R.; Mattoso, L.H.C.; McHugh, T.H. Recent Advances on Edible Films Based on Fruits and Vegetables-A Review: Fruit and Vegetable Edible Films. *Compr. Rev. Food Sci. Food Saf.* **2017**, *16*, 1151–1169. [CrossRef]
21. Trajkovska Petkoska, A.; Daniloski, D.; D'Cunha, N.M.; Naumovski, N.; Broach, A.T. Edible Packaging: Sustainable Solutions and Novel Trends in Food Packaging. *Food Res. Int.* **2021**, *140*, 109981. [CrossRef]
22. Paidari, S.; Zamindar, N.; Tahergorabi, R.; Kargar, M.; Ezzati, S.; Shirani, N.; Musavi, S.H. Edible Coating and Films as Promising Packaging: A Mini Review. *Food Meas.* **2021**, *15*, 4205–4214. [CrossRef]
23. Zambrano-Zaragoza, M.L.; González-Reza, R.; Mendoza-Muñoz, N.; Miranda-Linares, V.; Bernal-Couoh, T.F.; Mendoza-Elvira, S.; Quintanar-Guerrero, D. Nanosystems in Edible Coatings: A Novel Strategy for Food Preservation. *Int. J. Mol. Sci.* **2018**, *19*, 705. [CrossRef]
24. Valdés, A.; Ramos, M.; Beltrán, A.; Jiménez, A.; Garrigós, M.C. State of the Art of Antimicrobial Edible Coatings for Food Packaging Applications. *Coatings* **2017**, *7*, 56. [CrossRef]
25. Dhall, R.K. Advances in Edible Coatings for Fresh Fruits and Vegetables: A Review. *Crit. Rev. Food Sci. Nutr.* **2013**, *53*, 435–450. [CrossRef] [PubMed]
26. Han, J.H. Chapter 9—Edible Films and Coatings: A Review. In *Innovations in Food Packaging*, 2nd ed.; Han, J.H., Ed.; Food Science and Technology; Academic Press: San Diego, CA, USA, 2014; pp. 213–255, ISBN 978-0-12-394601-0.
27. Ungureanu, C.; Tihan, G.; Zgârian, R.; Pandelea (Voicu), G. Bio-Coatings for Preservation of Fresh Fruits and Vegetables. *Coatings* **2023**, *13*, 1420. [CrossRef]

28. Pham, T.T.; Nguyen, L.L.P.; Dam, M.S.; Baranyai, L. Application of Edible Coating in Extension of Fruit Shelf Life: Review. *AgriEngineering* **2023**, *5*, 520–536. [CrossRef]

29. Matloob, A.; Ayub, H.; Mohsin, M.; Ambreen, S.; Khan, F.A.; Oranab, S.; Rahim, M.A.; Khalid, W.; Nayik, G.A.; Ramniwas, S.; et al. A Review on Edible Coatings and Films: Advances, Composition, Production Methods, and Safety Concerns. *ACS Omega* **2023**, *8*, 28932–28944. [CrossRef] [PubMed]

30. Ezhilarasi, P.N.; Karthik, P.; Chhanwal, N.; Anandharamakrishnan, C. Nanoencapsulation Techniques for Food Bioactive Components: A Review. *Food Bioprocess Technol.* **2013**, *6*, 628–647. [CrossRef]

31. McClements, D.J.; Jafari, S.M. Chapter 1—General Aspects of Nanoemulsions and Their Formulation. In *Nanoemulsions*; Jafari, S.M., McClements, D.J., Eds.; Academic Press: Cambridge, MA, USA, 2018; pp. 3–20, ISBN 978-0-12-811838-2.

32. Tan, C.; McClements, D.J. Application of Advanced Emulsion Technology in the Food Industry: A Review and Critical Evaluation. *Foods* **2021**, *10*, 812. [CrossRef]

33. Chellaram, C.; Murugaboopathi, G.; John, A.A.; Sivakumar, R.; Ganesan, S.; Krithika, S.; Priya, G. Significance of Nanotechnology in Food Industry. *APCBEE Procedia* **2014**, *8*, 109–113. [CrossRef]

34. Imam, S.S.; Jahangir, M.A.; Gilani, S.J.; Zafar, A.; Alshehri, S. Nanoemulsions as Delivery Vehicle for Nutraceuticals and Improving Food Nutrition Properties. In *Nanoemulsions in Food Technology*; CRC Press: Boca Raton, FL, USA, 2021; ISBN 978-1-00-312112-1.

35. Wang, Q.; Chen, W.; Zhu, W.; McClements, D.J.; Liu, X.; Liu, F. A Review of Multilayer and Composite Films and Coatings for Active Biodegradable Packaging. *npj Sci. Food* **2022**, *6*, 18. [CrossRef]

36. Singh, Y.; Meher, J.G.; Raval, K.; Khan, F.A.; Chaurasia, M.; Jain, N.K.; Chourasia, M.K. Nanoemulsion: Concepts, Development and Applications in Drug Delivery. *J. Controll. Release* **2017**, *252*, 28–49. [CrossRef]

37. Acevedo-Fani, A.; Soliva-Fortuny, R.; Martín-Belloso, O. Nanoemulsions as Edible Coatings. *Curr. Opin. Food Sci.* **2017**, *15*, 43–49. [CrossRef]

38. Rashid, F.; Ahmed, Z.; Ameer, K.; Amir, R.M.; Khattak, M. Optimization of Polysaccharides-Based Nanoemulsion Using Response Surface Methodology and Application to Improve Postharvest Storage of Apple (*Malus domestica*). *Food Meas.* **2020**, *14*, 2676–2688. [CrossRef]

39. Jeevahan, J.; Chandrasekaran, M. Nanoedible Films for Food Packaging: A Review. *J. Mater. Sci.* **2019**, *54*, 12290–12318. [CrossRef]

40. Hernalsteens, S. Edible Films and Coatings Made up of Fruits and Vegetables. In *Biopolymer Membranes and Films*; Elsevier: Amsterdam, The Netherlands, 2020; pp. 575–588, ISBN 978-0-12-818134-8.

41. Zareie, Z.; Tabatabaei Yazdi, F.; Mortazavi, S.A. Development and Characterization of Antioxidant and Antimicrobial Edible Films Based on Chitosan and Gamma-Aminobutyric Acid-Rich Fermented Soy Protein. *Carbohydr. Polym.* **2020**, *244*, 116491. [CrossRef] [PubMed]

42. Xue, R.; Zhao, H.; An, Z.-W.; Wu, W.; Jiang, Y.; Li, P.; Huang, C.-X.; Shi, D.; Li, R.K.Y.; Hu, G.-H.; et al. Self-Healable, Solvent Response Cellulose Nanocrystal/Waterborne Polyurethane Nanocomposites with Encryption Capability. *ACS Nano* **2023**, *17*, 5653–5662. [CrossRef] [PubMed]

43. Guerreiro, A.C.; Gago, C.M.L.; Faleiro, M.L.; Miguel, M.G.C.; Antunes, M.D.C. The Effect of Edible Coatings on the Nutritional Quality of 'Bravo de Esmolfe' Fresh-Cut Apple through Shelf-Life. *LWT* **2017**, *75*, 210–219. [CrossRef]

44. Martínez-Hernández, G.B.; Amodio, M.L.; Colelli, G. Carvacrol-Loaded Chitosan Nanoparticles Maintain Quality of Fresh-Cut Carrots. *Innov. Food Sci. Emerg. Technol.* **2017**, *41*, 56–63. [CrossRef]

45. Kumar, R.; Ghoshal, G.; Goyal, M. Effect of Basil Leaves Extract on Modified Moth Bean Starch Active Film for Eggplant Surface Coating. *LWT* **2021**, *145*, 111380. [CrossRef]

46. Gomes, M.d.S.; Cardoso, M.d.G.; Guimarães, A.C.G.; Guerreiro, A.C.; Gago, C.M.L.; Vilas Boas, E.V.d.B.; Dias, C.M.B.; Manhita, A.C.C.; Faleiro, M.L.; Miguel, M.G.C.; et al. Effect of Edible Coatings with Essential Oils on the Quality of Red Raspberries over Shelf-Life. *J. Sci. Food Agric.* **2017**, *97*, 929–938. [CrossRef]

47. Sharma, S.; Ramana Rao, T.V. Xanthan Gum Based Edible Coating Enriched with Cinnamic Acid Prevents Browning and Extends the Shelf-Life of Fresh-Cut Pears. *LWT Food Sci. Technol.* **2015**, *62*, 791–800. [CrossRef]

48. Lima, Á.M.; Cerqueira, M.A.; Souza, B.W.S.; Santos, E.C.M.; Teixeira, J.A.; Moreira, R.A.; Vicente, A.A. New Edible Coatings Composed of Galactomannans and Collagen Blends to Improve the Postharvest Quality of Fruits—Influence on Fruits Gas Transfer Rate. *J. Food Eng.* **2010**, *97*, 101–109. [CrossRef]

49. Feng, Z.; Wu, G.; Liu, C.; Li, D.; Jiang, B.; Zhang, X. Edible Coating Based on Whey Protein Isolate Nanofibrils for Antioxidation and Inhibition of Product Browning. *Food Hydrocoll.* **2018**, *79*, 179–188. [CrossRef]

50. Li, J.; Sun, Q.; Sun, Y.; Chen, B.; Wu, X.; Le, T. Improvement of Banana Postharvest Quality Using a Novel Soybean Protein Isolate/Cinnamaldehyde/Zinc Oxide Bionanocomposite Coating Strategy. *Sci. Hortic.* **2019**, *258*, 108786. [CrossRef]

51. Boyacı, D.; Iorio, G.; Sozbilen, G.S.; Alkan, D.; Trabattoni, S.; Pucillo, F.; Farris, S.; Yemenicioğlu, A. Development of Flexible Antimicrobial Zein Coatings with Essential Oils for the Inhibition of Critical Pathogens on the Surface of Whole Fruits: Test of Coatings on Inoculated Melons. *Food Packag. Shelf Life* **2019**, *20*, 100316. [CrossRef]

52. Baswal, A.K.; Dhaliwal, H.S.; Singh, Z.; Mahajan, B.; Kalia, A.; Gill, K.S. Influence of Carboxy Methylcellulose, Chitosan and Beeswax Coatings on Cold Storage Life and Quality of Kinnow Mandarin Fruit. *Sci. Hortic.* **2020**, *260*, 108887. [CrossRef]

53. Ahmed, Z.F.R.; Palta, J.P. Postharvest Dip Treatment with a Natural Lysophospholipid plus Soy Lecithin Extended the Shelf Life of Banana Fruit. *Postharvest Biol. Technol.* **2016**, *113*, 58–65. [CrossRef]

54. Mellinas, C.; Valdés, A.; Ramos, M.; Burgos, N.; Garrigós, M.d.C.; Jiménez, A. Active Edible Films: Current State and Future Trends. *J. Appl. Polym. Sci.* **2016**, *133*, 42631. [CrossRef]
55. Saberi, B.; Golding, J.B. Postharvest Application of Biopolymer-Based Edible Coatings to Improve the Quality of Fresh Horticultural Produce. In *Polymers for Food Applications*; Gutiérrez, T.J., Ed.; Springer International Publishing: Cham, Switzerland, 2018; pp. 211–250, ISBN 978-3-319-94625-2.
56. Moeini, A.; Pedram, P.; Fattahi, E.; Cerruti, P.; Santagata, G. Edible Polymers and Secondary Bioactive Compounds for Food Packaging Applications: Antimicrobial, Mechanical, and Gas Barrier Properties. *Polymers* **2022**, *14*, 2395. [CrossRef]
57. Kouhi, M.; Prabhakaran, M.P.; Ramakrishna, S. Edible Polymers: An Insight into Its Application in Food, Biomedicine and Cosmetics. *Trends Food Sci. Technol.* **2020**, *103*, 248–263. [CrossRef]
58. Zubair, M.; Pradhan, R.A.; Arshad, M.; Ullah, A. Recent Advances in Lipid Derived Bio-Based Materials for Food Packaging Applications. *Macromol. Mater. Eng.* **2021**, *306*, 2000799. [CrossRef]
59. Kumar, L.; Ramakanth, D.; Akhila, K.; Gaikwad, K.K. Edible Films and Coatings for Food Packaging Applications: A Review. *Environ. Chem. Lett.* **2022**, *20*, 875–900. [CrossRef]
60. Kyei, S.K.; Eke, W.I.; Darko, G.; Akaranta, O. Natural Polyhydroxy Resins in Surface Coatings: A Review. *J. Coat. Technol. Res.* **2022**, *19*, 775–794. [CrossRef]
61. Roy, S.; Rhim, J.-W. Fabrication of Pectin/Agar Blended Functional Film: Effect of Reinforcement of Melanin Nanoparticles and Grapefruit Seed Extract. *Food Hydrocoll.* **2021**, *118*, 106823. [CrossRef]
62. Donsì, F. Chapter 11—Applications of Nanoemulsions in Foods. In *Nanoemulsions*; Jafari, S.M., McClements, D.J., Eds.; Academic Press: Cambridge, MA, USA, 2018; pp. 349–377, ISBN 978-0-12-811838-2.
63. Yaashikaa, P.R.; Kamalesh, R.; Senthil Kumar, P.; Saravanan, A.; Vijayasri, K.; Rangasamy, G. Recent Advances in Edible Coatings and Their Application in Food Packaging. *Food Res. Int.* **2023**, *173*, 113366. [CrossRef]
64. Kahramanoğlu, İ.; Panfilova, O.; Kesimci, T.G.; Bozhüyük, A.U.; Gürbüz, R.; Alptekin, H. Control of Postharvest Gray Mold at Strawberry Fruits Caused by Botrytis Cinerea and Improving Fruit Storability through *Origanum onites* L. and *Ziziphora clinopodioides* L. Volatile Essential Oils. *Agronomy* **2022**, *12*, 389. [CrossRef]
65. Shahbaz, M.U.; Arshad, M.; Mukhtar, K.; Nabi, B.G.; Goksen, G.; Starowicz, M.; Nawaz, A.; Ahmad, I.; Walayat, N.; Manzoor, M.F.; et al. Natural Plant Extracts: An Update about Novel Spraying as an Alternative of Chemical Pesticides to Extend the Postharvest Shelf Life of Fruits and Vegetables. *Molecules* **2022**, *27*, 5152. [CrossRef]
66. Tajkarimi, M.M.; Ibrahim, S.A.; Cliver, D.O. Antimicrobial Herb and Spice Compounds in Food. *Food Control* **2010**, *21*, 1199–1218. [CrossRef]
67. Sánchez-González, L.; Vargas, M.; González-Martínez, C.; Chiralt, A.; Cháfer, M. Use of Essential Oils in Bioactive Edible Coatings: A Review. *Food Eng. Rev.* **2011**, *3*, 1–16. [CrossRef]
68. Pandey, V.K.; Islam, R.U.; Shams, R.; Dar, A.H. A Comprehensive Review on the Application of Essential Oils as Bioactive Compounds in Nano-Emulsion Based Edible Coatings of Fruits and Vegetables. *Appl. Food Res.* **2022**, *2*, 100042. [CrossRef]
69. Mustafa, I.F.; Hussein, M.Z. Synthesis and Technology of Nanoemulsion-Based Pesticide Formulation. *Nanomaterials* **2020**, *10*, 1608. [CrossRef]
70. Espitia, P.J.P.; Fuenmayor, C.A.; Otoni, C.G. Nanoemulsions: Synthesis, Characterization, and Application in Bio-Based Active Food Packaging. *Compr. Rev. Food Sci. Food Saf.* **2019**, *18*, 264–285. [CrossRef] [PubMed]
71. Guerra-Rosas, M.I.; Morales-Castro, J.; Ochoa-Martínez, L.A.; Salvia-Trujillo, L.; Martín-Belloso, O. Long-Term Stability of Food-Grade Nanoemulsions from High Methoxyl Pectin Containing Essential Oils. *Food Hydrocoll.* **2016**, *52*, 438–446. [CrossRef]
72. Rao, J.; McClements, D.J. Food-Grade Microemulsions and Nanoemulsions: Role of Oil Phase Composition on Formation and Stability. *Food Hydrocoll.* **2012**, *29*, 326–334. [CrossRef]
73. Ribeiro, A.M.; Estevinho, B.N.; Rocha, F. Preparation and Incorporation of Functional Ingredients in Edible Films and Coatings. *Food Bioprocess Technol.* **2021**, *14*, 209–231. [CrossRef]
74. Kamal, I. Edible Films and Coatings: Classification, Preparation, Functionality and Applications—A Review. *Arch. Organ. Inorg. Chem. Sci.* **2020**, *4*, 501–509. [CrossRef]
75. Andrade, R.; Skurtys, O.; Osorio, F. Atomizing Spray Systems for Application of Edible Coatings. *Compr. Rev. Food Sci. Food Saf.* **2012**, *11*, 323–337. [CrossRef]
76. Priya, K.; Thirunavookarasu, N.; Chidanand, D.V. Recent Advances in Edible Coating of Food Products and Its Legislations: A Review. *J. Agric. Food Res.* **2023**, *12*, 100623. [CrossRef]
77. Miranda, M.; Sun, X.; Marín, A.; dos Santos, L.C.; Plotto, A.; Bai, J.; Benedito Garrido Assis, O.; David Ferreira, M.; Baldwin, E. Nano- and Micro-Sized Carnauba Wax Emulsions-Based Coatings Incorporated with Ginger Essential Oil and Hydroxypropyl Methylcellulose on Papaya: Preservation of Quality and Delay of Post-Harvest Fruit Decay. *Food Chem. X* **2022**, *13*, 100249. [CrossRef]
78. Tabassum, N.; Aftab, R.A.; Yousuf, O.; Ahmad, S.; Zaidi, S. Application of Nanoemulsion Based Edible Coating on Fresh-Cut Papaya. *J. Food Eng.* **2023**, *355*, 111579. [CrossRef]
79. Zillo, R.R.; da Silva, P.P.M.; de Oliveira, J.; da Glória, E.M.; Spoto, M.H.F. Carboxymethylcellulose Coating Associated with Essential Oil Can Increase Papaya Shelf Life. *Sci. Hortic.* **2018**, *239*, 70–77. [CrossRef]

80. Luciano, W.A.; Pimentel, T.C.; Bezerril, F.F.; Barão, C.E.; Marcolino, V.A.; de Siqueira Ferraz Carvalho, R.; dos Santos Lima, M.; Martín-Belloso, O.; Magnani, M. Effect of Citral Nanoemulsion on the Inactivation of Listeria Monocytogenes and Sensory Properties of Fresh-Cut Melon and Papaya during Storage. *Int. J. Food Microbiol.* **2023**, *384*, 109959. [CrossRef] [PubMed]

81. Gundewadi, G.; Rudra, S.G.; Sarkar, D.J.; Singh, D. Nanoemulsion Based Alginate Organic Coating for Shelf Life Extension of Okra. *Food Packag. Shelf Life* **2018**, *18*, 1–12. [CrossRef]

82. Bashir, O.; Amin, T.; Hussain, S.Z.; Naik, H.R.; Goksen, G.; Wani, A.W.; Manzoor, S.; Malik, A.R.; Wani, F.J.; Proestos, C. Development, Characterization and Use of Rosemary Essential Oil Loaded Water-Chestnut Starch Based Nanoemulsion Coatings for Enhancing Post-Harvest Quality of Apples Var. Golden Delicious. *Curr. Res. Food Sci.* **2023**, *7*, 100570. [CrossRef] [PubMed]

83. Jo, W.-S.; Song, H.-Y.; Song, N.-B.; Lee, J.-H.; Min, S.C.; Song, K.B. Quality and Microbial Safety of 'Fuji' Apples Coated with Carnauba-Shellac Wax Containing Lemongrass Oil. *LWT Food Sci. Technol.* **2014**, *55*, 490–497. [CrossRef]

84. Naqash, F.; Masoodi, F.a.; Ayob, O.; Parvez, S. Effect of Active Pectin Edible Coatings on the Safety and Quality of Fresh-Cut Apple. *Int. J. Food Sci. Technol.* **2022**, *57*, 57–66. [CrossRef]

85. Farina, V.; Passafiume, R.; Tinebra, I.; Palazzolo, E.; Sortino, G. Use of Aloe Vera Gel-Based Edible Coating with Natural Anti-Browning and Anti-Oxidant Additives to Improve Post-Harvest Quality of Fresh-Cut 'Fuji' Apple. *Agronomy* **2020**, *10*, 515. [CrossRef]

86. Gull, A.; Bhat, N.; Wani, S.M.; Masoodi, F.A.; Amin, T.; Ganai, S.A. Shelf Life Extension of Apricot Fruit by Application of Nanochitosan Emulsion Coatings Containing Pomegranate Peel Extract. *Food Chem.* **2021**, *349*, 129149. [CrossRef]

87. Das, S.; Vishakha, K.; Banerjee, S.; Mondal, S.; Ganguli, A. Sodium Alginate-Based Edible Coating Containing Nanoemulsion of Citrus Sinensis Essential Oil Eradicates Planktonic and Sessile Cells of Food-Borne Pathogens and Increased Quality Attributes of Tomatoes. *Int. J. Biol. Macromol.* **2020**, *162*, 1770–1779. [CrossRef]

88. Robledo, N.; López, L.; Bunger, A.; Tapia, C.; Abugoch, L. Effects of Antimicrobial Edible Coating of Thymol Nanoemulsion/Quinoa Protein/Chitosan on the Safety, Sensorial Properties, and Quality of Refrigerated Strawberries (*Fragaria* × *Ananassa*) Under Commercial Storage Environment. *Food Bioprocess Technol.* **2018**, *11*, 1566–1574. [CrossRef]

89. Das, S.K.; Vishakha, K.; Das, S.; Chakraborty, D.; Ganguli, A. Carboxymethyl Cellulose and Cardamom Oil in a Nanoemulsion Edible Coating Inhibit the Growth of Foodborne Pathogens and Extend the Shelf Life of Tomatoes. *Biocatal. Agric. Biotechnol.* **2022**, *42*, 102369. [CrossRef]

90. Gómez-Contreras, P.; Figueroa-Lopez, K.J.; Hernández-Fernández, J.; Cortés Rodríguez, M.; Ortega-Toro, R. Effect of Different Essential Oils on the Properties of Edible Coatings Based on Yam (*Dioscorea rotundata* L.) Starch and Its Application in Strawberry (*Fragaria vesca* L.) Preservation. *Appl. Sci.* **2021**, *11*, 11057. [CrossRef]

91. Oliveira Filho, J.G.d.; Albiero, B.R.; Calisto, Í.H.; Bertolo, M.R.V.; Oldoni, F.C.A.; Egea, M.B.; Bogusz Junior, S.; de Azeredo, H.M.C.; Ferreira, M.D. Bio-Nanocomposite Edible Coatings Based on Arrowroot Starch/Cellulose Nanocrystals/Carnauba Wax Nanoemulsion Containing Essential Oils to Preserve Quality and Improve Shelf Life of Strawberry. *Int. J. Biol. Macromol.* **2022**, *219*, 812–823. [CrossRef]

92. Sathiyaseelan, A.; Saravanakumar, K.; Mariadoss, A.V.A.; Ramachandran, C.; Hu, X.; Oh, D.-H.; Wang, M.-H. Chitosan-Tea Tree Oil Nanoemulsion and Calcium Chloride Tailored Edible Coating Increase the Shelf Life of Fresh Cut Red Bell Pepper. *Prog. Organ. Coat.* **2021**, *151*, 106010. [CrossRef]

93. Arabpoor, B.; Yousefi, S.; Weisany, W.; Ghasemlou, M. Multifunctional Coating Composed of *Eryngium Campestre* L. Essential Oil Encapsulated in Nano-Chitosan to Prolong the Shelf-Life of Fresh Cherry Fruits. *Food Hydrocoll.* **2021**, *111*, 106394. [CrossRef]

94. Basumatary, I.B.; Mukherjee, A.; Katiyar, V.; Dutta, J.; Kumar, S. Chitosan-Based Active Coating for Pineapple Preservation: Evaluation of Antimicrobial Efficacy and Shelf-Life Extension. *LWT* **2022**, *168*, 113940. [CrossRef]

95. Prakash, A.; Baskaran, R.; Vadivel, V. Citral Nanoemulsion Incorporated Edible Coating to Extend the Shelf Life of Fresh Cut Pineapples. *LWT* **2020**, *118*, 108851. [CrossRef]

96. Yang, R.; Miao, J.; Shen, Y.; Cai, N.; Wan, C.; Zou, L.; Chen, C.; Chen, J. Antifungal Effect of Cinnamaldehyde, Eugenol and Carvacrol Nanoemulsion against Penicillium Digitatum and Application in Postharvest Preservation of Citrus Fruit. *LWT* **2021**, *141*, 110924. [CrossRef]

97. Phothisuwan, S.; Koomhin, P.; Matan, N.; Matan, N. Quality Maintenance of Salacca Fruit with a Carnauba Wax Coating Containing Orange Oil and Detection of Sensory Perception Improvement with Electroencephalography to Appraise Brain Responses. *LWT* **2021**, *147*, 111628. [CrossRef]

98. Vilaplana, R.; Guerrero, K.; Guevara, J.; Valencia-Chamorro, S. Chitosan Coatings to Control Soft Mold on Fresh Blackberries (Rubus Glaucus Benth.) during Postharvest Period. *Sci. Hortic.* **2020**, *262*, 109049. [CrossRef]

99. López-Palestina, C.U.; Aguirre-Mancilla, C.L.; Raya-Pérez, J.C.; Ramírez-Pimentel, J.G.; Gutiérrez-Tlahque, J.; Hernández-Fuentes, A.D. The Effect of an Edible Coating with Tomato Oily Extract on the Physicochemical and Antioxidant Properties of Garambullo (*Myrtillocactus geometrizans*) Fruits. *Agronomy* **2018**, *8*, 248. [CrossRef]

100. Aboryia, M.S.; El-Gioushy, S.F.; Sami, R.; Aljumayi, H.; Alyamani, A.; Almasoudi, A.; Gawish, M.S. Synergistic Effect of Dipping in Aloe Vera Gel and Mixing with Chitosan or Calcium Chloride on the Activities of Antioxidant Enzymes and Cold Storage Potential of Peach (*Prunus persica* L.) Fruits. *Coatings* **2022**, *12*, 498. [CrossRef]

101. Sortino, G.; Inglese, P.; Farina, V.; Passafiume, R.; Allegra, A. The Use of Opuntia Ficus-Indica Mucilage and Aloe Arborescens as Edible Coatings to Improve the Physical, Chemical, and Microbiological Properties of 'Hayward' Kiwifruit Slices. *Horticulturae* **2022**, *8*, 219. [CrossRef]

102. Manzoor, S.; Gull, A.; Wani, S.M.; Ganaie, T.A.; Masoodi, F.A.; Bashir, K.; Malik, A.R.; Dar, B.N. Improving the Shelf Life of Fresh Cut Kiwi Using Nanoemulsion Coatings with Antioxidant and Antimicrobial Agents. *Food Biosci.* **2021**, *41*, 101015. [CrossRef]
103. Iftikhar, A.; Rehman, A.; Usman, M.; Ali, A.; Ahmad, M.M.; Shehzad, Q.; Fatim, H.; Mehmood, A.; Moiz, A.; Shabbir, M.A.; et al. Influence of Guar Gum and Chitosan Enriched with Lemon Peel Essential Oil Coatings on the Quality of Pears. *Food Sci. Nutr.* **2022**, *10*, 2443–2454. [CrossRef]
104. Moradinezhad, F.; Naeimi, A.; Farhangfar, H. Influence of Edible Coatings on Postharvest Quality of Fresh Chinese Jujube Fruits during Refrigerated Storage. *J. Hortic. Postharvest Res.* **2018**, *1*, 1–14. [CrossRef]
105. Utama, N.A.; Pranata, I.A.; Pramesi, P.C. Maintaining Physicochemical and Sensory Properties of Guava Var. Getas Merah Using Alginate and Cyclea Barbata Leaveas Powder as Edible Coating. *Adv. Hortic. Sci.* **2022**, *36*, 135–144. [CrossRef]
106. Shivangi, S.; Dorairaj, D.; Negi, P.S.; Shetty, N.P. Development and Characterisation of a Pectin-Based Edible Film That Contains Mulberry Leaf Extract and Its Bio-Active Components. *Food Hydrocoll.* **2021**, *121*, 107046. [CrossRef]
107. Arnon-Rips, H.; Porat, R.; Poverenov, E. Enhancement of Agricultural Produce Quality and Storability Using Citral-Based Edible Coatings; the Valuable Effect of Nano-Emulsification in a Solid-State Delivery on Fresh-Cut Melons Model. *Food Chem.* **2019**, *277*, 205–212. [CrossRef]
108. Bhargava, K.; Conti, D.S.; da Rocha, S.R.P.; Zhang, Y. Application of an Oregano Oil Nanoemulsion to the Control of Foodborne Bacteria on Fresh Lettuce. *Food Microbiol.* **2015**, *47*, 69–73. [CrossRef]
109. Suhag, R.; Kumar, N.; Petkoska, A.T.; Upadhyay, A. Film Formation and Deposition Methods of Edible Coating on Food Products: A Review. *Food Res. Int.* **2020**, *136*, 109582. [CrossRef]
110. Mahmud, N.; Islam, J.; Tahergorabi, R. Marine Biopolymers: Applications in Food Packaging. *Processes* **2021**, *9*, 2245. [CrossRef]
111. Chettri, S.; Sharma, N.; Mohite, A.M. Edible Coatings and Films for Shelf-Life Extension of Fruit and Vegetables. *Biomater. Adv.* **2023**, *154*, 213632. [CrossRef] [PubMed]
112. Maringgal, B.; Hashim, N.; Mohamed Amin Tawakkal, I.S.; Muda Mohamed, M.T. Recent Advance in Edible Coating and Its Effect on Fresh/Fresh-Cut Fruits Quality. *Trends Food Sci. Technol.* **2020**, *96*, 253–267. [CrossRef]
113. Sousa Gallagher, M.J.; Mahajan, P.V. 22—The Stability and Shelf Life of Fruit and Vegetables. In *Food and Beverage Stability and Shelf Life*; Kilcast, D., Subramaniam, P., Eds.; Woodhead Publishing Series in Food Science, Technology and Nutrition; Woodhead Publishing: Sawston, UK, 2011; pp. 641–656, ISBN 978-1-84569-701-3.
114. Abdalla, G.; Mussagy, C.U.; Sant'Ana Pegorin Brasil, G.; Scontri, M.; da Silva Sasaki, J.C.; Su, Y.; Bebber, C.; Rocha, R.R.; de Sousa Abreu, A.P.; Goncalves, R.P.; et al. Eco-Sustainable Coatings Based on Chitosan, Pectin, and Lemon Essential Oil Nanoemulsion and Their Effect on Strawberry Preservation. *Int. J. Biol. Macromol.* **2023**, *249*, 126016. [CrossRef] [PubMed]
115. McClements, D.J.; Rao, J. Food-Grade Nanoemulsions: Formulation, Fabrication, Properties, Performance, Biological Fate, and Potential Toxicity. *Crit. Rev. Food Sci. Nutr.* **2011**, *51*, 285–330. [CrossRef] [PubMed]

Article

The Influence of Biopolymer Coating Based on Pumpkin Oil Cake Activated with *Mentha piperita* Essential Oil on the Quality and Shelf-Life of Grape

Danijela Šuput [1], Lato Pezo [2,*], Biljana Lončar [1], Senka Popović [1], Aleksandra Tepić Horecki [1], Tatjana Daničić [1], Dragoljub Cvetković [1], Aleksandra Ranitović [1], Nevena Hromiš [1] and Jovana Ugarković [1]

[1]   Faculty of Technology Novi Sad, University of Novi Sad, Bulevar Cara Lazara 1, 21000 Novi Sad, Serbia
[2]   Institute of General and Physical Chemistry, University of Belgrade, Studentski Trg 12, 11000 Belgrade, Serbia
*   Correspondence: latopezo@yahoo.co.uk

Abstract: This work aimed to determine the influence of biopolymer coatings based on pumpkin oil cake, with and without the addition of *Mentha piperita* essential oil, on the quality and shelf-life of the Afus Ali variety of grapes, stored at room temperature and in the refrigerator. Furthermore, a 10% (w/w) aqueous solution of composite pumpkin oil cake (PuOC) with the addition of 30% glycerol was prepared at 60 °C and pH 10. The active biopolymer coating was prepared similarly by adding 1% (v/v) *Mentha piperita* essential oil. The quality of packed grapes was tested by determining the dry matter content, total sugar content, total acidity, alcohol content, total phenolic compounds content, and total flavonoid content, as well as by determining the antioxidant activity, through the application of the DPPH, FRAP and ABTS tests. Additionally, microbiological parameters were investigated: total aerobic microbial count, yeasts, and molds. The obtained results proved that in all tested samples, over a certain period of time, the content of dry matter, content of phenolic and flavonoids substances and sugar content decreased as a consequence of the spoilage of grapes, that is, the consumption of sugar for the production of alcohol, which consequently leads to the total acidity increasing. The application of lower storage temperatures and active coating (with *Mentha piperita* essential oil) had a positive effect on all inevitable reactions. Grapes' antioxidant potential may be enhanced or maintained by applying PuOC coating with or without *Mentha piperita* essential oil, which is best observed in the case of the DPPH test. The uncoated sample stored at room temperature had the largest decrease in DPPH values during storage, with changes ranging from 2.119 mg/g to 1.471 µmol mg/g. The samples, coated with PuOC and PuOC with the addition of essential oil, had uniform DPPH values throughout the entire storage period. Additionally, regarding phenolic content, at the end of storage period the highest phenolic content was observed in samples with active coating stored at room temperature (734.746 ± 2.462) and at refrigerator temperature (680.827 ± 0.448) compared with untreated samples and with samples with plain PuOC coating. The presence of active essential oil in the applied coating significantly affected the microbiological profile of grapes during the storage period. Besides the positive impact of the applied lower storage temperature, the effectiveness of the applied active packaging is even greater (microbiological results were in the order of PuOC+essential oil < PuOC < Control). The developed artificial neural networks were found to be adequate for modeling the microbiological profile, antioxidant activity, phenolic and flavonoid content.

Keywords: biopolymer; coatings; pumpkin oil cake; essential oil; *Mentha piperita*; grapes; artificial neural networks; sensitivity analysis

Citation: Šuput, D.; Pezo, L.; Lončar, B.; Popović, S.; Tepić Horecki, A.; Daničić, T.; Cvetković, D.; Ranitović, A.; Hromiš, N.; Ugarković, J. The Influence of Biopolymer Coating Based on Pumpkin Oil Cake Activated with *Mentha piperita* Essential Oil on the Quality and Shelf-Life of Grape. *Coatings* **2023**, *13*, 299. https://doi.org/10.3390/coatings13020299

Academic Editor: Jun Mei

Received: 30 December 2022
Revised: 21 January 2023
Accepted: 25 January 2023
Published: 28 January 2023

1. Introduction

The focus of researchers on creating natural bioplastics from biodegradable biopolymers for food packaging applications has expanded recently. Fossil fuels are a finite resource, and it is essential to discover not only new energy sources but also creative materials to replace plastics made from petroleum [1]. Additionally, the disposal of packaging made of petroleum-based materials has resulted in enormous waste issues and subsequent environmental degradation [2]. The biobased and biodegradable natural polymers are the most intriguing sustainable possible replacements for the fossil sources of plastic products [1,3,4]. Because of their accessibility, environmental friendliness, and capacity to disintegrate via direct consumption, biodegradable polymers, made from natural resources, are regarded as promising alternatives to non-biodegradable synthetic polymers [5].

Even if there is plenty of novel research on this topic, the number of bioplastics generated today is still a small portion compared to oil-based products [2]. On the other hand, the agricultural industry produces a large number of different by-products with biomacromolecules, such as proteins and polysaccharides, so biopolymers from agricultural sources are an intriguing option for producing biodegradable/edible polymers [3].

Oilseed crops, regarded as foods high in energy, are farmed worldwide primarily for edible oil production. Oilseed crops are rich in fibers, antioxidants, vitamins (vitamin E, niacin, and folate), minerals (phosphorus, iron, and magnesium), and monounsaturated and polyunsaturated fatty acids [3]. One of the main byproducts of the oil extraction from oilseeds is seed oil cake, accounting for about 50% of the original seed total weight. Traditionally, either a screw press or solvents are used to extract the oil from oilseeds. The co-product that comes directly from the expeller is referred to as "cake," but the co-product that has gone through an extra, mostly organic solvent-based de-oiling process is referred to as "meal." However, there is some ambiguity in how both names are used [6].

According to USDA [7], in 2021/22 world production of major oilseeds (copra, cottonseed, palm kernel, peanut, rapeseed, soybean and sunflower seed) was 604.61 million metric tons. However, despite the enormous amount of agricultural biomass produced worldwide, only a small portion of it is currently used for applications other than those related to human nutrition or animal feed.

The aim of this work was to contribute to the application of biopolymer coatings based on oil seed cakes for food packaging. So far, various biopolymers have been applied on grapes with the aim of testing their efficacy, keeping up their quality and/or prolonging the shelf-life of grapes. These uses include: coatings based on starch/gelatin [8], chitosan and polyvinyl alcohol blending with salicylic acid [9], multilayer films composed of chitosan, sodium alginate and carboxymethyl chitosan-ZnO nanoparticles [10], edible coatings composed of alginate, galactomannans, cashew gum, and gelatin [11], starch-based films reinforced with cellulosic nanocrystals and essential oil [12], natamycin-incorporated nano-TiO_2/poly(butylene adipate-co-terephthalate) (PBAT)/poly(lactic acid) (PLA) biodegradable active film [13], poly(lactic acid) nanofiber packaging containing essential oils [14], epigallocatechin gallate grafted with pectin [15], etc. The novelty of this work is the application of composite biopolymers obtained from waste. These products are also applied as active packaging with the aim to keep up the quality of grapes and prolong their shelf-life.

After cold-pressing the oil from pumpkin seeds (*Cucurbita pepo* L.), a byproduct known as pumpkin oil cake (PuOC) is produced. Unfortunately, the majority of this cake is either consumed as animal feed or exposed. The purest form of pumpkin seed protein, PuOC, contains 63% proteins, 12% carbohydrates, 4.5% crude fiber, 8.4% oils, and 13% of other components [16]. Pumpkin oil cake was examined by Popovic et al. [17] and Hromiš et al. [18] where the influence of process parameters on the properties of the obtained biopolymer films was examined. The possibility of forming pouches was the next area of investigation into potential uses for PuOC films. Because PuOC films do not have the ability to seal in heat, they were laminated using zoin, a material that possesses the aforementioned properties [19]. Earlier work [20] pointed at the antioxidant activity of PuOC films without incorporating any active components. It was further proven that such films can be used as active packaging material and that the addition of essential oils could improve their properties [21,22]. Since it has been proven that pouches made of pumpkin oil cake-based films can preserve modified atmosphere conditions [23], they have been used for flaxseed oil packaging. It was proven that pumpkin oil cake-based packaging could ensure good oxidative stability without inducing significant changes in oil composition [24].

In this work, the biopolymer coating based on the pumpkin oil cake, native and activated with *Mentha piperita* essential oil, was applied by the immersion method to grapes of the Afus Ali variety and the quality of the grapes was monitored during the storage under certain conditions. Additionally, the principal purpose of this study was to investigate the possibility of anticipating the microbiological features (number of aerobic bacteria—TNAB, number of yeasts and molds) and antioxidant parameters (DPPH, FRAP, ABTS, total phenols content—TPC—and total flavonoids content—TFC) of a coating according to time, data of treatment and spread type (introduced as categorical variables), thus developing an artificial neural network model (ANN).

2. Materials and Methods

2.1. Experimental Material

The following materials were procured for the experiment: umpkin oil cake (PuOC) obtained after cold pressing of the oil (Linum, Serbia), glycerol (>99%) (Fisher Chemicals, USA), demineralized (distilled) water (Alfapanon, Serbia), NaOH (Lach-Ner, Czech Republic), *Mentha piperita* oil extract (Kirka Corporation, Serbia), "Afus ali" grapes purchased from a local store one day before packing. The grapes were harvested in October 2022 from the local vineyard (Južnobački district, Serbia) one day prior to purchase. Damaged grains were discarded. Only grains of similar size, color, and shape were considered for this study.

2.2. Synthesis of a Biopolymer Coating Based on Pumpkin Seed Oil Cake (PuOC)

A 10% (w/w) PuOC aqueous solution was prepared by using distilled water and grounded PuOC, followed by the addition of 0.3 g glycerol/g PuOC. The pH value of the film-forming solution was adjusted to pH 10 by the gradual addition of 50% aqueous NaOH solution on a magnetic stirrer (IKA, Germany). After achieving a pH of 10, the film-forming solution was heated for 20 min in a water bath at 60 °C and finally filtered through nylon mesh. The active biopolymer coating was prepared in the same way with the addition of 1% (v/v) *Mentha piperita* essential oil at the very end, so that the obtained biopolymer coating was cooled to room temperature. The active biopolymer coating was homogenized using a homogenizer at 166.67 Hz for 1 min (The SilentCrusher M Homogenizer, Heidolph Instruments, Germany).

2.3. Preparation, Packaging and Storage of Grape Samples with and without a Biopolymer Coating Based on Oil Pumpkin Cake

Grapes of the "Afus Ali" variety were immersed in the prepared solutions: PuOC and PuOC with the addition of *Mentha piperita* essential oil. After soaking for two minutes, the grapes were left to drain off the excess biopolymer coating by using a sieve and then spread on plastic trays and left in room conditions for 2 h to dry. Grapes without an applied coating were used as a control sample. Grape samples were packed into polystyrene trays measuring 17.5 cm × 13 cm and covered with polyethylene stretch film. Half of the samples were stored at room temperature (23 °C), and the other half at refrigerator temperature (4 °C). The sampling of the dynamics of grape samples stored at room temperature was performed on the 2nd, 4th, 6th, 8th and 10th days, and the samples stored at refrigerator temperature were sampled on the 3rd, 6th, 9th, 12th and 15th days. Different sampling days were chosen based on the assumption that the shelf life of samples stored at refrigerator temperature will be longer, and that therefore it is not necessary to sample these samples as often as samples stored at room temperature. In this way, the experiment itself would be simplified. Table 1 explains the sample designations used, while Figure 1 shows the appearance of the samples at the beginning of storage.

Table 1. Sample labeling and experimental design.

Label	Control Grape Sample (Uncoated)	Grape Sample with Biopolymer Coating	Grape Sample with Active Biopolymer Coating (*Mentha Piperita* Essential Oil)	Storage Period (Days)									Storage Temperature (°C)	
				2	3	4	6	8	9	10	12	15	4	23
S2	+			+										+
SB2		+		+										+
SU2			+	+										+
S4	+					+								+
SB4		+				+								+
SU4			+			+								+
S6	+						+							+
SB6		+					+							+
SU6			+				+							+
S8	+							+						+
SB8		+						+						+
SU8			+					+						+
S10	+									+				+
SB10		+								+				+
SU10			+							+				+
F3	+				+								+	
FB3		+			+								+	
FU3			+		+								+	
F6	+						+						+	
FB6		+					+						+	
FU6			+				+						+	
F9	+								+				+	
FB9		+							+				+	
FU9			+						+				+	
F12	+										+		+	
FB12		+									+		+	
FU12			+								+		+	
F15	+											+	+	
FB15		+										+	+	
FU15			+									+	+	

S—control grape sample, stored at room temperature; SB—grape sample with biopolymer coating, stored at room temperature; SU—grape sample with active biopolymer coating (with the addition of essential oil), stored at room temperature; F—control grape sample stored at refrigerator temperature; FB—grape sample with biopolymer coating, stored at refrigerator temperature; FU—grape sample with active biopolymer coating (with the addition of essential oil), stored at refrigerator temperature.

Figure 1. The appearance of samples at the beginning of storage: (**1**) Control sample, (**2**) Grapes coated with PuOC, (**3**) Grapes coated with activated PuOC with *Mentha piperita* essential oil.

2.4. Examination of the Quality of Packaged Grapes

2.4.1. Preparation of Liquid Extract

The grape samples were turned into a mash using a stick mixer. About 5 g of the prepared sample was weighed into an Erlenmeyer flask and 50 mL of 95% methanol was poured as an extraction agent. The flasks were covered and placed on a laboratory mixer (Unimax 1010, Heidolph Instruments GmbH & CoKG, Germany) for 24 h in a dark place. After extraction, the samples were filtered into vials and stored in a refrigerator until analysis. Methanolic extracts were used to determine the content of total phenols, total flavonoids and DPPH.

2.4.2. Dry Matter Content

The dry matter content was determined gravimetrically by drying the samples at a temperature of 105 ± 0.5 °C to a constant mass. The percentage of dry matter content is equal to:

$$DM = \frac{M_2 - M_0}{M_1 - M_2} \cdot 100 \ (\%) \tag{1}$$

wherein:

M_0—mass of vessel and auxiliary material (filter paper, sand, glass rod, lid), in g;
M_1—mass of the same container with the tested sample before drying, in g;
M_2—mass of the same vessel with the tested sample after drying, in g.

2.4.3. Content of Total Sugars

Total sugars were determined by the Luff–Schoorl method [25]. The method is based on the principle that, under certain conditions, reducing sugars (natural inversion) will convert copper sulfate ($CuSO_4$) from Luff's solution into copper oxide (Cu_2O). The unused amount of cupric ions is retitrated with a thiosulfate solution. The amount of sugar from the table is read from the difference between the consumption for the blank and the test. The non-reducing disaccharide (sucrose) must first be inverted, that is, the reducing monosaccharides must be hydrolyzed with an acid, and then determined using Luff's solution. The difference between the obtained total invert and the natural invert gives the amount of reducing sugars formed by sucrose inversion.

2.4.4. Total Acidity

The total (titratable) acidity of the sample was determined according to ISO 750:1998. It is a volumetric method, using a sodium hydroxide (NaOH) standard solution, with phenolphthalein as an indicator. Total acidity was expressed in g of tartaric acid per 100 g sample [25].

2.4.5. Alcohol Content

Ethanol, separated by distillation, is oxidized with potassium bichromate in the presence of sulfuric acid, and excess potassium bichromate is retitrated with ammonium ferrous sulfate in the presence of the iron-ortho-phenanthroline indicator. The method is applied for the determination of ethanol in fruit and vegetable products where the amount of ethanol does not exceed 5% (*m*/*m*).

2.4.6. Content of Total Phenolic Compounds

The content of total phenols in liquid methanolic extracts was determined by the spectrophotometric method according to Folin–Ciocalteu [26]. The reaction mixture for determining the content of total phenols in the sample was prepared by mixing 2 mL of the sample, 2.5 mL of the Folin–Ciocalteu reagent, and 7.5 mL of Na_2CO_3 in a measuring vessel of 50 mL. The measuring vessel was filled up to the mark with distilled water. Gallic acid was used as a standard. Absorbances were measured at 750 nm. Based on the measured absorbance, the concentration (mg/mL) of phenolic compounds was read from the calibration curve of the standard gallic acid solution, and then the content of total phenolic compounds in the sample was expressed as the gallic acid equivalent (mg GAE/100 g).

2.4.7. Content of Total Flavonoids

The content of total flavonoids in extracts of fresh and dried grapes was determined by the colorimetric method, in accordance with Markham [27]. The reaction mixture was prepared by mixing 1 mL of the extract with 4 mL of the distilled water and 0.3 mL of 5% the $NaNO_2$ solution. The mixture was then incubated for five minutes at room temperature, after which 0.3 mL of a 10% $AlCl_3 \times 6H_2O$ solution was added. After six minutes, when the solution became intensely yellow, 2 mL of NaOH solution was added. The reaction mixture was supplemented with distilled water up to 10 mL and the absorbance was measured at 510 nm. The content of total flavonoids is expressed in catechin equivalents per unit mass of the sample (mg CAE/100 g). A standard catechin solution was used to create the calibration curve.

2.5. Determination of Antioxidant Activity

2.5.1. 2,2-Diphenyl-1-Picryl-Hydrazyl-Hydrate Assay (DPPH)

The DPPH assay was performed using a modified method originally presented in the study by Brand-Williams et al. [28]. A methanolic solution of the DPPH reagent (65 μM) was adjusted by adding methanol to obtain an absorbance of 0.70 (±0.01). The previously prepared extract and DPPH reagent were mixed (0.1 mL + 2.9 mL) in 10 mm glass cuvettes and incubated at room temperature for 60 min. Measurement of the neutralization of free radicals was carried out at 517 nm. A UV/VIS spectrophotometer (LLG-uniSPEC 2 Spectrophotometer) was used for the spectrophotometric measurement of absorbance. Results are expressed as mg Trolox equivalents per g (mg Trolox/g).

2.5.2. Ferric Reducing Antioxidant Power Assay (FRAP)

The FRAP test was performed by a modified method originally presented in the study by Benzie and Strain [29]. The FRAP reagent was prepared from 300 mM acetic buffer (pH 3.6), 10 mM 2,4,6-tripyridyl-s-triazine (TPCTZ) in 40 mmol/L HCl solution and 20 mM/L $FeCl_3$ aqueous solution in the ratio 10:1:1 ($v/v/v$). Previously prepared extracts and FRAP reagent were mixed (0.1 + 2.9 mL) and incubated in the dark at a temperature of 37 °C for 10 min. After incubation, the absorbance of the sample was measured at 593 nm. The results are expressed as mg equivalent of Fe^{2+} ions per g (mg Fe^{2+}/g).

2.5.3. ABTS Radical Scavenging Assay (ABTS)

The ABTS test was performed by a modified method described in the study by Re et al. [30]. The ABTS reagent solution was prepared by mixing a 7 mM aqueous solution of 2,2′-azino-bis(3-ethylbenzothiazoline-6-sulfonic acid) diammonium salt (ABTS) and 2.45 mM potassium persulfate in a 1:1 ratio (v/v) and incubating them in the dark at room temperature for 16 h. The ABTS reagent was diluted with 300 mM acetate buffer (pH 3.6) to adjust the absorbance at 0.734 to 0.70 (±0.01). The previously prepared extract and ABTS reagent were mixed (0.1 + 2.9 mL) and incubated in the dark at room temperature for 5 h. After incubation, the absorbance of the sample was measured at 734 nm. A UV/VIS spectrophotometer (LLG-uniSPEC 2 Spectrophotometer) was used for spectrophotometric measurement of absorbance. Results are expressed as mg Trolox equivalents per g (mg Trolox/g).

All measurements were performed in three replicates.

2.6. Microbiological Examination

The preparation of samples for microbiological testing was carried out in accordance with standard ISO 6887-1:2017 [31]. Standard methods were used to enumerate selected groups of microorganisms: total aerobic microbial count [32], total molds and yeasts count [33]. The diluent (buffered peptone water) and all culture media were acquired from HiMedia (Mumbai, India). The number of microorganisms in the tested samples is expressed as colony forming units (cfu) per gram.

The number of microorganisms in the tested samples is expressed as the mean of three measurements with standard deviation.

2.7. Statistical Analyses

Statistical processing of the data was presented by the STATISTICA 10.0 software package (StatSoft Inc., Tulsa, OK, USA). The results were exhibited as mean ± standard deviation of triplicate analyses for all measurements.

Principal component analysis (PCA) was applied to reveal the feasible correlations between measured parameters and employed to categorize objects.

2.7.1. Artificial Neural Network (ANN) Modeling

A multi-layer perceptron model (MLP), which consisted of three layers (input, hidden and output layers), was engaged in the establishment of a model building. Prior to the computation, experimental data were normalized to enhance the behaviour of the ANN. The Broyden–Fletcher–Goldfarb–Shanno (BFGS) algorithm was employed as an iterative method for solving unconstrained nonlinear optimization problems in ANN modeling [34].

The experimental database for ANN was randomly divided into training, cross-validation and testing data (with 60%, 20% and 20% of experimental data, respectively). The training data set was used for the learning cycle of ANN and also for the evaluation of the optimal number of neurons in the hidden layer and the weight coefficient of each neuron in the network [35].

Coefficients associated with the hidden layer were grouped in matrices W_1 and B_1, while coefficients associated with the output layer were grouped in matrices W_2 and B_2. The neural network is usually presented using matrix notation (Y is the matrix of the output variables, f_1 and f_2 are transfer functions in the hidden and output layers, respectively, and X is the matrix of input variables) [34]:

$$Y = f_1(W_2 \cdot f_2(W_1 \cdot X + B_1) + B_2) \tag{2}$$

The elements of matrices W_1 and W_2 were determined during the ANN learning cycle and during the iterative procedure, with an optimization algorithm being used to minimize the error between network outputs and experimental results. The coefficients of determination were used as parameters to check the performance of the obtained ANN model [36].

2.7.2. Global Sensitivity Analysis

Yoon's global sensitivity equation for the obtained ANN model was exploited to estimate the relative impact of the input parameters (time, temperature and applied coating) on output variables (microbiological profile, antioxidant activity, phenolic and flavonoid content), depending on the designed ANN model weight coefficients [37]:

$$RI_{ij}(\%) = \frac{\sum\limits_{k=0}^{n} (w_{ik} \cdot w_{kj})}{\sum\limits_{i=0}^{m} \left| \sum\limits_{k=0}^{n} (w_{ik} \cdot w_{kj}) \right|} \cdot 100\% \tag{3}$$

where: w—weight coefficient in ANN model, i—input variable, j—output variable, k—hidden neuron, n—number of hidden neurons, m—number of inputs.

2.7.3. The Accuracy of the Model

The numerical confirmation of the obtained ANN and RFR models was performed using statistical tests, such as coefficient of determination (r^2), reduced chi-square (χ^2), mean bias error (*MBE*), root mean square error (*RMSE*) and mean percentage error (*MPE*) methods. These commonly used parameters were calculated according to Puntarić et al., [38]:

$$\chi^2 = \frac{\sum\limits_{i=1}^{N} \left(x_{\exp,i} - x_{pre,i}\right)^2}{N - n} \tag{4}$$

$$RMSE = \left[\frac{1}{N} \cdot \sum\limits_{i=1}^{N} \left(x_{pre,i} - x_{\exp,i}\right)^2\right]^{1/2} \tag{5}$$

$$MBE = \frac{1}{N} \cdot \sum\limits_{i=1}^{N} \left(x_{pre,i} - x_{\exp,i}\right) \tag{6}$$

$$MPE = \frac{100}{N} \cdot \sum\limits_{i=1}^{N} \left(\frac{\left|x_{pre,i} - x_{\exp,i}\right|}{x_{\exp,i}}\right) \tag{7}$$

where $x_{\exp,i}$ were experimental values and $x_{pre,i}$ were the model predicted values, N and n are the number of observations and constants, accordingly.

3. Results and Discussion

3.1. Dry Matter Content

In all tested samples, over time, the content of dry matter decreased (Table 2) from an initial 20.440% to 17.169% for SB10 sample; 18.003% for SU10; 19.405% for FB15 and 19.790% for FU15 sample A greater reduction or loss in dry matter was observed in grapes stored at room temperature, while there were smaller changes in grapes stored at refrigerator (4 °C), which is in accordance with the findings of de Souza et al. [11]. According to Melo et al. [39], water evaporation in the fruit, which can be very large at higher temperatures during storage, is linked to weight loss. Additionally, the coating's porosity can let water evaporate and lead to weight loss [40], as can the fact that biopolymer structures tend to swell in the presence of humidity, with all these factors causing poor resistance to moisture and water loss [41]. On the other hand, there were higher values of dry matter observed in samples with an active PuOC coating compared to the applied biopolymer itself, which is a consequence of the hydrophobic structure of the added essential oil.

Table 2. The change in the dry matter content, total sugars/alcohol content and total acidity of packed grapes.

Sample	DM	Sugar	TA	ALC
0	20.440 ± 0.025 [h]	19.678	0.568	0.124
S2	19.690 ± 0.248 [fgh]	12.716	0.453	0.648
SB2	18.666 ± 0.089 [bcdef]	14.249	0.553	0.316
SU2	19.434 ± 0.346 [efgh]	14.150	0.571	0.303
S4	18.712 ± 0.353 [bcdef]	15.944	0.700	
SB4	18.232 ± 0.760 [abc]	15.938	0.801	
SU4	18.471 ± 0.533 [bcde]	17.648	0.786	
S6	19.312 ± 0.175 [defg]	15.640	0.723	0.114
SB6	19.112 ± 0.142 [cdefg]	16.287	1.041	0.194
SU6	19.079 ± 0.374 [cdefg]	17.497	0.970	0.093
S8	18.722 ± 0.143 [bcdef]	14.515	0.943	
SB8	18.423 ± 0.329 [bcde]	13.503	1.021	
SU8	18.675 ± 0.066 [bcdef]	17.969	1.228	
S10	18.286 ± 0.169 [bcd]	17.142	1.197	0.040
SB10	17.169 ± 0.190 [a]	10.698	1.382	0.031
SU10	18.003 ± 0.048 [ab]	15.667	1.401	0.688

Table 2. *Cont.*

Sample	DM	Sugar	TA	ALC
F3	19.189 ± 0.201 [cdefg]	12.707	0.452	1.684
FB3	19.072 ± 0.400 [cdefg]	17.332	0.467	0.054
FU3	19.255 ± 0.118 [cdefg]	12.141	0.480	0.685
F6	19.315 ± 0.205 [defg]	12.040	0.663	
FB6	20.096 ± 0.142 [gh]	15.244	0.842	
FU6	19.598 ± 0.223 [fgh]	14.057	0.993	
F9	19.640 ± 0.258 [fgh]	16.980	0.736	0.644
FB9	19.382 ± 0.141 [efgh]	13.828	0.661	0.654
FU9	19.887 ± 0.081 [gh]	19.197	0.700	1.354
F12	20.120 ± 0.034 [gh]	17.572	0.671	0.720
FB12	19.593 ± 0.273 [fgh]	16.948	0.692	0.725
FU12	19.932 ± 0.023 [gh]	13.566	1.937	0.732
F15	19.730 ± 0.048 [fgh]	8.945	1.495	
FB15	19.405 ± 0.007 [efgh]	16.608	2.014	
FU15	19.790 ± 0.015 [gh]	11.977	1.969	

Means in the same column with different superscript are statistically different ($p \leq 0.05$).

3.2. Content of Total Sugars/Alcohol Content and Total Acidity

It is noticeable that the sugar content changes during storage in both temperature regimes. The decrease in sugar content compared to day 0 is a consequence of the spoilage of grapes, that is, the consumption of sugar for the production of alcohol (Table 2).

Table 2 shows data related to the change in total acidity content, which is expressed in relation to the dry matter of the sample. A trend of increasing total acidity can be observed in samples stored at room temperature (23 °C), as well as those stored in a refrigerator (4 °C). The increase in total acidity was expected due to the grape spoilage during storage. The application of lower storage temperatures contributed to better values of total acidity of the grape samples. Previous research has demonstrated that organic acids and soluble solids are used by respiratory mechanisms to preserve the fruit's normal activity during storage [42]. Cold storage is considered an additional process that may slowdown the physiological processes in grapes, which in turn may slow the consumption of organic acids [39].

Higher values of total acidity are observed in the sample with biopolymer coating (both with and without the addition of essential oil) compared to the untreated samples in both temperature regimes. On the 10th day of storage for samples stored at room temperature, total acidity was 1.197%, for samples with a coating of 1.382% and for samples with an active coating of 1.401%. The same trend was observed for samples stored at refrigerator temperature: uncoated samples had a total acidity of 1.495%, coated samples 2.014%, and active coated samples 1.969% (whose increase in total acidity value occurred on the 12th day of storage (1.937%)). An increase in total acidity is a consequence of the appearance of spoilage and, given that the grape samples were also visually assessed by the presence of color changes, of spots with molds.

3.3. Determination of Antioxidant Activity

Results of the antioxidant potential of packed grapes are presented in Table 3 and they differ in all evaluated methods (ABTS, DPPH, and FRAP). These differences result from the various mechanisms used in radical antioxidant responses [43]. Although these differences are undeniable, for all antioxidant responses it can be noted that the values are higher in grape samples stored at a lower temperature, regardless of whether an (active) coating based on PuOC was applied.

Table 3. The change of antioxidant activity, total phenolic compounds and total flavonoids of packed grapes.

Sample	DPPH	FRAP	ABTS	TPC	TFC
0	2.268 ± 0.018 [P]	0.666 ± 0.005 [hi]	4.227 ± 0.034 [r]	640.100 ± 1.301 [j]	1.583 ± 0.005 [o]
S2	2.119 ± 0.010 [m]	0.775 ± 0.009 [lm]	3.473 ± 0.059 [m]	659.981 ± 1.350 [l]	1.470 ± 0.015 [m]
SB2	1.895 ± 0.013 [k]	0.776 ± 0.004 [lm]	3.388 ± 0.025 [l]	683.697 ± 0.475 [op]	1.450 ± 0.015 [m]
SU2	1.708 ± 0.008 [h]	0.655 ± 0.005 [gh]	3.066 ± 0.008 [i]	639.622 ± 1.368 [j]	1.252 ± 0.002 [j]
S4	1.812 ± 0.010 [i]	0.807 ± 0.009 [no]	3.642 ± 0.008 [o]	689.813 ± 1.421 [pq]	1.316 ± 0.012 [k]
SB4	1.862 ± 0.015 [jk]	0.916 ± 0.005 [q]	4.015 ± 0.042 [P]	749.782 ± 1.458 [u]	1.392 ± 0.015 [l]
SU4	1.605 ± 0.015 [g]	0.666 ± 0.013 [hi]	2.981 ± 0.042 [h]	698.297 ± 2.879 [r]	1.005 ± 0.005 [de]
S6	1.408 ± 0.010 [cd]	0.513 ± 0.009 [a]	2.524 ± 0.059 [b]	560.532 ± 0.918 [e]	0.897 ± 0.005 [ab]
SB6	1.441 ± 0.018 [de]	0.628 ± 0.011 [efg]	3.219 ± 0.042 [k]	577.991 ± 2.319 [fg]	1.054 ± 0.015 [fg]
SU6	1.320 ± 0.008 [a]	0.495 ± 0.002 [a]	2.897 ± 0.042 [efg]	615.128 ± 1.393 [h]	0.888 ± 0.015 [a]
S8	1.532 ± 0.008 [f]	0.577 ± 0.000 [bc]	2.820 ± 0.051 [cd]	650.627 ± 1.420 [k]	1.032 ± 0.007 [ef]
SB8	1.292 ± 0.015 [a]	0.549 ± 0.002 [b]	2.507 ± 0.042 [b]	552.976 ± 1.924 [d]	1.069 ± 0.020 [fg]
SU8	1.620 ± 0.010 [g]	0.749 ± 0.013 [kl]	2.905 ± 0.017 [fg]	693.767 ± 2.372 [qr]	0.995 ± 0.005 [de]
S10	1.471 ± 0.018 [e]	0.611 ± 0.002 [de]	2.778 ± 0.059 [c]	622.801 ± 2.907 [i]	0.932 ± 0.015 [bc]
SB10	1.365 ± 0.013 [b]	0.591 ± 0.007 [cd]	2.481 ± 0.034 [b]	575.685 ± 2.064 [f]	1.005 ± 0.005 [de]
SU10	1.630 ± 0.015 [g]	0.689 ± 0.011 [ij]	3.134 ± 0.042 [j]	734.746 ± 2.462 [t]	1.091 ± 0.012 [g]
F3	2.316 ± 0.005 [q]	1.083 ± 0.013 [t]	4.142 ± 0.051 [q]	721.047 ± 0.924 [s]	1.832 ± 0.015 [q]
FB3	2.225 ± 0.015 [o]	0.956 ± 0.013 [r]	4.312 ± 0.051 [s]	726.979 ± 0.465 [s]	1.744 ± 0.015 [P]
FU3	2.182 ± 0.013 [n]	0.834 ± 0.011 [o]	3.651 ± 0.034 [o]	657.112 ± 3.683 [l]	1.538 ± 0.015 [n]
F6	1.902 ± 0.010 [k]	0.640 ± 0.005 [fgh]	2.956 ± 0.051 [gh]	526.757 ± 3.211 [b]	1.208 ± 0.012 [i]
FB6	1.393 ± 0.015 [bc]	0.686 ± 0.011 [ij]	2.854 ± 0.017 [def]	522.188 ± 0.441 [b]	0.966 ± 0.015 [cd]
FU6	2.278 ± 0.008 [Pq]	0.793 ± 0.009 [mn]	3.490 ± 0.025 [mn]	675.671 ± 0.452 [mn]	1.450 ± 0.015 [m]
F9	1.817 ± 0.010 [i]	0.736 ± 0.007 [k]	2.905 ± 0.034 [fg]	534.419 ± 2.256 [c]	1.250 ± 0.010 [j]
FB9	1.844 ± 0.013 [ij]	0.735 ± 0.002 [k]	2.930 ± 0.025 [gh]	584.114 ± 1.829 [g]	1.167 ± 0.015 [h]
FU9	1.532 ± 0.018 [f]	0.617 ± 0.007 [def]	2.320 ± 0.025 [a]	524.583 ± 0.891 [b]	0.897 ± 0.005 [ab]
F12	2.961 ± 0.005 [t]	1.239 ± 0.009 [u]	5.820 ± 0.051 [t]	694.766 ± 0.881 [qr]	$2.219 + 0.005$ [r]
FB12	1.549 ± 0.010 [f]	0.698 ± 0.013 [j]	2.490 ± 0.025 [b]	493.260 ± 3.167 [a]	0.966 ± 0.015 [cd]
FU12	1.948 ± 0.015 [l]	0.775 ± 0.013 [lm]	2.837 ± 0.017 [cde]	620.019 ± 2.223 [hi]	1.074 ± 0.015 [g]
F15	2.618 ± 0.015 [s]	1.001 ± 0.007 [s]	4.100 ± 0.042 [q]	582.735 ± 2.695 [g]	1.707 ± 0.022 [P]
FB15	2.396 ± 0.015 [r]	0.936 ± 0.007 [qr]	3.549 ± 0.051 [n]	673.958 ± 1.827 [m]	1.450 ± 0.015 [m]
FU15	2.094 ± 0.015 [m]	0.873 ± 0.002 [P]	3.693 ± 0.042 [o]	680.827 ± 0.448 [no]	1.245 ± 0.010 [ij]

Means in the same column with different superscript are statistically different ($p \leq 0.05$).

The uncoated samples stored at room temperature had the highest decrease in DPPH values during storage, ranging from 2.119 mg/g to 1.471 μmol mg/g. The samples coated with PuOC and PuOC with the addition of essential oil had uniform DPPH values through the entire storage period. The uncoated grape sample stored at refrigerator temperature and samples on the 12th day (sample labeled as F12) had the highest DPPH value (2.961 mg Trolox/g). The same sample also had the highest ABTS (5.82 mg Trolox/g) and FRAP (1.239 mg Fe^{2+}/g) values.

In both groups of samples, there was a slight decrease in capacity for ABTS radical scavenging, although in some samples, the decrease is not linear, which may be a consequence of sample inhomogeneity. The ABTS values on the second and third days of storage are uniform for all samples and range from 3.066 mg Trolox/g to 4.142 mg Trolox/g. Later, during storage, significantly lower values were obtained for samples stored at room temperature, compared to samples stored at refrigerator temperature.

For the FRAP method, the Fe^{3+} reducing power in all treatments undergoes a slight decline until halfway throughout the storage period, followed by an increase in values, so that the final values are the same as the initial values or slightly higher.

The findings indicate that grapes' antioxidant potential may be enhanced or maintained by applying PuOC coatings, with or without *Mentha piperita* essential oil. In all of the samples, the coating application of PuOC (with or without the addition of *Mentha piperita* essential oil) preserved the antioxidant potential of the grapes, which is in the agreement with Tahir et al. [44]. The reason is that biopolymer coatings regulate the ripening process as well as the hydrolysis reactions and reduce changes in phenolic compounds, effects which can have an impact on grapes' antioxidant potential [45].

3.4. Content of Total Phenolic Compounds and Total Flavonoids

Table 3 shows data related to the change in the composition of total phenolic compounds. The presence of phenolic compounds was found in all tested samples. Phenol substances are positively correlated to the quality of grapes and with the antioxidant activity of grapes. As can be noted from Table 3, the values for the content of phenolic substances in grapes varied during storage with a slight decreasing trend. The underlying mechanism is related to the activity of polyphenol oxidase and peroxidase present in grapes [46]. Polyphenol oxidase can oxidize polyphenols into quinones in an aerobic environment, which has the effect of promoting grape browning and polyphenols content reductions [47]. At the end of storage period, the highest phenolic content was observed for samples with active coating stored at room temperature (734.746 ± 2.462) and at refrigerator temperature (680.827 ± 0.448) compared with untreated samples and with samples with plain PuOC coating.

The presence of flavonoids was also found in all tested samples. Similar to the phenolic content, a discrete decrease in the flavonoid content can be observed during the storage period, which is in agreement with the findings of Lo'ay et al. [9]. In general, higher values were obtained for samples stored at refrigerator temperature compared to samples stored at room temperature. According to the obtained results presented in Table 3, the contribution of PuOC coating to the preservation of flavonoid content is significant compared to that of untreated samples and samples with an active PuOC coating with the addition of *Mentha piperita* essential oil.

3.5. Microbiological Examination

Table 4 provides results related to the microbiological profile of grape samples. High initial values for total aerobic microbial count were observed, which affected all other values during storage. Although the obtained results are uneven, it can be noted that the application of an active PuOC coating with the addition of *Mentha piperita* essential oil is the most important factor for the microbiological stability of grapes packaged at room temperature. SU10 sample had 0.777×10^7 cfu/g, compared with SB10 sample (1.157×10^7 cfu/g) and compared with the uncoated S10 sample (1.52×10^7 cfu/g). The grape quality preservation implies the application of lower temperatures, which was confirmed because the values on 21st day for samples stored at refrigerator temperature were significantly lower (in the range 0.065×10^7–0.237×10^7 cfu/g) than those for samples stored on 10th day at room temperature (0.777×10^7–1.52×10^7 cfu/g).

The results of yeast determination showed that the influence of storage temperature is negligible. On the other hand, the application of an active coating based on PuOC is more significant because, for all tested samples, on each sampling day, lower values were obtained compared to untreated samples, as well as for samples with only a coating based on PuOC. This would mean that the biggest contribution to the low presence of mold is the application of *Mentha piperita* essential oil. The total number of yeasts in each sample group was in the order of PuOC+essential oil < PuOC < Control.

The same results were obtained when determining molds. In each tested sample group, the lowest values were obtained for samples coated with added essential oil, i.e., for which an active coating was applied. The first significant increase in the mold value of the samples stored at room temperature was observed on the sixth day and for the untreated sample (S6) was 0.627×10^5 cfu/g. In samples stored at refrigerator temperature, the increase in the number of molds was observed at the 12th day, and it was 0.93×10^5 cfu/g for sample FB12. This fact favors the use of lower storage temperatures. By the 21st day of storage, there was a significant increase in the presence of mold, 3.167×10^5 cfu/g for sample F21, 3.6×10^5 cfu/g for sample FB and 0.837×10^5 cfu/g for sample FU21.

Table 4. The change of microbiological profile of packed grapes.

Sample	TNAB [$\times 10^7$]	Yeasts [$\times 10^7$]	Molds [$\times 10^5$]
0	0.013 ± 0.004 [a]	0.002 ± 0.001 [a]	0.002 ± 0.002 [a]
S2	1.203 ± 0.454 [cdef]	0.507 ± 0.210 [abcdef]	0.048 ± 0.016 [a]
SB2	0.477 ± 0.087 [abcd]	0.480 ± 0.288 [abcde]	0.045 ± 0.010 [a]
SU2	2.133 ± 0.757 [g]	0.103 ± 0.032 [abc]	0.008 ± 0.002 [a]
S4	0.413 ± 0.101 [abcd]	0.217 ± 0.076 [abc]	0.037 ± 0.015 [a]
SB4	0.433 ± 0.076 [abcd]	0.533 ± 0.076 [abcdef]	0.038 ± 0.020 [a]
SU4	0.163 ± 0.032 [ab]	0.203 ± 0.127 [abc]	0.029 ± 0.015 [a]
S6	0.247 ± 0.050 [ab]	0.190 ± 0.050 [abc]	0.056 ± 0.017 [a]
SB6	0.330 ± 0.125 [abc]	0.520 ± 0.075 [abcdef]	0.034 ± 0.006 [a]
SU6	0.230 ± 0.062 [ab]	0.213 ± 0.078 [abc]	0.011 ± 0.001 [a]
S8	0.217 ± 0.057 [ab]	0.132 ± 0.010 [abc]	0.627 ± 0.110 [ab]
SB8	0.163 ± 0.057 [ab]	0.303 ± 0.135 [abcd]	0.320 ± 0.080 [a]
SU8	0.133 ± 0.031 [ab]	0.217 ± 0.031 [abc]	0.055 ± 0.015 [a]
S10	1.520 ± 0.495 [efg]	0.150 ± 0.040 [abc]	0.473 ± 0.110 [ab]
SB10	1.157 ± 0.319 [cdef]	0.393 ± 0.179 [abcde]	0.467 ± 0.125 [ab]
SU10	0.777 ± 0.493 [abcde]	0.627 ± 0.583 [abcdef]	0.120 ± 0.010 [a]
F3	0.713 ± 0.090 [abcde]	0.026 ± 0.019 [a]	0.006 ± 0.003 [a]
FB3	1.253 ± 0.647 [defg]	0.220 ± 0.060 [abc]	0.010 ± 0.001 [a]
FU3	0.250 ± 0.100 [ab]	0.002 ± 0.000 [a]	0.003 ± 0.001 [a]
F6	0.213 ± 0.099 [ab]	0.173 ± 0.050 [abc]	0.017 ± 0.006 [a]
FB6	1.467 ± 0.569 [efg]	0.793 ± 0.555 [bcdefg]	0.016 ± 0.006 [a]
FU6	0.500 ± 0.100 [abcd]	0.160 ± 0.061 [abc]	0.009 ± 0.001 [a]
F9	0.391 ± 0.494 [abcd]	0.068 ± 0.014 [ab]	0.043 ± 0.012 [a]
FB9	0.157 ± 0.038 [ab]	1.267 ± 0.473 [fg]	0.019 ± 0.006 [a]
FU9	1.800 ± 0.361 [fg]	1.467 ± 0.723 [g]	0.010 ± 0.001 [a]
F12	0.163 ± 0.047 [ab]	0.127 ± 0.040 [abc]	0.867 ± 0.084 [ab]
FB12	1.020 ± 0.164 [bcdef]	1.090 ± 0.271 [efg]	0.930 ± 0.200 [ab]
FU12	1.187 ± 0.359 [cdef]	1.007 ± 0.261 [defg]	0.423 ± 0.090 [ab]
F15	0.053 ± 0.013 [a]	0.723 ± 0.357 [abcdefg]	1.800 ± 0.458 [bc]
FB15	0.540 ± 0.053 [abcd]	0.833 ± 0.176 [cdefg]	1.000 ± 0.092 [ab]
FU15	0.433 ± 0.123 [abcd]	0.253 ± 0.084 [abcd]	0.600 ± 0.075 [ab]
F18	0.063 ± 0.007 [a]	0.133 ± 0.042 [abc]	1.300 ± 0.265 [ab]
FB18	0.390 ± 0.066 [abcd]	0.530 ± 0.108 [abcdef]	1.027 ± 0.155 [ab]
FU18	0.145 ± 0.048 [ab]	0.283 ± 0.126 [abcd]	0.683 ± 0.070 [ab]
F21	0.065 ± 0.023 [a]	0.113 ± 0.035 [abc]	3.167 ± 1.795 [cd]
FB21	0.207 ± 0.116 [ab]	0.450 ± 0.087 [abcde]	3.600 ± 1.709 [d]
FU21	0.237 ± 0.146 [ab]	0.225 ± 0.158 [abc]	0.837 ± 0.111 [ab]

Means in the same column with different superscript are statistically different ($p \leq 0.05$).

Figure 2 displays the appearance of the samples at the end of the experiment (18th and 21st day), when microbiological damage is already visible.

Mentha piperita has various biological activities: antioxidant activities, cytotoxicity activities, anti-inflammatory properties, as well as antimicrobial activities [48]. According to [49], *Mentha piperita* common major components are menthol (oxygenated monoterpene), menthone (oxygenated monoterpene), carvone (oxygenated monoterpene), anethole (phenylproprenoid), 1,8-cineole (oxygenated monoterpene) and common minor components are menthyl acetate, limonene (monoterpene hydrocarbon), α-pinene (monoterpene hydrocarbon), β-pinene (monoterpene hydrocarbon) and myrcene (monoterpene hydrocarbon).

The structural functional groups of major components play an important role in the biological activity of essential oils. Menthol and menthone are cyclic and oxygenated monoterpenes that play essential roles in the disorganization of cell membrane structures, causing depolarization and physical or chemical alterations, thereby disrupting metabolic activities [50]. These major active components penetrate the cell membrane and target the ergosterol biosynthesis pathway, thus impairing its biosynthesis. Simultaneously, they react with the membrane itself with their reactive hydroxyl moiety, and the extensive lesion on the membrane is a combined effect of the two events [51]. Minor components also significantly influence the antimicrobial properties of the *Mentha piperita* essential oil through synergistic interactions [52].

Figure 2. Samples appearance with visible microbiological decay at 18th and 21st day.

According to the available literature, *Mentha piperita* has very strong antimicrobial potential against various bacteria, yeasts and molds [51,53–55]. As such, numerous applications in the food industry have been conducted [50,56,57]. The results of this research support the fact about the antimicrobial effect of the *Mentha piperita* essential oil when it is incorporated into a biopolymer coating.

3.6. PCA Analysis

The points displayed in the PCA graphic, which are numerically in close vicinity to each other, demonstrate the similarity of patterns that portray these data. The direction of the vector explaining the variable in factor space discloses a rising trend of these variables, and the longitude of the vector is relative to the square of the correlation values among the fitting value for the variable and the variable itself. The angles, amidst corresponding variables, denote the degree of their correlations (minor angles corresponding to elevated correlations).

The PCA of the microbiological data explained that the first two components accounted for 81.63% of the total variance (47.85 and 33.78%, respectively) in the three-variable factor space (microbiological parameters). Considering the mapping of the PCA performed on the data, molds (which contributed 19.4% of the total variance, based on correlations) exhibited positive scores according to the first principal component, whereas TNAB (50.8%) and yeasts (29.8%) showed negative score values according to the first principal component (Figure 3a). A positive contribution to the second principal component calculation was observed for: yeasts (40.0% of the total variance, based on correlations) and molds (60.0%).

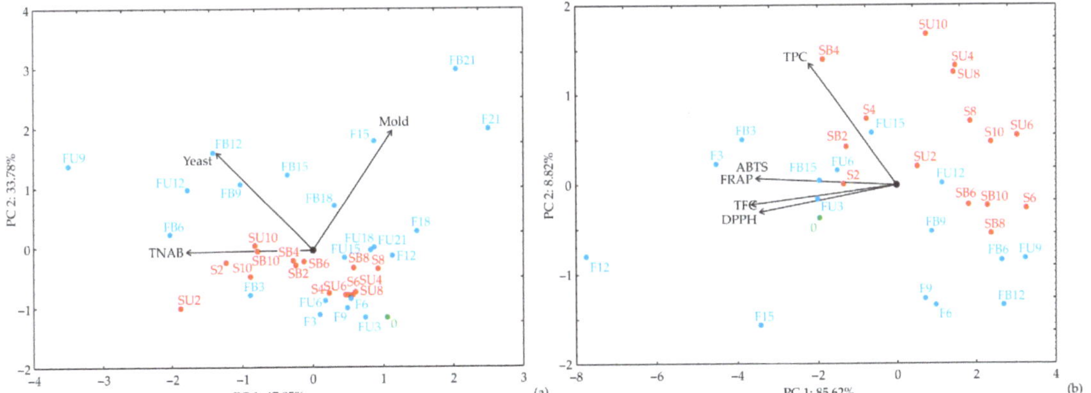

Figure 3. PCA ordination of variables based on component correlations for: (**a**) microbiological features (TNAB, number of yeasts and molds) and (**b**) antioxidant parameters (DPPH, FRAP, ABTS, TPC and TFC).

The PCA of the antioxidant data explained that the first two components accounted for 94.44% of the total variance (85.62% and 8.82%, respectively) in the six-variable factor space (antioxidant data). Considering the mapping of the PCA performed on the data, DPPH (which contributed 12.6% of the total variance, based on correlations), FRAP (12.3%), ABTS (13.2%), and TFC (14.2%) exhibited negative scores according to the first principal component (Figure 3b). A positive contribution to the second principal component calculation was observed for TPC (90.4% of the total variance, based on correlations).

According to Figure 4a, there is a positive correlation between total aerobic microbial count and yeast count (r = 0.342). On the other hand, correlation between TNAB and mold content is negative. There are positive correlations between DPPH, FRAP, ABTS, TPC and TFC (Figure 4b). The highest positive correlations were found between ABTS and TFC (r = 0.928), DPPH and TFC (r = 0.920), DPPH and FRAP (r = 0.857), (Figure 4b).

3.7. ANN Model

The calculated optimal neural network model for microbiological parameters, such as the number of aerobic bacteria (TNAB), yeasts and molds count showed adequate generalization capabilities for the modeling of experimental results: The optimum number of neurons in the hidden layer of ANN model was 10 (network MLP 7-10-3) (Table 5), while the r^2 values were: 0.742; 0.659; and 0.792, accordingly, during the training, testing and validation cycles for output variables, for the training, testing and validation cycles for output variables.

Table 5. Artificial neural network model summary (performance and errors), for training, testing and validation cycles.

Network	Performance			Error			Training Algorithm	Error Function	Activation	
	Train.	Test.	Valid.	Train.	Test	Valid.			Hidden	Output
MLP 7-10-3	0.742	0.659	0.792	$5.4 \cdot 10^7$	$2.1 \cdot 10^7$	$8.8 \cdot 10^7$	BFGS 106	SOS	Tanh	Logistic
MLP 7-10-5	0.982	0.956	0.960	38.339	96.566	79.667	BFGS 226	SOS	Tanh	Tanh

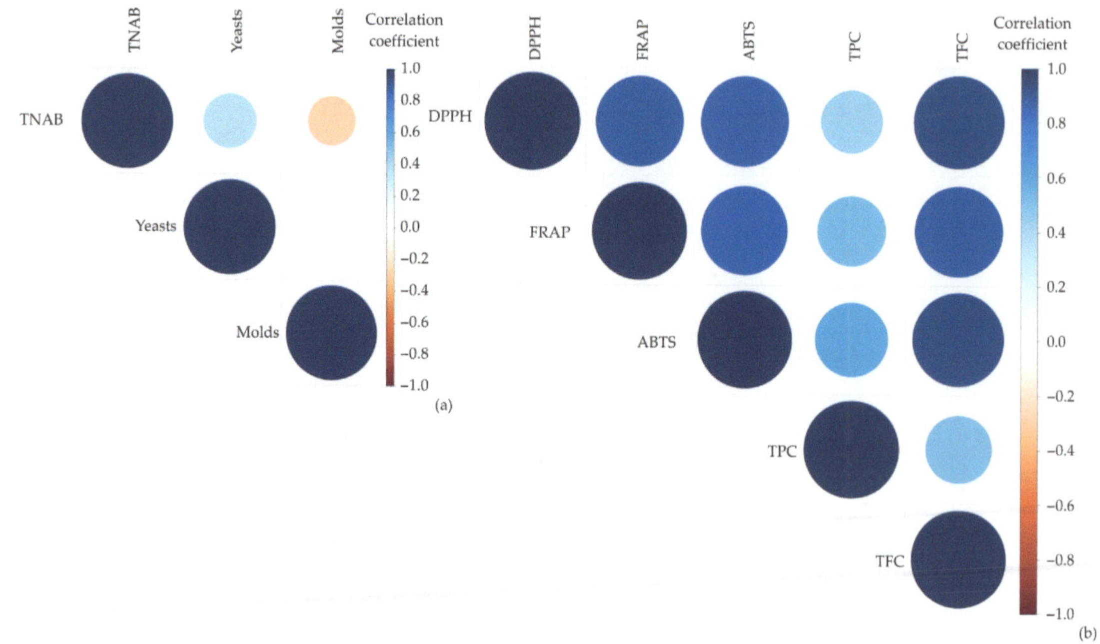

Figure 4. Color correlation diagram between: (**a**) total aerobic microbial count, yeasts, and molds count and (**b**) DPPH, FRAP, ABTS, total phenolic compounds content, and total flavonoid content.

The optimal neural network model for antioxidant parameters, such as DPPH, FRAP, ABTS, TPC and TFC, showed quite good generalization capabilities for the modeling of experimental results: the optimum number of neurons in the hidden layer of ANN model was 10 (network MLP 7-10-5) (Table 5), while the r^2 values were: 0.982; 0.956 and 0.960, accordingly, during the training, testing and validation cycles for output variables, for the training, testing and validation cycles for output variables.

The obtained r^2 values during the testing cycle were: 0.652; 0.799 and 0.780 for TNAB, yeasts and molds count modeling, while the obtained r^2 values for DPPH, FRAP, ABTS, TPC and TFC were: 0.978; 0.967; 0.954; 0.984 and 0.995.

The goodness of fit between experimental results and model-calculated outputs, represented as ANN performance (sum of r^2 between measured and calculated TNAB, Yeasts, Molds, DPPH, FRAP, ABTS, TPC and TFC), observed during training, testing and validation steps, are shown in Table 6.

Table 6. The "goodness of fit" tests for the developed ANN model.

	χ^2	*RMSE*	*MBE*	*MPE*	r^2	**Skew**	**Kurt**	**Mean**	**StDev**	**Var**
TNAB	1.8×10^7	4.2×10^3	-4.0×10^2	0.266	0.685	-1.1×10	1.1×10^2	-4.0×10^2	4.2×10^3	1.8×10^7
Yeasts	1.8×10^6	1.3×10^3	1.3×10^2	0.505	0.702	1.1×10	1.1×10^2	1.3×10^2	1.3×10^3	1.8×10^6
Molds	7.273	2.685	−0.084	0.084	0.805	−0.492	0.172	−0.084	2.695	7.266
DPPH	7.091	2.648	−0.035	117.584	0.976	−0.378	0.155	−0.035	2.663	7.090
FRAP	7.091	2.648	−0.035	290.328	0.962	−0.379	0.155	−0.035	2.663	7.090
ABTS	7.091	2.649	−0.033	65.939	0.942	−0.380	0.155	−0.033	2.663	7.090
TPC	7.130	2.656	−0.055	0.337	0.979	−0.356	0.112	−0.055	2.670	7.127
TFC	7.091	2.648	−0.035	173.988	0.995	−0.378	0.155	−0.035	2.663	7.090

r^2—coefficient of determination, χ^2- reduced chi-square, *MBE*—mean bias error, *RMSE*—root mean square error and *MPE*—mean percentage error.

The ANN model predicted experimental variables (TNAB, yeasts and molds, DPPH, FRAP, ABTS, TPC and TFC) reasonably well for a broad range of the process variables (as seen in Figure 5, where the experimentally measured and ANN model predicted values of TNAB, Yeasts and Molds are presented).

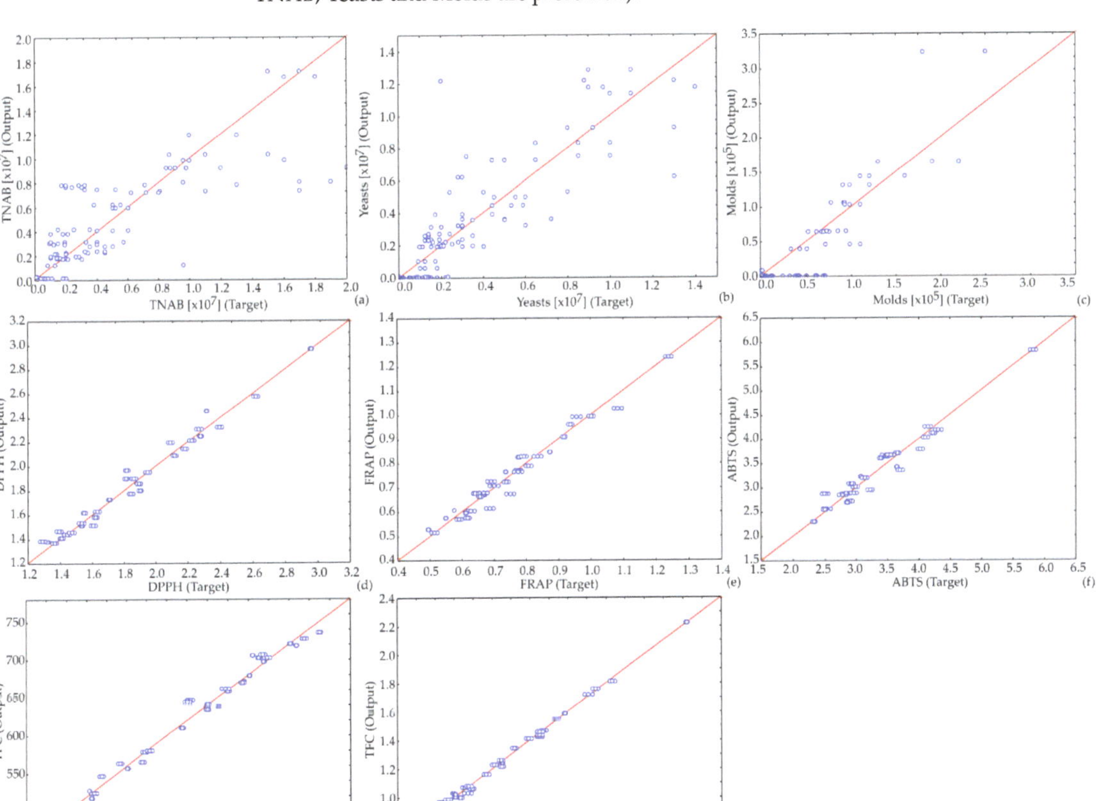

Figure 5. Comparison between experimentally obtained and model-predicted values of TNAB (**a**), Yeasts (**b**), Molds (**c**), DPPH (**d**), FRAP (**e**), ABTS (**f**), TPC (**g**) and TFC (**h**).

The efficiency of the ANN model in modeling TNAB, yeasts and molds is graphically illustrated by scatter plots (Figure 5). In most scatter plots, data are distributed with large dispersion, indicating low prediction accuracy.

The results obtained from the database were fitted to the developed ANN model. Reduced chi-square (χ^2), root mean square error (*RMSE*), mean bias error (*MBE*), mean percentage error (*MPE*), and coefficient of determination (r^2) were calculated statistical parameters applied for the determination of fitting quality between database and the developed model. The particularly high values of r^2 and low values of χ^2, *RMSE*, *MBE* and *MPE* suggested adequate fit (Table 6). The ANN model showed better fit to DPPH, FRAP, ABTS, TPC and TFC data, according to relatively low χ^2, *RMSE*, *MBE*, and *MPE*, as well as the high r^2 values (Table 6).

The ANN models satisfactorily modelled experimental variables for various process variables.

For the ANN model, the model calculated *TNAB*, *Yeasts* and *Molds*, were not too close to the experimental values in most cases in terms of r^2 values, while the sum of squares (SOS) values acquired using the ANN model were of the same order of magnitude as experimental errors for the outputs mentioned in the literature [58–60].

The ANN model predicted experimental variables (DPPH, FRAP, ABTS, TPC and TFC) reasonably well for a broad range of the process variables (as seen in Figure 1, where the experimentally measured and ANN model predicted values of DPPH, FRAP, ABTS, TPC and TFC are presented).

The efficiency of the ANN model in modeling DPPH, FRAP, ABTS, TPC and TFC is graphically illustrated by scatter plots (Figure 5). In most scatter plots, data are distributed with large dispersion, indicating low prediction accuracy.

The developed ANN model for TNAB, yeast and mold modeling consisted of 113 weights-bias coefficients, while the developed ANN model for DPPH, FRAP, ABTS, TPC and TFC modeling consisted of 168 weights-bias coefficients showing the high nonlinearity of the system [61–63].

Table 7 presents the elements of matrix W_1 and vector B_1, while Table 8 presents the elements of matrix W_2 and vector B_2. These were derived during the ANN model development using Equation (1). The goodness of fits between experimental and model-calculated results were shown in Table 3.

Table 7. Elements of matrix W_1 and vector B_1 (presented in the bias column).

	1	2	3	4	5	6	7	8	9	10
Time	−18.323	0.258	−14.374	−22.471	−14.094	−10.103	8.658	3.175	−3.196	−1.702
Treatment(C)	3.538	8.000	−6.033	−12.017	−4.536	−2.369	−1.394	4.059	6.748	−4.653
Treatment(F)	1.833	5.212	1.067	−0.556	−4.657	2.060	−5.744	−2.246	−2.978	−0.275
Treatment(S)	3.561	−7.635	11.311	13.295	10.044	−4.592	−2.174	−0.839	0.714	7.602
Spread(B)	−0.760	1.646	0.453	1.468	−14.473	4.619	5.329	0.063	−3.376	−0.170
Spread(C)	12.301	1.096	5.841	11.777	5.212	−5.080	−7.528	5.368	2.277	−1.214
Spread(U)	−2.669	2.745	0.004	−12.523	10.037	−4.370	−7.103	−4.475	5.645	4.143
Bias	8.892	5.623	6.253	0.725	0.857	−4.826	−9.264	0.861	4.466	2.752

Table 8. Elements of matrix W_2 and vector B_2 (presented in the bias column).

	1	2	3	4	5	6	7	8	9	10	Bias
TNAB	−9.068	3.236	5.956	10.256	3.307	−3.085	5.517	−7.085	−6.790	2.167	5.501
Yeasts	−7.647	−14.211	8.626	−10.375	−3.223	−1.412	−3.777	0.832	−0.424	1.886	−6.377
Molds	0.799	−0.104	−1.480	−3.937	−2.262	−7.018	8.270	2.259	1.484	1.906	−8.626

Table 9 presents the elements of matrix W_1 and vector B_1 (presented in the bias column), and Table 10 presents the elements of matrix W_2 and vector B_2 (bias) for the hidden layer, used for calculation in Equation (1).

Table 9. Elements of matrix W_1 and vector B_1 (presented in the bias column).

	1	2	3	4	5	6	7	8	9	10
Time	7.100	0.257	−17.721	−0.075	8.369	4.650	14.590	−6.199	−0.817	19.669
Treatment(C)	−1.072	1.133	−1.250	0.815	−1.449	0.740	0.173	0.885	1.067	0.115
Treatment(F)	−2.672	−0.859	4.899	0.275	−0.666	−3.011	−7.300	−0.007	−0.370	−2.419
Treatment(S)	2.284	0.237	0.461	−0.230	0.524	2.230	5.004	0.983	−1.201	−1.500
Spread(B)	−0.307	0.350	0.425	−3.578	−0.255	0.318	0.697	4.467	−2.723	−1.019
Spread(C)	−1.359	−0.638	2.212	5.292	−1.463	−0.651	−4.015	−2.191	0.386	−1.378
Spread(U)	0.288	0.525	1.566	−0.860	0.144	0.249	1.175	−0.303	1.641	−1.371
Bias	−1.343	0.394	4.114	0.897	−1.678	−0.084	−2.205	1.913	−0.549	−3.760

The quality of the model fit was tested, and the residual analysis of the developed model was presented in Table 8. The ANN model had an insignificant lack of fit tests, which means the model satisfactorily predicted the pig carcass compositions. A high r^2 is indicative that the variation was accounted for and that the data fitted the proposed model.

Global Sensitivity Analysis—Yoon's Interpretation Method

The effects of analytical method parameters (time, temperature, and applied coating) on the determination of output variables (microbiological profile, antioxidant activity, phenolic, and flavonoid content) was analyzed by employing Yoon's global sensitivity equation corresponding to the weight coefficients of the obtained ANN model [64,65]. Following the global sensitivity analysis of a displayed ANN model, the graphical illustration of Yoon's interpretation method results was shown in Figure 6. Time was the most positively influential parameter influencing yeasts and molds count, with an approximately relative importance of +39.83% and +46.52%, respectively. On the other hand, the time influence on the TNAB count was quite the opposite −16.71%. The most negative effect on yeasts and olds count was observed for spread (c) (−23.52% and −12.21%, accordingly), as shown in Figure 6a–c.

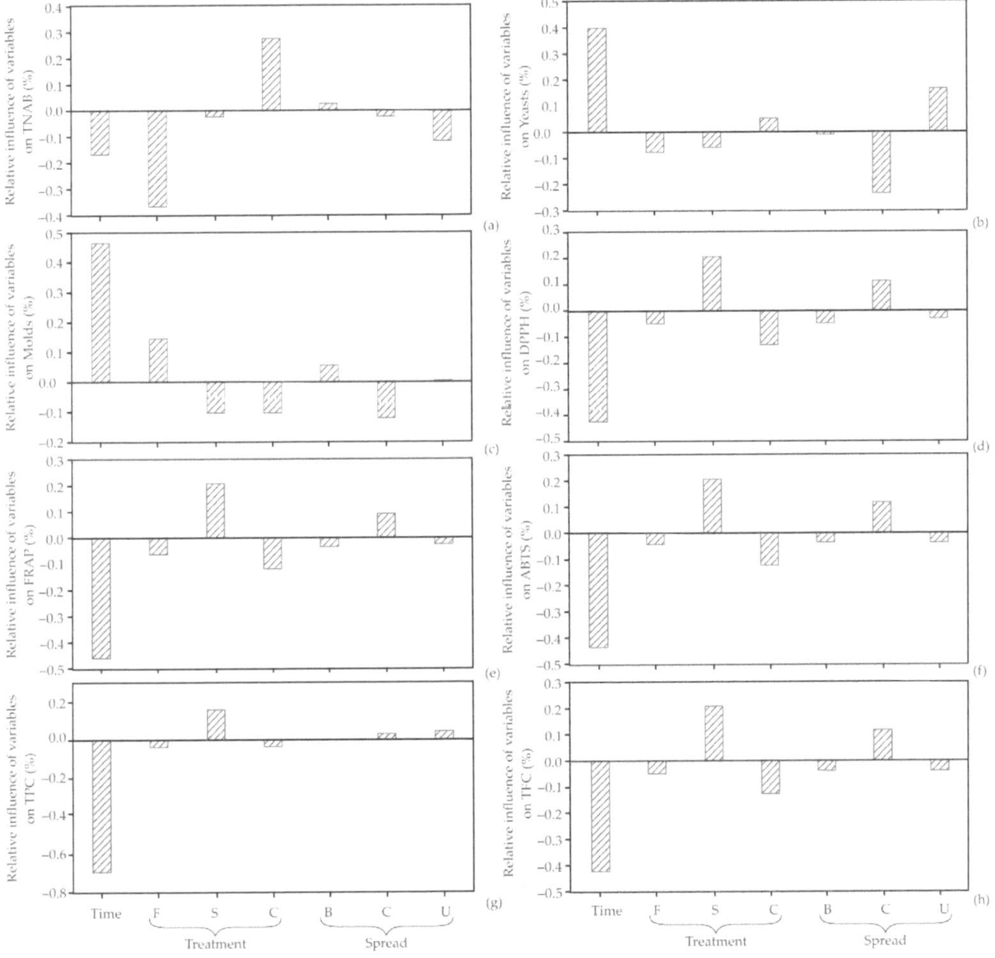

Figure 6. The relative importance of the time, treatment and coatings on: (**a**) TNAB, (**b**) yeasts, (**c**) molds count, (**d**) DPPH, (**e**) FRAP, (**f**) ABTS, (**g**) TPC and (**h**) TFC determined using Yoon's interpretation method. Treatments: C—control (untreated), F—sample stored at refrigerator temperature, S—sample kept under room temperature, Spread type: C—control (uncoated), B—biopolymer coating; U—biopolymer coating with the addition of essential oil.

Table 10. Elements of matrix W_2 and vector B_2 (presented in the bias column).

	1	2	3	4	5	6	7	8	9	10	Bias
DPPH	2.078	−0.827	−0.268	−0.452	−0.234	0.866	−2.572	−0.200	0.377	−0.334	0.782
FRAP	1.303	−0.621	−0.297	−0.479	−0.041	0.526	−1.655	−0.289	0.139	−0.492	0.621
ABTS	1.432	−0.245	−0.241	−0.216	−0.144	0.432	−1.760	−0.131	0.054	−0.310	0.439
TPC	−0.470	0.307	−0.525	−0.656	0.502	0.692	−0.553	−0.966	−0.345	−1.278	0.580
TFC	2.325	−0.439	−0.303	−0.280	−0.321	0.393	−2.533	−0.114	0.113	−0.341	0.588

Furthermore, time was the most negatively influential parameter for antioxidant parameters (DPPH, FRAP, and ABTS), total phenols content and total flavonoids content, with approximate relative importance of −42.24%; −45.89%; −43.56%; −69.45% and −42.29%, respectively. On the other hand, sample stored at refrigerator temperature generated the enhanced antioxidant parameters (DPPH, FRAP and ABTS), and total phenols content and total compared to other treatments, expressing the positive influence of storage at refrigerated temperatures with the following relative influences: +20.79%; +20.90%; +20.70%; +15.96% and +20.85%, respectively, Figure 6d–h.

According to the global sensitivity analysis, it can be concluded that the most influential analytical method parameter was time.

4. Conclusions

The results of this paper prove that pumpkin oil cake can be used for the synthesis of biopolymer coating, but also as a carrier for the active component (+essential oil) in order to obtain active packaging. The positive impact of the pumpkin oil cake-based applied coating based on the sustainability of the treated grapes was confirmed, but the effectiveness of the applied active packaging was even greater, especially in terms of microbiological stability.

The exposed results are inspiring because as the approaches using mathematical modeling microbiological and antioxidant parameters during storage were found to be effective tool. The influence of time, treatment and coatings on microbiological and antioxidant parameters was assessed in the sensitivity analysis. The outcomes of this study reveal that number of aerobic bacteria, number of yeasts and molds, as well as antioxidant parameters (DPPH, FRAP and ABTS), total phenols content and total flavonoids content of biopolymers coating based on pumpkin oil cake, activated with *Mentha piperita*, can be modeled based on time, treatment and coatings, and during storage of packed grapes. This modeling should take place according to reasonably depleted χ^2, *RMSE*, *MBE*, and *MPE*, and the increased r^2 values. It was confirmed that the artificial neural network is appropriate for the modeling of output variables.

Author Contributions: Conceptualization, D.Š., S.P. and A.T.H.; methodology, D.C., N.H. and J.U.; software, L.P.; validation, A.R., T.D. and S.P.; formal analysis, D.Š., D.C. and T.D.; investigation, A.T.H. and A.R.; resources, D.Š.; data curration, B.L. and L.P.; writing—original draft preparation, D.Š., L.P. and B.L.; writing—review and editing, D.Š., B.L. and A.T.H.; visualization, S.P. and N.H.; supervision, A.T.H.; project administration, D.Š.; funding acquisition, L.P. All authors have read and agreed to the published version of the manuscript.

Funding: This research was funded by Serbian program of the Ministry of Education, Science and Technological Development, grant numbers: 451-03-68/2022-14/200051 and 451-03-68/2022-14/200134.

Institutional Review Board Statement: Not applicable.

Informed Consent Statement: Not applicable.

Data Availability Statement: Not applicable.

Conflicts of Interest: The authors declare no conflict of interest.

References

1. Letcher, T.M. (Ed.) *Plastic Waste and Recycling: Environmental Impact, Societal Issues*; Academic Press: Cambridge, MA, USA, 2020; pp. 1–664.
2. Porta, R. The plastics sunset and the bio-plastics sunrise. *Coatings* **2019**, *9*, 526. [CrossRef]
3. Mirpoor, S.F.; Valeria, C.; Giosafatto, L.; Porta, R. Biorefining of seed oil cakes as industrial co-streams for production of innovative bioplastics. A review. *Trends Food Sci. Technol.* **2021**, *109*, 259–270. [CrossRef]
4. Ghanbarzadeh, B.; Almasi, H. Boidegradable polymers. In *Biodegradation-life of Science*; Chamy, R., Rosenkranz, F., Eds.; Tech Publications: Zagreb, Croatia, 2013; pp. 141–186.
5. Tian, H.; Guo, G.; Fu, X.; Yao, Y.; Yuan, L.; Xiang, A. Fabrication, properties and applications of soy-protein-based materials: A review. *Int. J. Biol. Macromol.* **2018**, *120*, 475–490. [CrossRef] [PubMed]
6. Arrutia, F.; Binner, E.; Williams, P.; Waldron, W. Oilseeds beyond oil: Press cakes and meals supplying global protein requirements. *Trends Food Sci. Technol.* **2020**, *100*, 88–102. [CrossRef]
7. USDA Oilseeds. *World Markets and Trade*; Circular: Washington, DC, USA, 2022. Available online: https://downloads.usda.library.cornell.edu/usda-esmis/files/tx31qh68h/hq37wz129/wd377597b/oilseeds.pdf (accessed on 22 November 2022).
8. Fakhouri, F.M.; Martelli, S.M.; Caon, T.; Velasco, J.I.; Mei, L.H.I. Edible films and coatings based on starch/gelatin: Film properties and effect of coatings on quality of refrigerated Red Crimson grapes. *Postharvest Biol. Technol.* **2015**, *109*, 57–64. [CrossRef]
9. Lo'ay, A.A.; Taha, N.A.; EL-Khateeb, Y.A. Storability of 'Thompson Seedless' grapes: Using biopolymer coating chitosan and polyvinyl alcohol blending with salicylic acid and antioxidant enzymes activities during cold storage. *Sci. Hortic.* **2019**, *249*, 314–321. [CrossRef]
10. Wang, H.; Gong, X.; Miao, Y.; Guo, X.; Liu, C.; Fan, Y.-Y.; Zhang, J.; Niu, B.; Li, W. Preparation and characterization of multilayer films composed of chitosan, sodium alginate and carboxymethyl chitosan-ZnO nanoparticles. *Food Chem.* **2019**, *283*, 397–403. [CrossRef]
11. De Souza, W.F.C.; de Lucena, F.A.; da Silva, K.G.; Martins, L.P.; de Castro, R.J.S.; Sato, H.H. Influence of edible coatings composed of alginate, galactomannans, cashew gum, and gelatin on the shelf-life of grape cultivar 'Italia': Physicochemical and bioactive properties. *LWT* **2021**, *152*, 112315. [CrossRef]
12. Bangar, S.P.; Whiteside, W.S.; Ozogul, F.; Dunno, K.D.; Cavender, G.A.; Dawson, P. Development of starch-based films reinforced with cellulosic nanocrystals and essential oil to extend the shelf life of red grapes. *Food Biosci.* **2022**, *47*, 101621. [CrossRef]
13. Zheng, Y.; Jia, X.; Zhao, Z.; Ran, Z.; Du, M.; Ji, H.; Pan, Y.; Li, Z.; Ma, X.; Liu, Y.; et al. Innovative natural antimicrobial natamycin incorporated titanium dioxide (nano-TiO$_2$)/ poly(butylene adipate-co-terephthalate) (PBAT) /poly (lactic acid) (PLA) biodegradable active film (NTP@PLA) and application in grape preservation. *Food Chem.* **2023**, *400*, 134100. [CrossRef]
14. Brandão, R.M.; Batista, L.R.; de Oliveira, J.E.; Barbosa, R.B.; Nelson, D.L.; Cardoso, M.G. In vitro and in vivo efficacy of poly (lactic acid) nanofiber packaging containing essential oils from *Ocimum basilicum* L. and *Ocimum gratissimum* L. against Aspergillus carbonarius and Aspergillus niger in table grapes. *Food Chem.* **2023**, *400*, 134087. [CrossRef] [PubMed]
15. Huang, X.; Hong, M.; Wang, L.; Meng, Q.; Ke, Q.; Kou, X. Bioadhesive and antibacterial edible coating of EGCG-grafted pectin for improving the quality of grapes during storage. *Food Hydrocoll.* **2023**, *136*, 108255. [CrossRef]
16. Peričin, D.; Radulović, L.J.; Mađarev, S.; Dimić, E. Bioprocess for value added products from oil cakes. *J. Edible Oil Ind.* **2007**, *38*, 35–40.
17. Popović, S.; Peričin, D.; Vaštag, Ž.; Popović, L.J.; Lazić, V. Evaluation of edible film-forming ability of pumpkin oil cake; effect of pH and temperature. *Food Hydrocoll.* **2011**, *25*, 470–476. [CrossRef]
18. Hromiš, N.; Popović, S.; Šuput, D.; Bulut, S.; Lazić, V. Composite films based on pumpkin oil cake obtained by different filtration process. *Food Feed Res.* **2019**, *46*, 1–10. [CrossRef]
19. Bulut, S.; Lazić, V.; Popović, S.; Hromiš, N.; Šuput, D. Influence of different concentrations of glycerol and guar xanthan on properties of pumpkin oil cake-zein bi-layer film. *Ratar. I Povrt.* **2017**, *54*, 19–24. [CrossRef]
20. Popović, S.; Lazić, V.; Popović, L.J.; Vaštag, Ž.; Peričin, D. Effect of the addition of pumpkin oil cake to gelatine to produce biodegradable composite films. *Int. J. Food Sci. Technol.* **2010**, *45*, 1184–1190. [CrossRef]
21. Bulut, S.; Popović, S.; Hromiš, N.; Šuput, D.; Lazić, V.; Kocić-Tanackov, S.; Dimić, G.; Kravić, S. Antibacterial activity of biopolymer composite materials obtained from pumpkin oil cake and winter savory or basil essential oil against various pathogenic bacteria. *J. Food Nutr. Res.* **2020**, *59*, 250–258.
22. Bulut, S.; Popović, S.; Hromiš, N.; Šuput, D.; Adamović, D.; Lazić, V. Incorporation of essential oils into pumpkin oil cake-based materials in order to improve their properties and reduce water sensitivity. *Hem. Ind.* **2020**, *74*, 313–325. [CrossRef]
23. Bulut, S.; Lazić, V.; Popović, S.; Šuput, D.; Hromiš, N.; Popović, L.J. Possibility to maintain modified atmosphere in pouches made from biopolymer materials. In Proceedings of the 3rd International Congress, Food Technology, Quality and Safety, Novi Sad, Serbia, 25–27 October 2016; pp. 122–127.
24. Hromiš, N.; Lazić, V.; Popović, S.; Šuput, D.; Bulut, S.; Kravić, S.; Romanić, R. The possible application of edible pumpkin oil cake film as pouches for flaxseed oil protection. *Food Chem.* **2022**, *371*, 131197. [CrossRef]
25. ISO /50:1998; Fruit and Vegetable Products—Determination of Titratable Acidity. ISO: Geneva, Switzerland, 1988.
26. Singleton, V.L.; Rossi, J.A. Colorimetry of total phenolics with phosphomolybdic-phosphotungstic acid reagents. *Am. J. Enol. Vitic.* **1965**, *16*, 144–158.

27. Harborne, J.B. Methods in plant biochemistry. In *Plant Phenolics*; Dey, P.M., Harborne, J.B., Eds.; Academic Press: London, UK, 1989; pp. 283–323.

28. Brand-Williams, W.; Cuvelier, M.E.; Berset, C.L.W.T. Use of a free radical method to evaluate antioxidant activity. *Food Sci. Technol.* **1995**, *28*, 25–30. [CrossRef]

29. Benzie, I.; Strain, J. The ferric reducing ability of plasma (FRAP) as a measure of "antioxidant power: The FRAP assay". *Anal. Biochem.* **1996**, *239*, 70–76. [CrossRef] [PubMed]

30. Re, R.; Pellegrini, N.; Proteggente, A.; Pannala, A.; Yang, M.; Rice-Evans, C. Antioxidant activity applying an improved ABTS radical cation decolorization assay. *Free Radic. Biol. Med.* **1999**, *26*, 1231–1237. [CrossRef]

31. *ISO 6887-1:2017*; Microbiology of the Food Chain—Preparation of Test Samples, Initial Suspension and Decimal Dilutions for Microbiological Examination—Part 1: General Rules for the Preparation of the Initial Suspension and Decimal Dilutions. ISO: Geneva, Switzerland, 2017.

32. *ISO 4833-1:2013*; Microbiology of the Food Chain—Horizontal Method for the Enumeration of Microorganisms—Part 1: Colony Count at 30 Degrees C by the Pour Plate Technique. ISO: Geneva, Switzerland, 2013.

33. *ISO 21527-2:2008*; Microbiology of Food and Animal Feeding Stuffs—Horizontal Method for the Enumeration of Yeasts and Moulds—Part 2: Colony Count Technique in Products with Water Activity Greater Than 0.95. ISO: Geneva, Switzerland, 2008.

34. Rajković, D.; Jeromela, A.M.; Pezo, L.; Lončar, B.; Grahovac, N.; Špika, A.K. Artificial neural network and random forest regression models for modelling fatty acid and tocopherol content in oil of winter rapeseed. *J. Food Compos. Anal.* **2023**, *115*, 105020. [CrossRef]

35. Pezo, L.; Lončar, B.; Šovljanski, O.; Tomić, A.; Travičić, V.; Pezo, M.; Aćimović, M. Agricultural parameters and essential oil content composition prediction of aniseed, based on growing year, locality and fertilization type—An artificial neural network approach. *Life* **2022**, *12*, 1722. [CrossRef] [PubMed]

36. Vojnov, B.; Jaćimović, G.; Šeremešić, S.; Pezo, L.; Lončar, B.; Krstić, Đ.; Vujić, S.; Ćupina, B. The effects of winter cover crops on maize yield and crop performance in semiarid conditions—Artificial neural network approach. *Agronomy* **2022**, *12*, 2670. [CrossRef]

37. Yoon, Y.; Uysal, M. An examination of the effects of motivation and satisfaction on destination loyalty: A structural model. *Tour. Manag.* **2005**, *26*, 45–56. [CrossRef]

38. Puntarić, E.; Pezo, L.; Zgorelec, Ž.; Gunjača, J.; Kučić Grgić, D.; Voća, N. Prediction of the production of separated municipal solid waste by artificial neural networks in Croatia and the European Union. *Sustainability* **2022**, *14*, 10133. [CrossRef]

39. Melo, N.F.C.B.; de Mendonça, B.L.S.; Diniz, K.M.; Leal, C.F.; Canto, D.; Flores, M.A.P.; da Costa Tavares-Filho, J.H.; Galembeck, A.; Stamford, T.L.M.; Stamford-Arnaud, T.M.; et al. Effects of fungal chitosan nanoparticles as eco-friendly edible coatings on the quality of postharvest table grapes. *Postharvest Biol. Technol.* **2018**, *139*, 56–66. [CrossRef]

40. Meighani, H.; Ghasemnezhad, M.; Bakhshi, D. Effect of different coatings on post-harvest quality and bioactive compounds of pomegranate (*Punica granatum* L.) fruits. *J. Food Sci. Technol.* **2014**, *52*, 4507–4514. [CrossRef] [PubMed]

41. Farris, S.; Schaich, K.M.; Liu, L.; Piergiovanni, L.; Yam, K.L. Development of polyion-complex hydrogels as an alternative approach for the production of bio-based polymers for food packaging applications: A review. *Trends Food Sci. Technol.* **2009**, *20*, 316–332. [CrossRef]

42. Zhou, R.; Mo, Y.; Li, Y.F.; Zhao, Y.Y.; Zhang, G.X.; Hu, Y.S. Quality and internal characteristics of huanghua pears (*Pyrus pyrifolia nakai* cv. Huanghua) treated with different kinds of coatings during storage. *Postharvest Biol. Technol.* **2008**, *49*, 171–179. [CrossRef]

43. Zhu, J.J.; Yang, J.J.; Wu, G.J.; Jiang, J.G. Comparative antioxidant, anticancer and antimicrobial activities of essential oils from Semen Platycladi by different extraction methods. *Ind. Crops Prod.* **2020**, *146*, 112206. [CrossRef]

44. Tahir, H.E.; Xiaobo, Z.; Mahunu, G.K.; Arslan, M.; Abdalhai, M.; Zhihua, L. Recent developments in gum edible coating applications for fruits and vegetables preservation: A review. *Carbohydr. Polym.* **2019**, *224*, 115141. [CrossRef]

45. Nair, M.S.; Tomar, M.; Punia, S.; Kukula-Koch, W.; Kumar, M. Enhancing the functionality of chitosan- and alginate-based active edible coatings/films for the preservation of fruits and vegetables: A review. *Int. J. Biol. Macromol.* **2020**, *164*, 304–320. [CrossRef]

46. Kim, I.-H.; Oh, Y.A.; Lee, H.; Song, K.B.; Min, S.C. Grape berry coatings of lemongrass oil-incorporating nanoemulsion. *LWT Food Sci. Technol.* **2014**, *58*, 1–10. [CrossRef]

47. Roux, E.; Billaud, C.; Maraschin, C.; Brun-Merimee, S.; Nicolas, J. Inhibitory effect of unheated and heated D-glucose, D-fructose and L-cysteine solutions and Maillard reaction product model systems on polyphenoloxidase from apple. II. Kinetic study and mechanism of inhibition. *Food Chem.* **2003**, *81*, 51–60. [CrossRef]

48. Eftekhari, A.; Khusro, A.; Ahmadian, E.; Dizaj, S.M.; Hasanzadeh, A.; Cucchiarini, M. Phytochemical and nutra-pharmaceutical attributes of *Mentha* spp.: A comprehensive review. *Arab. J. Chem.* **2021**, *14*, 103106. [CrossRef]

49. Singh, I.R.; Pulikkal, A.K. Preparation, stability and biological activity of essential oil-based nano emulsions: A comprehensive review. *OpenNano* **2022**, *8*, 100066. [CrossRef]

50. Guerra, I.C.D.; de Oliveira, P.D.L.; de Souza Pontes, A.L.; Lúcio, A.S.S.C.; Tavares, J.F.; Barbosa-Filho, J.M.; Madruga, M.S.; de Souza, E.L. Coatings comprising chitosan and *Mentha piperita* L. or Mentha × villosa Huds essential oils to prevent common postharvest mold infections and maintain the quality of cherry tomato fruit. *Int. J. Food Microbiol.* **2015**, *214*, 168–178. [CrossRef] [PubMed]

51. Samber, N.; Khan, A.; Varma, A.; Manzoor, N. Synergistic anti-candidal activity and mode of action of *Mentha piperita* essential oil and its major components. *Pharm. Biol.* **2015**, *53*, 1496–1504. [CrossRef] [PubMed]

52. Riahi, L.; Elferchichi, M.; Ghazghazi, H.; Jebali, J.; Ziadi, S.; Aouadhi, C.; Chograni, H.; Zaouali, Y.; Zoghlami, N.; Mliki, A. Phytochemistry, antioxidant and antimicrobial activities of the essential oils of *Mentha rotundifolia* L. in tunisia. *Ind. Crops Prod.* **2013**, *49*, 883–889. [CrossRef]

53. Singh, R.; Shushni, M.A.; Belkheir, A. Antibacterial and antioxidant activities of *Mentha piperita* L. *Arabian J. Chem.* **2015**, *8*, 322–328. [CrossRef]

54. Shalayel, M.H.F.; Asaad, A.M.; Qureshi, M.A.; Elhussein, A.B. Anti-bacterial activity of peppermint (*Mentha piperita*) extracts against some emerging multi-drug resistant human bacterial pathogens. *J. Herb. Med.* **2017**, *7*, 27–30. [CrossRef]

55. Khusro, A.; Aarti, C.; Agastian, P. Production and purification of anti-tubercular and anticancer protein from *Staphylococcushominis* under mild stress condition of *Mentha piperita* L. *J. Pharm. Biomed. Anal.* **2020**, *182*, 113136. [CrossRef]

56. Ed-dra, A.; Filai, F.; Bou-idra, M.; Zekkori, B.; Bouymajane, A.; Moukrad, N.; Benhallam, F.; Bentayeb, A. Application of Mentha suaveolens essential oil as an antimicrobial agent in fresh turkey sausages. *J. Appl. Biol. Biotechnol.* **2018**, *6*, 7–12.

57. Talebi, F.; Misaghi, A.; Khanjari, A.; Kamkar, A.; Gandomi, H.; Rezaeigolestani, M. Incorporation of spice essential oils into poly-lactic acid film matrix with the aim of extending microbiological and sensorial shelf life of ground beef. *LWT* **2018**, *96*, 482–490. [CrossRef]

58. Doumpos, M.; Zopounidis, C. Preference disaggregation and statistical learning for multicriteria decision support: A review. *Eur. J. Oper. Res.* **2011**, *209*, 203–214. [CrossRef]

59. Kollo, T.; von Rosen, D. Distribution expansions. In *Advanced Multivariate Statistics with Matrices*; Springer: Berlin/Heidelberg, Germany, 2005; pp. 277–354.

60. Ćurčić, L.J.; Lončar, B.; Pezo, L.; Stojić, N.; Prokić, D.; Filipović, V.; Pucarević, M. Chemometric approach to pesticide residue analysis in surface water. *Water* **2022**, *14*, 4089. [CrossRef]

61. Chattopadhyay, P.B.; Rangarajan, R. Application of ANN in sketching spatial nonlinearity of unconfined aquifer in agricultural basin. *Agric. Water Manag.* **2014**, *133*, 81–91. [CrossRef]

62. Montgomery, R. Canonical formulations of a classical particle in a Yang-Mills field and Wong's equations. *Lett. Math. Phys.* **1984**, *8*, 59–67. [CrossRef]

63. Demir, H.; Demir, H.; Lončar, B.; Nićetin, M.; Pezo, L.; Yilmaz, F. Artificial neural network and kinetic modeling of capers during dehydration and rehydration processes. *J. Food Process Eng.* 2022, *ahead of print*. [CrossRef]

64. Filipović, V.; Ugrenović, V.; Popović, V.; Dimitrijević, S.; Popović, S.; Aćimović, M.; Dragumilo, A.; Pezo, L. Productivity and flower quality of different pot marigold (*Calendula officinalis* L.) varieties on the compost produced from medicinal plant waste. *Ind. Crop. Prod.* **2023**, *192*, 116093. [CrossRef]

65. Erceg, T.; Šovljanski, O.; Stupar, A.; Ugarković, J.; Aćimović, M.; Pezo, L.; Tomić, A.; Todosijević, M. A comprehensive approach to chitosan-gelatine edible coating with β-cyclodextrin/lemongrass essential oil inclusion complex —Characterization and food application. *Int. J. Biol. Macromol.* **2022**, *228*, 400–410. [CrossRef]

Article

Antifungal Properties of Ozone Treatment against *P. citrinum* and *R. stolonifera* in Fresh-Peeled Garlic

Hong Liu [1], Xiangfeng Fan [1], Lin Cao [1], Xiaomin Wang [1], Xiaocui Liu [1], Jie Huang [1], Yage Xing [1,*] and Wanmin Luo [1,2]

[1] Food Microbiology Key Laboratory of Sichuan Province, College of Food and Bioengineering, Xihua University, Chengdu 610039, China; gyliuhong@126.com (H.L.); fanxiangfeng828@163.com (X.F.); cao552354250@163.com (L.C.); xiaominwang99@126.com (X.W.); xiaocuiliu777@126.com (X.L.); homer1024@163.com (J.H.); lwm17746776427@outlook.com (W.L.)

[2] Technology Research Center of Food Non-Thermal Processing, Yibin Xihua University Research Institute, Yibin 644004, China

*   Correspondence: xingyage1@163.com

Abstract: Garlic is susceptible to decay and presents a potential vehicle for foodborne disease transmission. Ozone treatment has proven to be an effective and innoxious method to provide bacterial resistance in this globally popular pungent vegetable. This study was conducted to evaluate the effects of different ozone treatments (differing in terms of concentration, treatment time and temperature) against *Penicillium citrinum* and *Rhizopus stolonifer* in spoiled fresh-peeled garlic. The results have shown that the most inhibitory conditions for in vitro treatments were achieved with an ozone concentration, treatment time and temperature of 6 ppm, for 20 min at 20 °C, respectively, on the *P. citrinum* and 8 ppm, for 20 min at 25 °C, respectively, on the *R. stolonifer*. The optimum in vivo ozone treatment conditions for fresh-peeled garlic inoculated with the same two kinds of spoilage molds remained the same for both, consisting of an ozone concentration of 6 ppm, a time of 15 min and a temperature of 20 °C. Following these ozone treatments, the total number of colonies of yeast and mold, as well as the incidence, lesion diameter and depth of spoilage in the fresh-peeled garlic was significantly reduced during storage, with improved bactericidal inhibition effects. In conclusion, this study showed that ozone treatment effectively inhibits the growth of spoilage molds, destroys cell structures, and affects the metabolic and physiological processes of *P. citrinum* and *R. stolonifer*. Thus, it provides a protective shield and extends the shelf life of fresh-peeled garlic.

Keywords: fresh-peeled garlic; ozone; antibacterial activity; mechanism

Citation: Liu, H.; Fan, X.; Cao, L.; Wang, X.; Liu, X.; Huang, J.; Xing, Y.; Luo, W. Antifungal Properties of Ozone Treatment against *P. citrinum* and *R. stolonifera* in Fresh-Peeled Garlic. *Coatings* **2023**, *13*, 1931. https://doi.org/10.3390/coatings13111931

Academic Editor: Simona Liliana Iconaru

Received: 11 October 2023
Revised: 6 November 2023
Accepted: 9 November 2023
Published: 11 November 2023

1. Introduction

Garlic (*Allium sativum* L.) stands as not only one of the most globally consumed vegetables, but also an enduring ingredient in traditional folk medicines [1]. Fresh garlic is rich in nutrients, including organosulfur compounds [1–5], amino acids [6], polyphenols [7] and vitamins [8]. Thus, fresh and fresh-peeled garlic has not only great commercial value but also significant research value. Fresh-peeled garlic can be described as a minimally processed vegetable (MPV). However, its vulnerability to decay, mildew and germination during storage due to the loss of its protective epidermis diminishes its commercial competitiveness significantly. *Penicillium citrinum* and *Rhizopus stolonifer* are the main infesting microorganisms. Among them, *Penicillium citrinum* colonies are mainly yellow-green in color with white edges. When it infests the garlic, it covers the surface with green molds, causing notable changes, such as reduced firmness of the garlic and loss of food value [9]. Spores of *Rhizopus stolonifer* are ubiquitous in the atmosphere and spread rapidly from infected fruits and vegetables to neighboring produce, particularly when the temperature exceeds 5 °C. These spores, susceptible to infestation of surrounding fruits and vegetables, contribute to spoilage during both pre-harvest and storage stages [10].

Fresh produce and MPVs hold a prominent position in global consumption patterns, serving as vital natural sources of essential nutrients [11]. The fresh-cut produce sector has been experiencing rapid growth, evolving into a multibillion-dollar industry [12] in which fresh-peeled garlic has huge commercial potential. Unfortunately, fresh produce is easily polluted through water, air, soil, insect vectors, processing equipment and even improper handling by employees in the food chain [13]. Multiple studies have highlighted that fresh-cut produce may carry a higher risk of foodborne illness compared to their unprocessed counterparts [14]. Herman [15] et al. analyzed the data on outbreaks of foodborne illness associated with fresh and fresh-cut vegetables between 1973 and 2012 and found that 73.6% were linked to fresh-cut leafy salad. Sirsat [16] et al. found that bagged fresh-cut romaine lettuce has equal if not higher levels of bacteria and spoilage microorganisms to that of whole lettuce. Processing (cutting, peeling, and shredding) can destroy cell surfaces, thereby exposing cytoplasm which provides an excellent source of nutrients for microorganisms [17]. The fresh-cut industry typically uses hypochlorous acid and hypochlorite for disinfection [13]; however, the reaction of these chemicals with organic molecules produces unhealthy by-products, including carcinogenic and mutagenic chlorinated compounds such as chloroform and other trihalomethanes, chloramines and haloacetic acids [18,19]. Therefore, the development of an efficient, yet safe and healthy, preservation method is imperative to prevent decay in fresh-peeled garlic.

Ozone (O_3), also known as triatomic oxygen, is naturally produced by oxygen via lightning or the action of ultraviolet (UV) light [20]. It is a strong oxidizing agent and powerful broad-spectrum antimicrobial agent [21] capable of combating bacteria, fungi, viruses and protozoa, as well as bacterial and fungal spores [22]. Molecular ozone or decomposed ozone products can efficiently and swiftly eliminate microorganisms, leaving no residue [23]. Gibson [24] et al. found that ozone is an effective sanitizer when applied to microorganisms on fresh produce. Zou [23] et al. reported that 10 to 15 min ozone treatments retarded mold spots and leakage in commercially packaged and processed produce. In an earlier study, Khadre [22] et al. reported that 0.3 ppm ozone treatments for 12 days at a storage temperature of 2 °C successfully prevented mold development in blackberries, without causing damage to the ozone-treated fruits. Moreover, ozone treatment was granted GRAS status (Generally Recognized As Safe) by the United States Food and Drug Administration (FDA) in 2001 [25]. Interest in ozone has steadily grown in the fruit and vegetable industry [26] and it is now widely applied in the treatment of fresh and dried fruit and vegetables, as well as frozen and processed products. Compared with other garlic preservation techniques, such as irradiation, air conditioning, and chemical preservatives, ozone sterilization has the advantages of high efficiency, affordability and high safety [23]. Among currently available technologies, ozone treatment provides a good alternative to chlorine and has been shown to result in a substantial reduction in microbiomes on various foods [27], such as the microflora and foodborne pathogens present on button mushrooms [28], microflora of dried figs [29], *Botrytis cinerea* on strawberries [30], green mold on tangerines [31], *Rhizopus stolonifer* on table grapes [32], molds on peaches and table grapes [33] and anthracnose rot on tomatoes [34]. However, no notable studies addressing the decontamination of fresh-peeled garlic using ozone have yet been reported.

After preliminary studies, we isolated and identified the molds responsible for spoilage in fresh-peeled garlic, among which *Penicillium citrinum* and *Rhizopus stolonifer* were subsequently determined to be the primary strains [35]. The main objective of this study was to investigate the antibacterial properties and mechanisms of ozone treatment specifically against *P. citrinum* and *R. stolonifer* isolated from fresh-peeled garlic. The exosmosis ratio (ER), protein dissolution rate (PDR), mycelial growth inhibition rate (MGIR) and lethality rate (LR) of *P. citrinum* and *R. stolonifer* were determined in vitro, while the total plate count (TPC), yeast and mold (YAM) counts and disease incidence rate (DIR), as well as the depth and diameter of the disease spot, were determined in vivo.

2. Materials and Methods

2.1. Materials

P. citrinum and *R. stolonifer* (SICC3.977 and SICC3.978, stored in the Southwest Center of Industrial Culture Collection in China, and isolated from spoiled and moldy fresh-peeled garlic) were used throughout the study. The isolates were maintained on potato dextrose agar (PDA, Beijing Aobox Biotechnology Co., Ltd., Beijing, China) at 4 °C until needed.

The *P. citrinum* and *R. stolonifer* samples were activated in potato dextrose broth (PDB, Beijing Aobox Biotechnology Co., Ltd., Beijing, China) and placed in a shaker at 28 °C for 72 h, according to a previously reported method [33,36,37], with some modifications. The *P. citrinum* and *R. stolonifer* were subsequently cultured on PDA at 28 °C for 72 h in an incubator (SKP-02, Huangshi Hengfeng Medical Apparatus and Instruments Co., Ltd., Huangshi, China). Finally, the *P. citrinum* and *R. stolonifer* conidia were collected from the PDA by adding 10 mL sterile water to the Petri dish, and the conidial suspensions were adjusted to a concentration of 10^6 conidia mL^{-1}.

2.2. Ozone Exposure

Ozone was produced using an ozone generator (YS-MJCB-S17, Hangzhou Yishi Technology Co., Ltd., Hangzhou, China), with an oxygen flow of 2 $L{\cdot}min^{-1}$ from the Mark 5 Plus 95 Concentrator Oxygen Concentrator (Nidek Medical Products Inc., Birmingham, AL, USA).

2.3. Sample Preparation and Ozone Treatment

To determine the ER and PDR of *P. citrinum* and *R. stolonifer*, their spores were collected from a PDA medium via an inoculation loop and transferred to the PDB medium, according to Diao [38] et al. and Yin [39]. The cell concentrations in the PDB medium were adjusted to approximately 10^6 conidia mL^{-1}, after which 20 mL of the diluted PDB medium was pipetted into empty Petri dishes (five dishes per treatment). For the determination of MGIR, mycelial plugs (5 mm in diameter) obtained from the periphery of actively growing three-day-old *P. citrinum* and *R. stolonifer* cultures were introduced mycelium-down at the center of Petri dishes containing 20 mL PDA medium (5 dishes per treatment). Furthermore, the LR of the two molds was determined according to the method described by Xu [40] et al. and Xing [41] et al., with slight modifications. Samples (1 mL) of each of the diluted spore suspensions were inoculated onto Petri dishes (five dishes per treatment) containing 20 mL PDA medium, after which the dish lids were removed to allow free air flow. The dishes were then placed in an ozone treatment room and the following three investigations were undertaken: (1) After 15 min ozone treatment at a temperature of 25 °C and a humidity level of 90%, the effects of different concentrations of ozone (0 (control), 2, 4, 6, 8, 10 ppm) on the two molds were explored; (2) When the concentration of ozone was 6 ppm, the temperature was 25 °C and the humidity was 90%, the effects of different treatment times (0 (control), 5, 10, 15, 20, 25, 30 min) on the two molds were explored; (3) When the concentration of ozone was 6 ppm, the treatment time was 25 min and the humidity was 90%, the effects of different temperatures (20, 25, 30 °C) on the two molds were explored.

The design and setup of the ozone fumigation system were established as previously described by Palou [33] et al., with some modifications. Pest-free fresh-peeled garlic of the same size was washed, dried and then divided into three groups of 2 kg each. A wound (diameter 2 mm, depth 5 mm) was introduced into the middle of each garlic using a sterile inoculation needle, and the wound surfaces were disinfected with 75% alcohol followed by UV irradiation for 30 min. Thereafter, 5 µL of a suspension of 1×10^6 colony forming unit (CFU)/mL spore concentration of *P. citrinum* and *R. stolonifer* were injected into each group of garlic wounds. The groups were then placed in the ozone treatment room and the following investigations were undertaken: In Group 1, after ozone treatment of 15 min at a temperature of 25 °C and a humidity level of 90%, the effects of different concentrations of ozone (0 (control), 1, 2, 3, 4, 5, 6 ppm) on the two molds were explored. In Group 2, when the concentration of ozone was 6 ppm, the temperature was 25 °C and the humidity was 90%, the effects of different treatment times (0 (control), 3, 6, 9, 12, 15, 18 min) on the two molds were studied. In Group 3, when the concentration of ozone was 6 ppm, the treatment time was 15 min and the humidity was 90%, the effects of different temperatures (15, 20, 25 °C) on the two molds were investigated. At the end of each exposure, the dishes were transferred for incubation at 28 °C, and the TPC, YAM, DIR, diameter and depth of the disease spots were determined. Each different treatment was evaluated at intervals of 2 days over 10 days of ambient storage.

2.4. Exosmosis Ratio (ER) and Protein Dissolution Rate (PDR) Determination

The ER of the fungi cells was measured for each ozone treatment group according to the method described by Xing [42] et al. Cells were cultured for 72 h in PDB medium, and then centrifuged at 4 °C and 4000 r/min for 10 min. The obtained precipitate was washed with deionized water 1–2 times, then weighed and divided into two groups, Group A and Group B, which were placed into small beakers and immersed with 20 mL of deionized water. The samples of Group A were placed in a vacuum oven in which vacuum gas drainage was repeated 3–4 times. The pressure was controlled at 450–500 mm Hg and the vacuum infiltration was restored to normal pressure after 30 min. The treated samples were then shaken at 28 °C for another 2–3 h. The samples of Group B were placed in a boiling water bath (100 °C) for 15 min to completely illuminate the electrolytes in the tissue. Finally, conductivity was measured in both Group A and Group B at a constant temperature of 25 °C using a conductivity meter (DBS-11A, Shanghai Yidian Scientific Instrument Co., Ltd., Shanghai, China). The ER was calculated as follows: ER (%) = (the conductivity of Group A/the conductivity of Group B) × 100%.

PDR was measured with reference to the method reported by Yin [39] et al. The bacterial suspension before and after ozone treatment was placed in a refrigerated centrifuge and centrifuged at 4000 r/min for 10 min at 4 °C. The supernatant was subsequently taken to measure the absorbance at 280 nm. Sterile water was used as a reference solution. The PDR was ultimately expressed as absorbance.

2.5. Mycelial Growth Inhibition Rate (MGIR) and Lethality Rate (LR) Determination

At the end of each exposure, the dishes were transferred for incubation at 28 °C for up to 72 h, after which the colony diameter of each treatment was measured with a Vernier caliper, recorded, and the MGIR was estimated by expressing the mean of the diameter at each exposure to ozone concentration. Finally, MGIR was estimated using the following equation:

$$\text{MGIR}(\%) = \frac{CD_0(\text{cm}) - CD_t(\text{cm})}{CD_0(\text{cm})} \times 100\%$$

Colony diameter = total colony diameter (cm)—0.5 cm, where MGIR is the mycelial growth inhibition rate, CD_0 is the colony diameter of control and CD_t is the colony diameter of the ozone treatment. At the end of each exposure, the dishes were transferred for incubation at 28 °C for up to 72 h, after which the total colonies of *P. citrinum* and *R. stolonifer* in the ozone treatment groups and the blank group were counted separately as follows:

LR (%) = [(total colonies of blank group—total colonies of ozone treatment)/total colonies of blank group] × 100%

2.6. Mycelial Morphology

Appropriate ozone treatment was adopted based on the above experimental results. One mL of diluted spore suspension was inoculated onto Petri dishes (five dishes per treatment) containing 20 mL PDA medium. Dish lids were removed to allow air flow and the dishes were placed in an ozone treatment library. At the end of exposure, the dishes were transferred for incubation at 28 °C for 72 h, after which the mycelial morphology of each replication dish was assessed via microscopic observation (10 × 40).

2.7. Microbiological Counts

The TPC and YAM counts in the fresh-peeled garlic were determined using the method described by Salve [43] et al. and Martiñon [44] et al. The estimations of TPC on plate count agar and of the YAM on potato dextrose agar were carried out using the spread plate method. TPC was counted after incubation at 30 °C for 24 h, while yeasts were counted at 30 °C for 5 d.

2.8. Disease Incidence Rate (DIR) and Disease Spot Diameter and Depth

Disease incidence in the fungi cells was measured according to the method described by Ong [45] et al., with some modifications. The effect of ozone on the disease incidence in the garlic from each treatment was evaluated at two-day intervals during 10 days of ambient storage by weighing the diseased garlic against the weight of total garlic as follows:

$$\text{Disease incidence rate (\%)} = [\text{weight of fresh-peeled garlic (g)} / \text{weight of total fresh-peeled garlic (g)}] \times 100\%$$

Ten garlic samples were taken each time for measurement. A Vernier caliper was used to measure the diameter of the disease spots via the cross method in the direction of the wound. The average value was calculated as the final result. Each treatment was evaluated at 2-day intervals throughout a 10-day period of ambient storage.

2.9. Statistical Analysis

The tests in this investigation were carried out in triplicate. The test results were analyzed using SPSS 20.0 software (SPSS Inc., Chicago, IL, USA) and expressed as mean ± standard deviations. One-way analysis of variance (ANOVA) was performed followed by the Student–Newman–Keuls test to determine the significant difference ($p < 0.05$) between the various means of treatments.

3. Results and Discussion

3.1. Results

3.1.1. Changes in ER and PDR

As shown in Figure 1A, a statistically significant increase in ER ($p < 0.05$) was observed when the ozone concentration increased from 0 ppm to 6 ppm. When the ozone concentration increased above 6 ppm, the ER of *P. citrinum* stabilized, whereas the ER of *R. stolonifer* continued to increase significantly until the ozone concentration reached 8 ppm. Similarly, when the treatment time lengthened from 5 min to 20 min, the ER of both *P. citrinum* and *R. stolonifer* increased significantly; however, the difference was no longer statistically significant after 20 min (Figure 1B). The highest ER observed in the *P. citrinum* and *R. stolonifer* were 88.21% and 90.56%, respectively. As shown in Figure 1C, the change in treatment temperature had little effect on the ER of *P. citrinum* and *R. stolonifer*. With the increase in temperature, the ER of *P. citrinum* showed neither a steady, increasing, nor decreasing trend, mainly because the decomposition rate increased but the inhibition rate of mold decreased, and the conductivity of the bacterial suspension also gradually declined. Thus, the increase in the permeability of the mold cell membranes of both *P. citrinum* and *R. stolonifer* indicated the destruction of the two fungi by the ozone treatment. However, trends differed according to the different ozone concentrations, treatment times and temperatures.

When the ozone concentration increased from 0 ppm to 10 ppm, the PDR of *P. citrinum* and *R. stolonifer* were both found to increase significantly, attaining a maximum of 0.234 and 0.394, respectively, at 10 ppm (Figure 2A). In Figure 2B it is evident that as the ozone treatment time increased from 0 min to 25 min the PDR of both molds increased, stabilizing for *P. citrinum* thereafter. Furthermore, in Figure 2C, the PDR of *P. citrinum* dropped significantly with rising temperatures, while that of the *R. stolonifer* did not show a trend change. Overall, as can be seen in Figure 2, the PDR of *P. citrinum* was significantly higher than that of the *R. stolonifer* after the same ozone treatments and, thus, ozone significantly increased the absorbance value of the mold suspension and increased the protein dissolution rate.

3.1.2. Changes in MGIR and LR

Ozone was found to have a significant inhibitory effect on the mycelial growth of both *P. citrinum* and *R. stolonifer* (Figure 3). In Figure 3A,B, it is evident that, at low ozone concentration or treatment time, the inhibitory effects on the same mold were different. In addition, the inhibitory effects of the same ozone concentration or time on *P. citrinum* were always higher than those on the *R. stolonifer*. As shown in Figure 3A, when the concentration was 6 ppm, the MGIR of *P. citrinum* reached 46.04% while that of the *R. stolonifer* reached only 18.14%. Nevertheless, when the concentration was 10 ppm, the inhibition rate of *R. stolonifer* reached 35.32%. As shown in Figure 3B, when the treatment time was 5 min, the MGIR of *P. citrinum* and *R. stolonifer* were 34.04% and 9.43%, respectively. The MGIR of *P. citrinum* reached its maximum value of 46.01% at 20 min, while that of *R. stolonifer* reached its maximum of 31.10% at 25 min. The MGIR of both fungi did not, however, change significantly with the increase in treatment time after 25 min. As can be seen in Figure 3C, the inhibition rate of *P. citrinum* and *R. stolonifer* did not change significantly as the temperature changed. At 20 °C, the MGIR of *P. citrinum* reached its maximum at 46.32%; however, the inhibition rate of *R. stolonifer* reached its maximum of 31.56% at 30 °C. These results, therefore, indicate that ozone treatment can significantly inhibit the growth of *P. citrinum* and *R. stolonifer*.

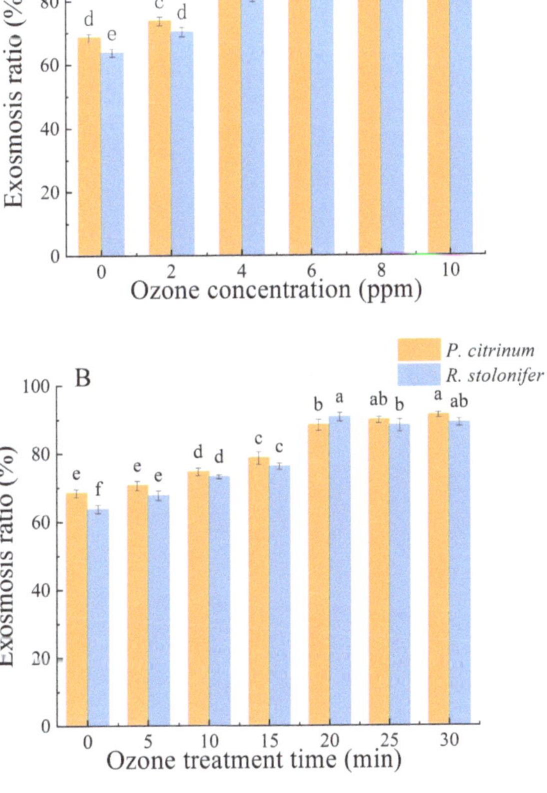

Figure 1. *Cont.*

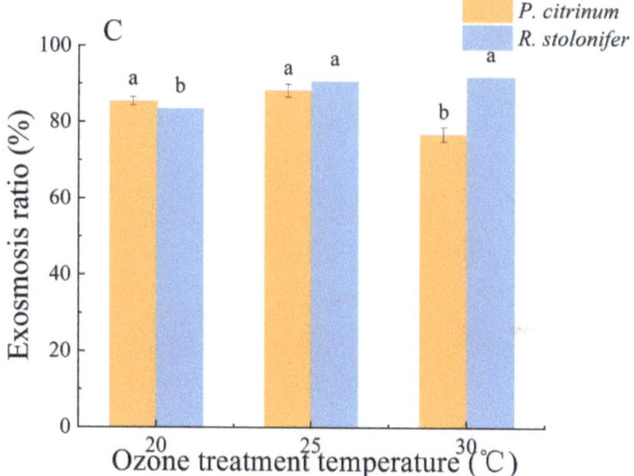

Figure 1. Effects of ozone treatments on the ER of *P. citrinum* and *R. stolonifer.* (**A**): Ozone concentration (ppm); (**B**): ozone treatment time (min); (**C**): ozone treatment temperature (°C). For *P. citrinum* and *R. stolonifer*, mean bars with different letters (a–d), (a–e) differed significantly ($p < 0.05$) for the same time and temperature at different ozone concentrations in (**A**) Mean bars with different letters (a–e), (a–f) within the same ozone concentration and temperature at different times differed significantly ($p < 0.05$) in (**B**). Mean bars with different letters (a,b), (a,b) differ significantly ($p < 0.05$) for the same ozone concentration and time at different temperatures in (**C**).

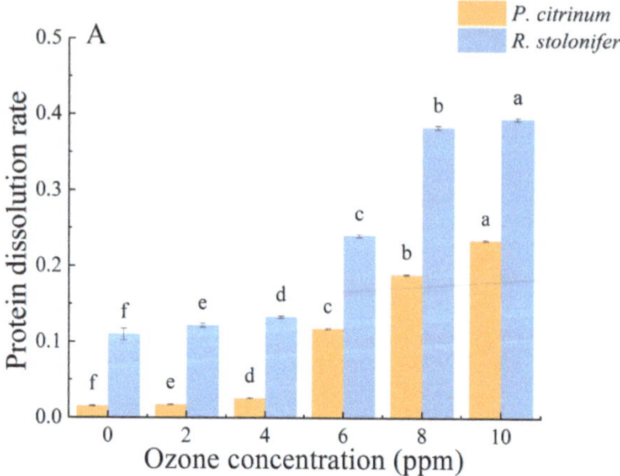

Figure 2. *Cont.*

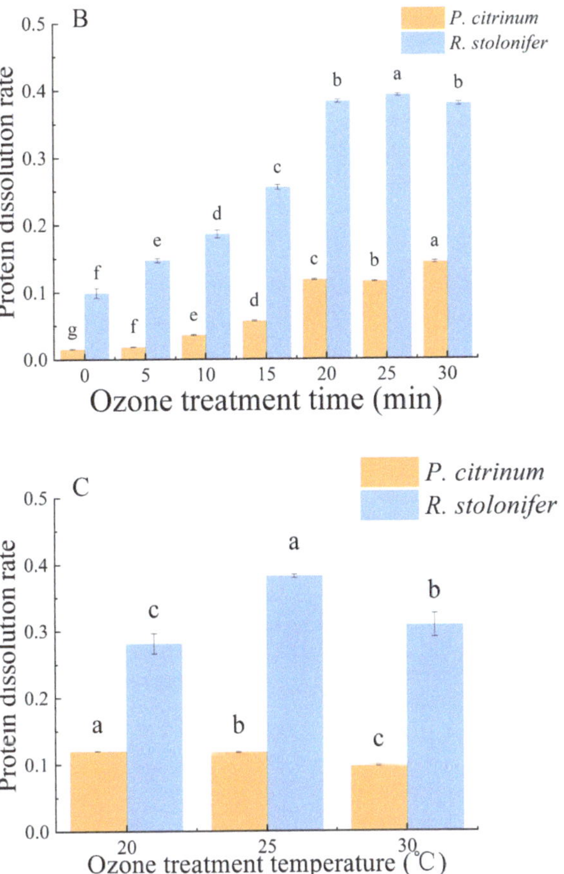

Figure 2. Effects of ozone treatments on the PDR of *P. citrinum* and *R. stolonifer*. (**A**): Ozone concentration (ppm); (**B**): ozone treatment time (min); (**C**): ozone treatment temperature (°C). For *P. citrinum* and R. stolonifer, mean bars with different letters (a–f), (a–f) differed significantly ($p < 0.05$) for the same time and temperature at different ozone concentrations in (**A**). Mean bars with different letters (a–g), (a–f) within the same ozone concentration and temperature at different times differed significantly($p < 0.05$) in (**B**). Mean bars with different letters (a–c), (a–c) differ significantly ($p < 0.05$) for the same ozone concentration and time at different temperatures in (**C**).

In order to understand the effect of ozone treatment on fungal morphology more intuitively, both *P. citrinum* and *R. stolonifer* were subjected to a 6 ppm exposure for 15 min. The spore stalks in the control group of *P. citrinum* were larger and more numerous than those of the treatment group (Figure 4b). Moreover, the conidiophore stem surface of the control group of *R. stolonifer* (Figure 4c) was smoother than those of the treatment groups (Figure 4d). These results showed that ozone treatment can effectively destroy the morphology of *P. citrinum* and *R. stolonifer*, reducing their size and, ultimately, killing the fungi.

The LR is one of the most direct indicators of the effect of ozone on mold. Here, LR was determined by measuring the total colonies. As shown in Figure 5A,B, the LR of the two molds continued to increase directly after ozone treatment; however, at 6 ppm exposure, the LR of *P. citrinum* was 55.45% and tending towards stability, while at 8 ppm exposure, the LR of *R. stolonifer* was 46.57% and also becoming stable. Before reaching stability, the LR of each mold was significantly different, possibly indicating that *P. citrinum* and *R. stolonifer* are tolerant of high concentrations of ozone. Furthermore, as the ozone treatment time extended, the LR of each mold continued to increase significantly up until 20 min of treatment. The LR of *P. citrinum* reached its highest peak of 55.45% at 20 min, while *R. stolonifer* reached its highest peak of 48.93% at 30 min. The effects of ozone treatment temperature on the two molds are shown in Figure 5C. The LR of *P. citrinum* and *R. stolonifer* reached their maximum of 55.45% and 46.57%, respectively, at 25 °C, and thereafter decreased as the treatment temperature continued to rise. This phenomenon may have been due to the low stability of ozone, as it decomposes into oxygen more rapidly at elevated temperatures, thereby reducing the bactericidal effectiveness.

3.1.3. Effects of Ozone Treatment on the Microbiological Counts of *P. citrinum* and *R. stolonifer*

As can be seen in Figure 6, the TPC and YAM counts in the fresh-peeled garlic inoculated with *P. citrinum* and *R. stolonifer* increased during their entire storage period, even though ozone is widely accepted to be an antimicrobial agent. During the storage process, the TPC in the treatment groups was significantly lower than that in the blank group ($p < 0.05$). In the *P. citrinum* (Figure 6(A1)), TPC dropped sharply when the ozone concentration increased from 2 ppm to 3 ppm and, in the fresh-peeled garlic treated at 1 ppm exposure, the TPC was not significantly different compared to that of the blank group by the 8th day of storage. Considering the observed variations, the most effective ozone treatment concentration for fresh-peeled garlic inoculated with *P. citrinum* was considered to be 6 ppm. TPC in the *R. stolonifera* (Figure 6(A2)) was found to drop sharply during storage when the ozone concentration increased from 3 ppm to 4 ppm. However, on the 2nd and 8th days, no significant difference was observed between these samples and the fresh-peeled garlic treated at 5 ppm and 6 ppm. The optimal concentration of ozone treatment for the fresh-peeled garlic inoculated by *R. stolonifer* was determined to be 5 ppm. Figure 6B shows variations in TPC at different times but in samples subjected to the same concentration of ozone treatment. In the *P. citrinum*, there was no significant difference between the treatment groups (3 min and 6 min) and the blank group. However, when the treatment time was extended to 15 min, the TPC was significantly lower than that at 6 min treatment ($p < 0.05$). Therefore, a 15-min treatment proved to be more effective in inhibiting *P. citrinum*. In *R. stolonifer*, when the treatment time was increased to 12 min, the TPC was significantly lower. After six days, there was no significant difference between the treatment groups at 15 min and 18 min. After 10 days of storage the TPC of the 15 min treatment group was 3.91 lg CFU/g, which was much less than that of the blank group, at 6.12 lg CFU/g. As shown in Figure 6(C1,C2), under the same storage time, there was no significant difference in the TPC on the surface of the garlic treated at 15 °C, 20 °C and 25 °C ($p > 0.05$).

As shown in Figure 7, the YAM trends were similar to those of the TCP. In the *P. citrinum* (Figure 7(A1)), the YAM in the 6 ppm ozone treatment group was 0.09 lg CFU/g after 2 days of storage, while the YAM in the blank group was much higher, at 2.97 lg CFU/g. Moreover, after 10 days of storage, the 6 ppm treatment group exhibited the lowest YAM count, indicating the best ozone treatment effect. In the *R. stolonifer* (Figure 7(A2)), when the ozone concentrate was increased from 3 ppm to 4 ppm, the YAM of the fresh-peeled garlic plummeted, consistently reaching its minimum during storage after 6 ppm ozone treatment. In addition, the YAM decreased continuously (Figure 7B) as the treatment time continued to increase. The YAM of the fresh-peeled garlic inoculated with *P. citrinum* (Figure 7(B1)) was, however, significantly different in the treatment and blank groups after 9 min. Overall, the optimal ozone treatment time was determined to be 15 min. In the *R. stolonifer* (Figure 7(B2)), when the ozone treatment time was increased from 15 min to 18 min, the counts of YAM were not significantly different ($p > 0.05$) at the same amount of storage time; however, they were significantly lower than the blank group. In conclusion, the most effective ozone treatment time was found to be 15 min for both the fungi in this study. Moreover, the change in the ozone treatment temperature did not affect the YAM counts in the fresh-peeled garlic samples (Figure 7C) and, therefore, similar to the TPC,

20 °C was selected as the optimal temperature condition for subsequent ozone sterilization experiments.

3.1.4. Effects of Ozone Treatment on DIR

Ozone treatment significantly inhibited the DIR in the fresh-peeled garlic. As shown in Figure 8, after four days of storage, in the garlic samples inoculated by *P. citrinum*, the DIR of the 6 ppm treatment group was only 4.25%, while that of the blank group was as high as 61.23%. By the 10th day, the DIR of the blank group and the 1 ppm treatment group had reached 100%. In the *R. stolonifer*, after exposures of 3 ppm and 4 ppm, the difference in DIR was the highest during each storage period. When increased to 6 ppm, the DIR was consistently the lowest throughout the storage period, and the antibacterial effect was, thus, the best. The changes during storage in the DIR in the fresh-peeled garlic inoculated with *P. citrinum* and *R. stolonifer* treated with ozone for different time are shown in Figure 8(B1,B2), respectively. After 10 days of storage, the DIR of the blank group had reached 100%, while the 15 min treatment groups had reached only 63.95% in the *P. citrinum*-inoculated samples and 56.03% in the *R. stolonifer*-inoculated samples. After 15 min ozone treatments, however, the changes in DIR were not significant, remaining at similarly low values. In Figure 8C, the coincidence degree of the curves in the graph indicates no difference in the DIR of the fresh-peeled garlic treated at the three different ozone treatment temperatures.

3.1.5. Changes in Diameter and Depth of Disease Spot

As shown in Figure 9, with the increase in storage time, the disease spot diameters in the fresh-peeled garlic after ozone treatment also increased, while the diameter of the spot in the blank group remained consistently at the highest value. Furthermore, the diameters of disease spots in the fresh-peeled garlic samples almost reflected the three ozone treatment temperatures of 15, 20 and 25 °C. After ozone treatment at various concentrations, the 6 ppm ozone treatment was found to have exerted the best bacteriostatic effect during storage, resulting in the smallest lesion diameters. When stored until the 10th day, the lesion diameters of the fresh-peeled garlic inoculated with *P. citrinum* and *R. stolonifer* were 3.47 mm and 3.17 mm, respectively, both much lower than that of the blank group. However, at ozone treatment times of 3 min and 6 min, under the same storage time, the disease spot diameters in the treated samples were not significantly different from that of the blank group. As treatment time continuously increased, however, the diameters continued to decrease, reaching a small value at 15 min and then stabilizing despite continuing treatment time.

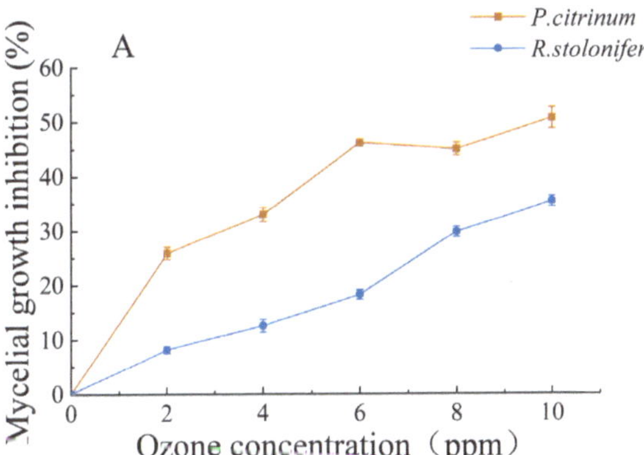

Figure 3. *Cont.*

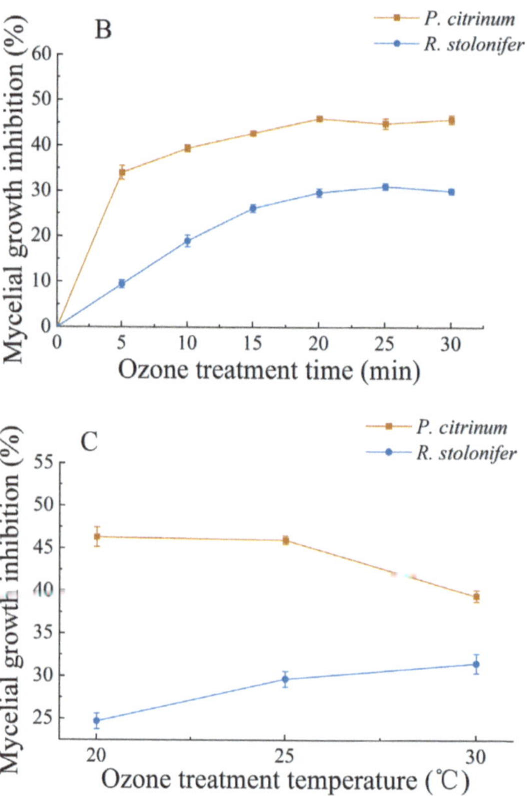

Figure 3. Effects of ozone treatments on the MGIR of *P. citrinum* and *R. stolonifer*. (**A**): Ozone concentration (ppm); (**B**): ozone treatment time (min); (**C**): ozone treatment temperature (°C).

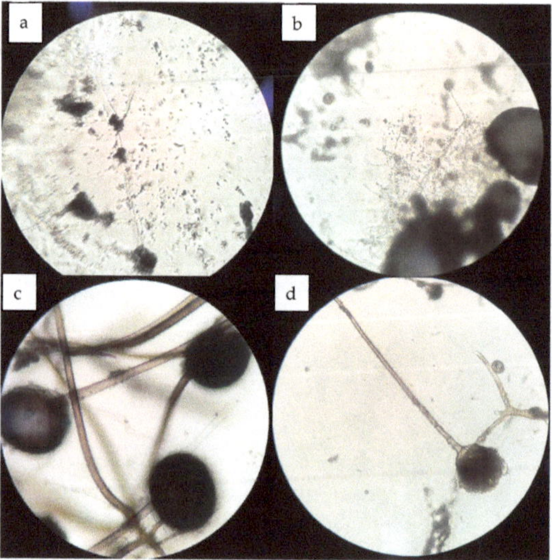

Figure 4. Effects of ozone treatments on mycelial morphology of *P. citrinum* and *R. stolonifer* in 10 × 40 field. (**a**): Control group of *P. citrinum*; (**b**): ozone treatment of *P. citrinum*; (**c**): control group of *R. stolonifer*; (**d**): ozone treatment of *R. stolonifer*.

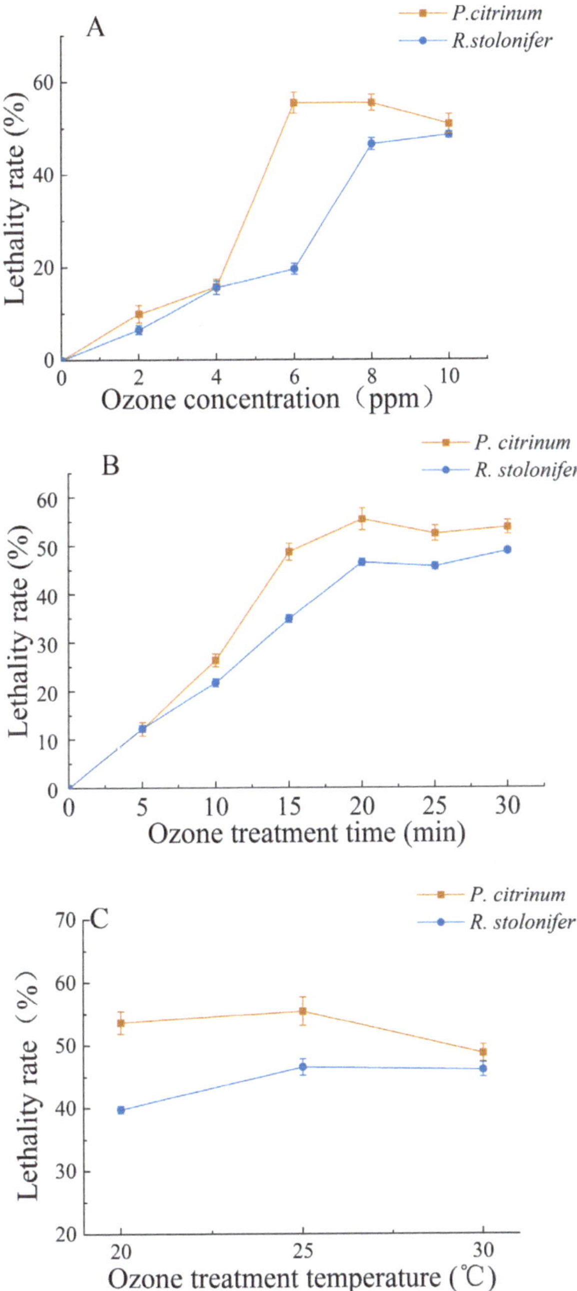

Figure 5. Effects of ozone treatments on the LR of *P. citrinum* and *R. stolonifer*. (**A**): Ozone concentration (ppm); (**B**); ozone treatment time (min); (**C**): ozone treatment temperature (^C).

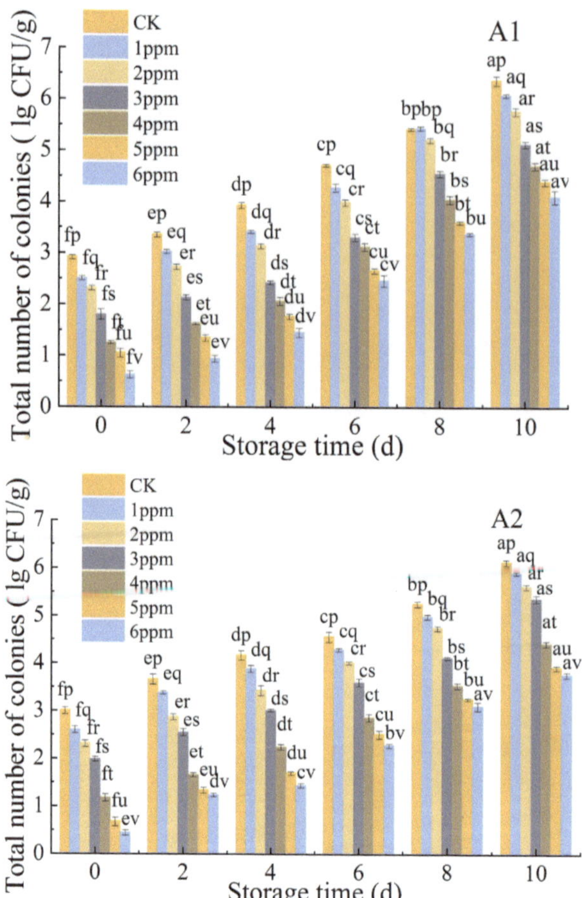

Figure 6. *Cont.*

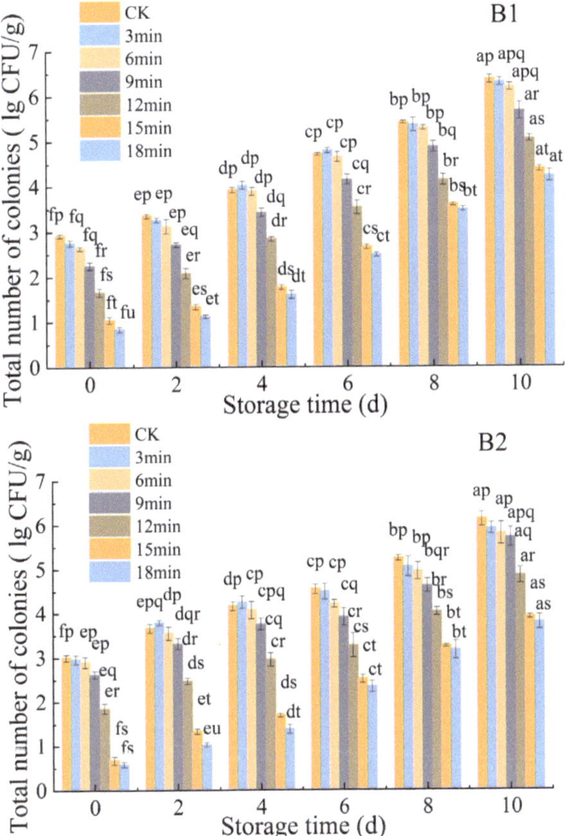

Figure 6. *Cont.*

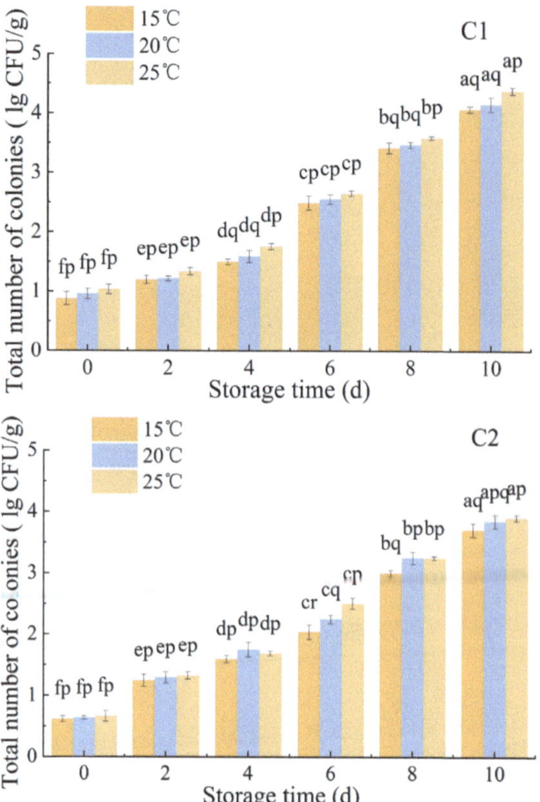

Figure 6. Effects of ozone treatments on the TPC counts of *P. citrinum* and *R. stolonifer* during storage. CK: control; (**A**): ozone concentration (ppm); (**B**): ozone treatment time (min); (**C**): ozone treatment temperature (°C); 1: *P. citrinum*; 2: *R. stolonifer*. For *P. citrinum* and *R. stolonifer*, mean bars with different letters (a–f) were significantly different ($p < 0.05$) for the same treatment, different storage days in (**A**,**B**,**C1**). Mean bars with different letters (p–v) differed significantly ($p < 0.05$) across treatments for the same number of storage days in (**A**,**B**).

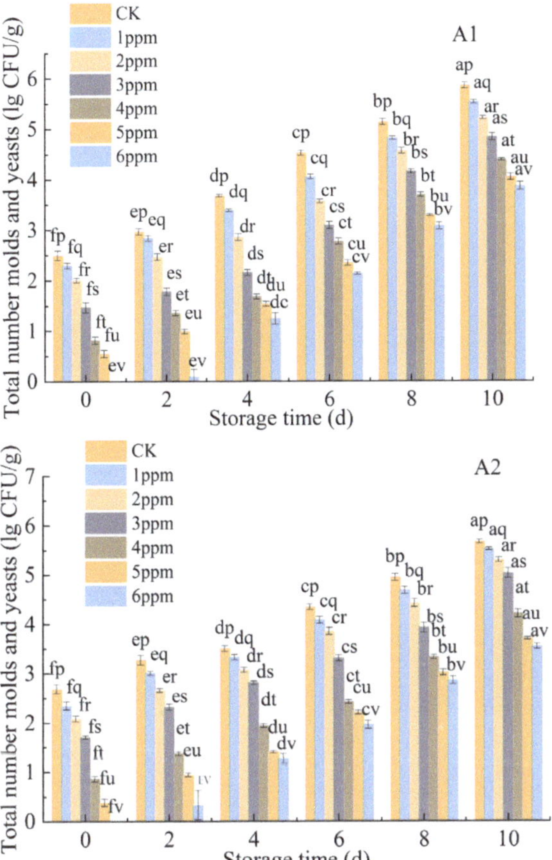

Figure 7. *Cont.*

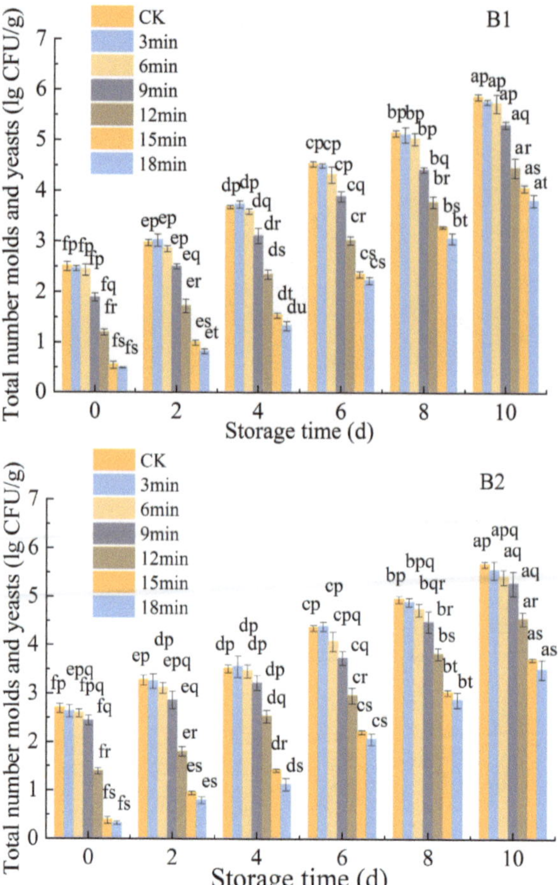

Figure 7. *Cont.*

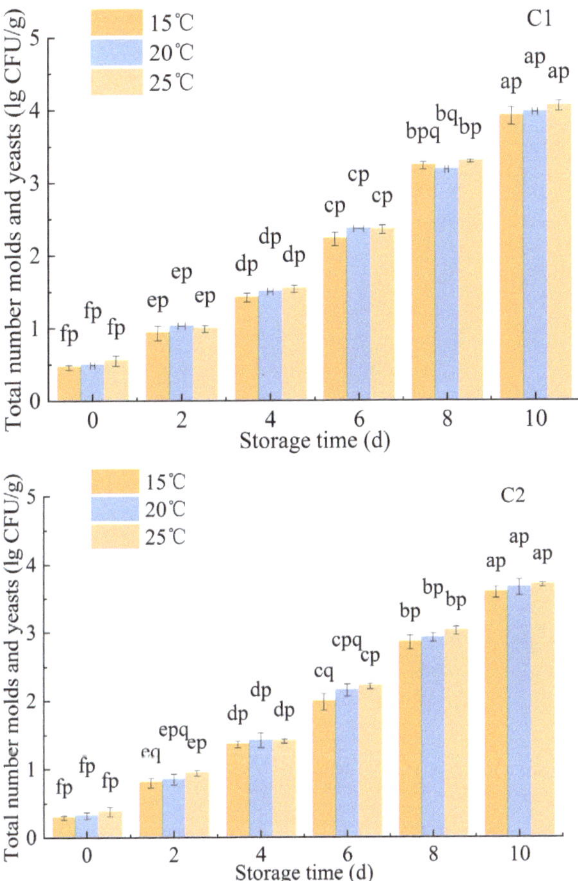

Figure 7. Effects of ozone treatments on the YAM counts of *P. citrinum* and *R. stolonifer* during storage. CK: control; (**A**): ozone concentration (ppm); (**B**): ozone treatment time (min); (**C**): ozone treatment temperature (°C); 1: *P. citrinum*; 2: *R. stolonifer*. For *P. citrinum* and *R. stolonifer*, mean bars with different letters (a–f) were significantly different ($p < 0.05$) for the same treatment, different storage days in (**A**,**B**). Mean bars with different letters (p–v) differed significantly ($p < 0.05$) across treatments for the same number of storage days in (**A**,**B1**).

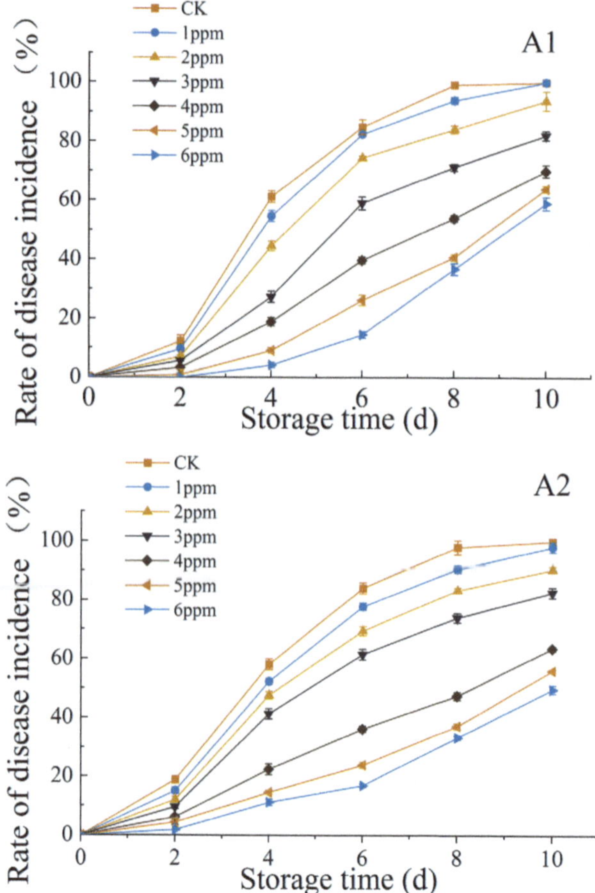

Figure 8. *Cont.*

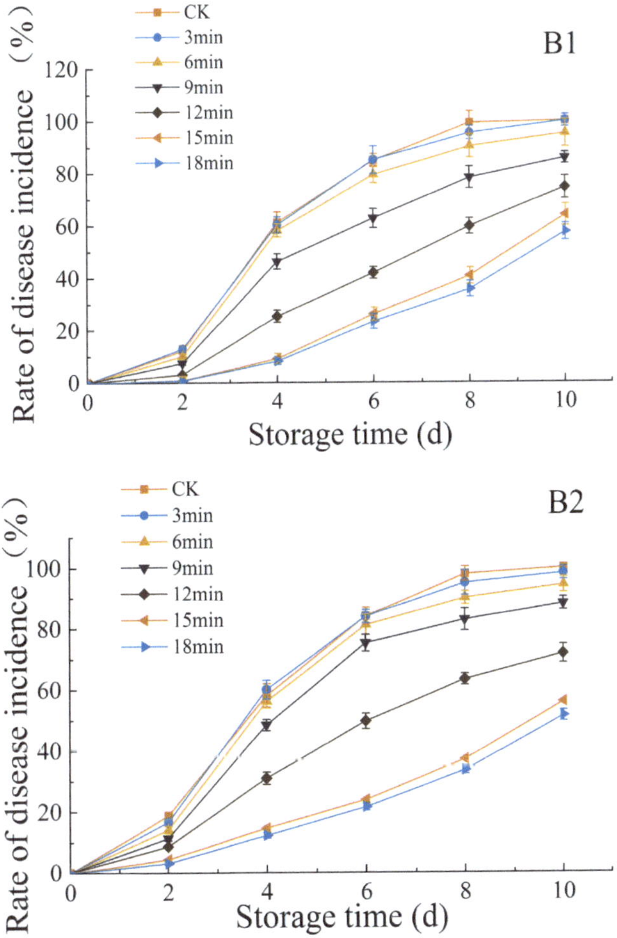

Figure 8. *Cont.*

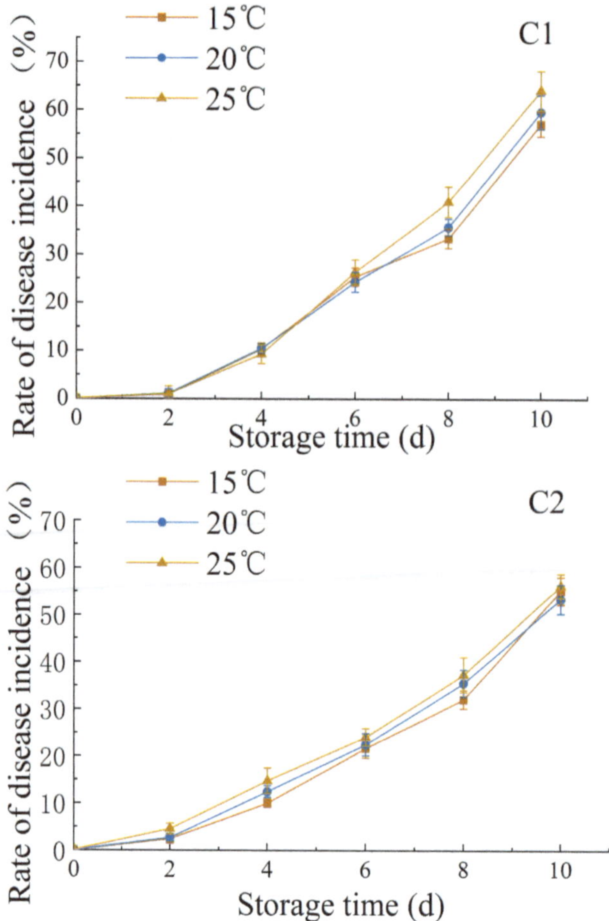

Figure 8. Effects of ozone treatments on the DIR. CK: control; (**A**): ozone concentration (ppm); (**B**): ozone treatment time (min); (**C**): ozone treatment temperature (°C); 1: *P. citrinum*; 2: *R. stolonifer*.

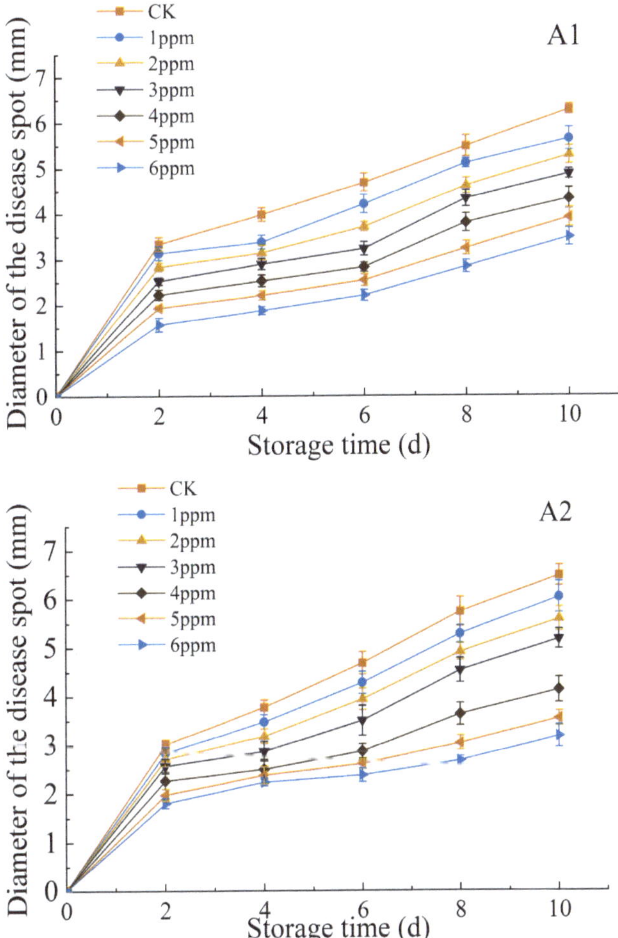

Figure 9. *Cont.*

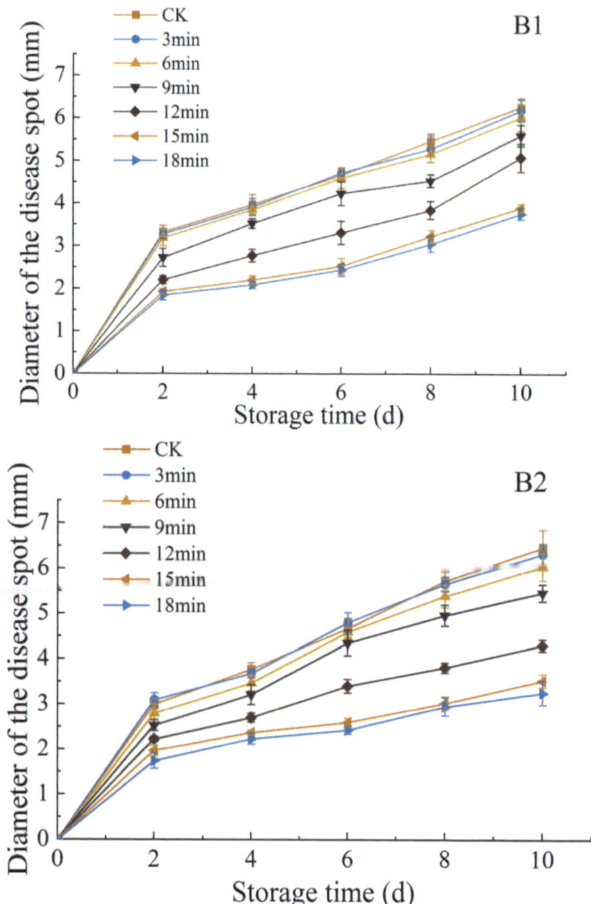

Figure 9. *Cont.*

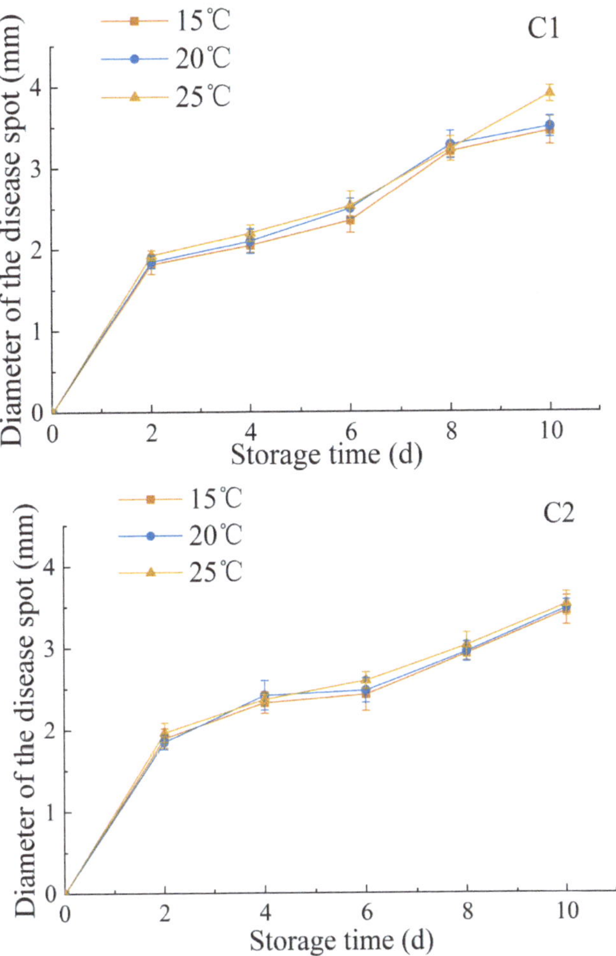

Figure 9. Effects of ozone treatments on the diameter of the disease spot. CK: control; (**A**): ozone concentration (ppm); (**B**): ozone treatment time (min); (**C**): ozone treatment temperature (°C); 1: *P. citrinum*; 2: *R. stolonifer*.

The depth of the disease spots in the fresh-peeled garlic continuously increased with increasing concentration and exposure time (Figure 10A,B). When the storage time was kept constant, no difference was observed in the depths of disease spots in the fresh-peeled garlic treatment groups of 15, 20 and 25 °C (Figure 10C).

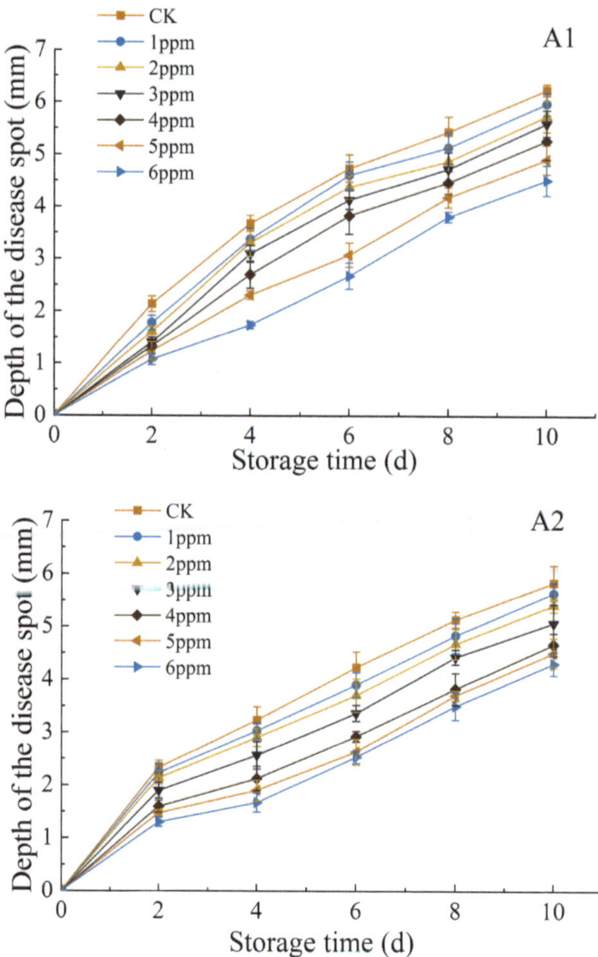

Figure 10. *Cont.*

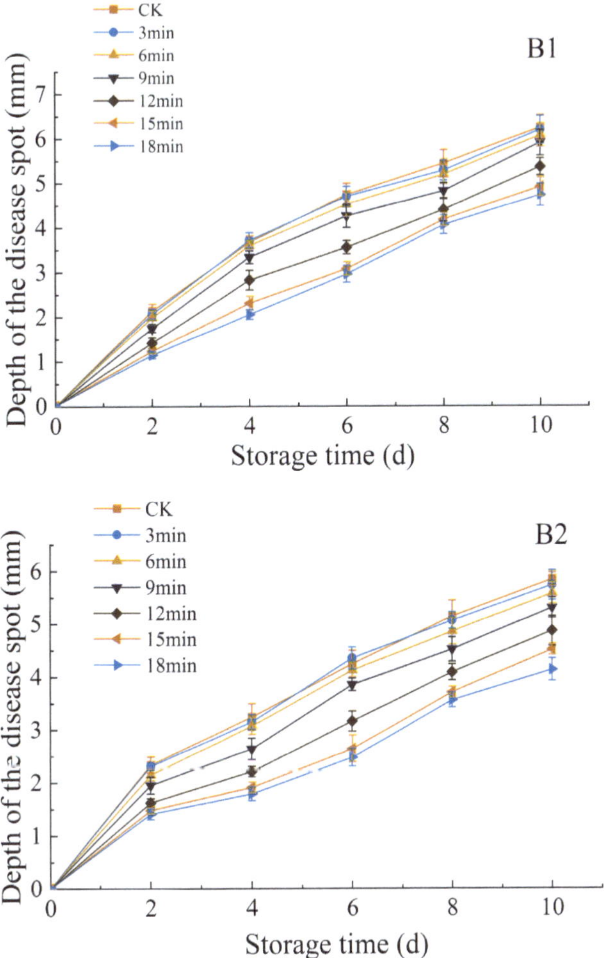

Figure 10. *Cont.*

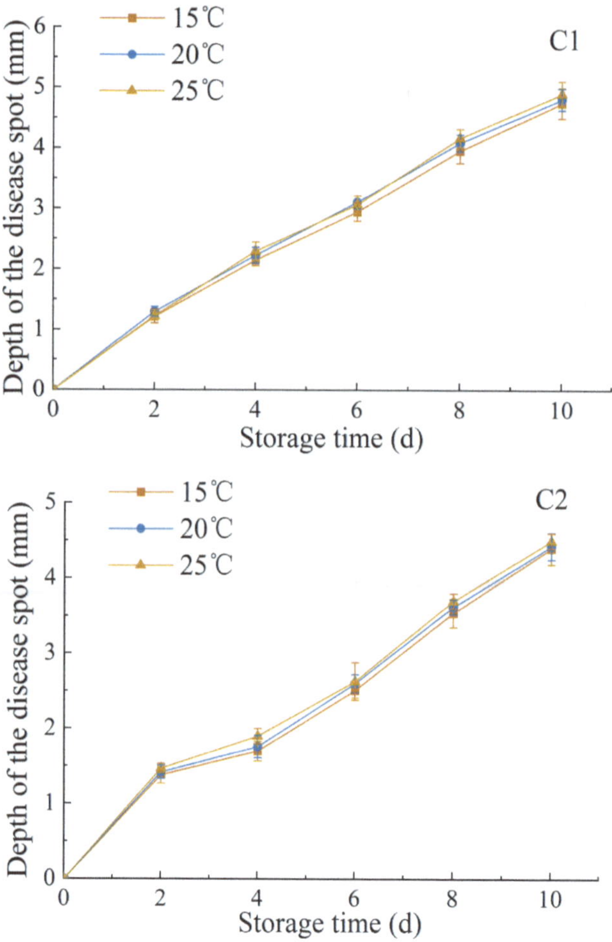

Figure 10. Effects of ozone treatments on the depth of the disease spot. CK: control; (**A**): ozone concentration (ppm); (**B**): ozone treatment time (min); (**C**): ozone treatment temperature (°C); 1: *P. citrinum*; 2: *R. stolonifer*.

3.2. Discussion

In the in vitro experiment, with increased ozone concentration, exposure time and temperature, the ER, PDR, MGIR and LR of the two molds changed (Figures 1–5). Ozone effectively inhibited the growth of *P. citrinum* and *R. stolonifer* in vitro, by disrupting spore development and inhibiting fungal cell growth through oxygenation. The result was consistent with that reported by Antony-Babu [46] et al., who found that ozone exposure significantly reduced the number of asexual spores formed in media. Another study reported that ozone did not kill all the spores in the in vitro tests [33], while Ozkan [47] et al. and Palou [48] et al. noted that different molds exhibit different levels of ozone tolerance. These results correspond with those of this study, shown in Figure 5, in which the trends of the two molds were similar but different, indicating that the ozone had similar inhibitory effects on *P. citrinum* and *R. stolonifer*, but could not attain an LR of 100%. Tzortzakis [49] et al. concluded that a low concentration of ozone could inhibit the growth of mold mycelium, while ozone treatment suppressed the germination of conidia but its inhibitory action was dependent on the concentration and duration of exposure. Therefore, the sensitivity of mold spores to the oxidizing action of ozone is considered to be dependent on fungal species, ozone dosage, exposure time and temperature. Further research is essential to explore the impacts of ozone on the ultramorphology, physiological metabolism and gene expression of molds, especially when examining its inhibitory properties on decay-causing molds in vitro.

In the in vivo experiment, as the storage period increased, the TPC, which predicts food shelf life resistance, YAM, which is one of the indicators for evaluating food hygiene, and DIR, as well as the diameter and depth of the two mold disease spots in the fresh-peeled garlic samples all increased (Figures 6–10). These results indicate that while ozone treatment can inhibit *P. citrinum* and *R. stolonifer* in vivo, it cannot completely eradicate all fungi in garlic, which are potentially affected by various factors. The effect of temperature on the two molds was relatively small in comparison to those of the ozone concentration and exposure time. These findings concur with those of Wang [35] et al., in which ozone delayed disease incidence and reduced its severity in fruit. Their study also found that prolonged ozone exposure decreased the germination of fungal spores on fruit peel. Similarly, Yeoh [50] et al. discovered that ozone effectively controlled fungal sporulation on citrus fruit. It also slightly delayed disease incidence and significantly lowered disease severity. Palou [33] et al. found that fungal structures within wounds remained protected from the oxidizing effect of ozone because of limited ozone penetration. Consequently, while exposure to gaseous ozone delayed disease incidence and reduced disease severity on wound-inoculated fruit, it did not totally prevent fruit decay. Overall, however, previous studies have found that ozone does exert a significantly beneficial effect on the preservation of fresh-peeled garlic and other MPV, such as table grapes [32,47,50,51]. Additionally, investigating the preservation effect of ozone treatment on fresh-peeled garlic remains a crucial area for future studies.

It is worth noting that due to the mechanism of action of ozone, its application in food disinfection and sterilization basically does not compromise food quality and safety. However, strict control over the ozone concentration is imperative. Exceeding prescribed limits may result in harmful substances, especially if ozone reacts with bromide. In the disinfection process using ozone, it is also necessary to strictly control the ozone quantity. Ozone has a stimulating effect on the human body, and excessively high concentration can cause respiratory diseases, headaches, tachycardia, and other health threats, posing risks to human safety [52].

4. Conclusions

In this study, ozone treatment was found to partially damage the integrity of the fungi cell membrane, causing the leakage of intracellular proteins, polysaccharides and other substances, which led to an increase in the conductivity of the fungi suspension and, ultimately, the death of the mold. Furthermore, the exposure of fresh-peeled garlic samples to ozone exerted a significant inhibitory effect on *P. citrinum* and *R. stolonifer*, two of the main strains causing spoilage in fresh-peeled garlic. It was also evident that ozone treatment under appropriate conditions will provide optimal treatment results. In vitro, the best inhibitory effect of the ozone treatment on *P. citrinum* was evidenced at 6 ppm, for 20 min at 20 °C, while on *R. stolonifer* the optimal conditions were 8 ppm, for 20 min at 25 °C. In vivo, the best inhibitory effect of ozone treatment on both of the molds was evidenced at 6 ppm, for 15 min at 20 °C. Overall, the application of appropriate ozone treatment was found to provide a feasible solution for the inhibition of *P. citrinum* and *R. stolonifer* and may prolong the shelf life of peeled garlic.

Author Contributions: Conceptualization, H.L. and X.W.; methodology, L.C.; software, X.F.; validation, Y.X. and X.F.; formal analysis, W.L.; investigation, J.H.; resources, X.L.; data curation, X.L.; writing—original draft preparation, H.L.; writing—review and editing, X.F.; visualization, J.H. and L.C.; supervision, Y.X.; project administration, W.L. and J.H.; funding acquisition, Y.X. and X.L. All authors have read and agreed to the published version of the manuscript.

Funding: This work was supported by Science and technology support program of Sichuan (2021YFQ0071 and 2020YFN0151), the Open Research Subject of International Science and Technology Cooperation (Australia and New Zealand) Institute of Sichuan (AXYJ2022-012 and AXYJ2022-013) and the College Students Innovation and Entrepreneurship Training Program of Sichuan Province (S202110650007).

Institutional Review Board Statement: Not applicable.

Informed Consent Statement: Not applicable.

Data Availability Statement: Data are contained within the article.

Conflicts of Interest: The authors declare no conflict of interest.

References

1. Martins, N.; Petropoulos, S.; Ferreira, I.C.F.R. Chemical Composition and Bioactive Compounds of Garlic (*Allium Sativum* L.) as Affected by Pre- and Post-Harvest Conditions: A Review. *Food Chem.* **2016**, *211*, 41–50. [CrossRef] [PubMed]
2. Beato, V.M.; Orgaz, F.; Mansilla, F.; Montaño, A. Changes in Phenolic Compounds in Garlic (*Allium Sativum* L.) Owing to the Cultivar and Location of Growth. *Plant Foods Hum. Nutr.* **2011**, *66*, 218–223. [CrossRef] [PubMed]
3. González, R.E.; Soto, V.C.; Sance, M.M.; Camargo, A.B.; Galmarini, C.R. Variability of Solids, Organosulfur Compounds, Pungency and Health-Enhancing Traits in Garlic (*Allium Sativum* L.) Cultivars Belonging to Different Ecophysiological Groups. *J. Agric. Food Chem.* **2009**, *57*, 10282–10288. [CrossRef]
4. Ichikawa, M.; Mizuno, I.; Yoshida, J.; Ide, N.; Ushijima, M.; Kodera, Y.; Hayama, M.; Ono, K. Pharmacokinetics of Cycloalliin, an Organosulfur Compound Found in Garlic and Onion, in Rats. *J. Agric. Food Chem.* **2006**, *54*, 9811–9819. [CrossRef] [PubMed]
5. Jabbes, N.; Arnault, I.; Auger, J.; Al Mohandes Dridi, B.; Hannachi, C. Agro-Morphological Markers and Organo-Sulphur Compounds to Assess Diversity in Tunisian Garlic Landraces. *Sci. Hortic.* **2012**, *148*, 47–54. [CrossRef]
6. Lee, J.; Harnly, J.M. Free Amino Acid and Cysteine Sulfoxide Composition of 11 Garlic (*Allium Sativum* L.) Cultivars by Gas Chromatography with Flame Ionization and Mass Selective Detection. *J. Agric. Food Chem.* **2005**, *53*, 9100–9104. [CrossRef] [PubMed]
7. Hirata, S.; Abdelrahman, M.; Yamauchi, N.; Shigyo, M. Diversity Evaluation Based on Morphological, Physiological and Isozyme Variation in Genetic Resources of Garlic (*Allium Sativum* L.) Collected Worldwide. *Genes Genet. Syst.* **2016**, *91*, 161–173. [CrossRef]
8. Fratianni, F.; Riccardi, R.; Spigno, P.; Ombra, M.N.; Cozzolino, A.; Tremonte, P.; Coppola, R.; Nazzaro, F. Biochemical Characterization and Antimicrobial and Antifungal Activity of Two Endemic Varieties of Garlic (*Allium Sativum* L.) of the Campania Region, Southern Italy. *J. Med. Food* **2016**, *19*, 686–691. [CrossRef]
9. Li, R.; Wu, W.; Fang, X.; Chen, H.; Han, Y.; Niu, B.; Chen, H.; Gao, H. Antifungal Activity and Mechanism of Perillaldehyde against *Penicillium citrinum*, a Major Fungal Pathogen of *Myrica rubra*. *Food Sci.* **2023**, *44*, 196–203.
10. Kong, J.; Zhang, Y.; Ju, J.; Xie, Y.; Guo, Y.; Cheng, Y.; Qian, H.; Quek, S.Y.; Yao, W. Antifungal Effects of Thymol and Salicylic Acid on Cell Membrane and Mitochondria of Rhizopus Stolonifer and Their Application in Postharvest Preservation of Tomatoes. *Food Chem.* **2019**, *285*, 380–388. [CrossRef]
11. Meireles, A.; Giaouris, E.; Simões, M. Alternative Disinfection Methods to Chlorine for Use in the Fresh-Cut Industry. *Food Res. Int.* **2016**, *82*, 71–85. [CrossRef]
12. Rico, D.; Martín-Diana, A.B.; Barat, J.M.; Barry-Ryan, C. Extending and Measuring the Quality of Fresh-Cut Fruit and Vegetables: A Review. *Trends Food Sci. Technol.* **2007**, *18*, 373–386. [CrossRef]
13. Martínez-Vaz, B.M.; Fink, R.C.; Diez-Gonzalez, F.; Sadowsky, M.J. Enteric Pathogen-Plant Interactions: Molecular Connections Leading to Colonization and Growth and Implications for Food Safety. *Microb. Environ.* **2014**, *29*, 123–135. [CrossRef] [PubMed]
14. Yu, H.; Neal, J.A.; Sirsat, S.A. Consumers' Food Safety Risk Perceptions and Willingness to Pay for Fresh-Cut Produce with Lower Risk of Foodborne Illness. *Food Control* **2018**, *86*, 83–89. [CrossRef]
15. Herman, K.M.; Hall, A.J.; Gould, L.H. Outbreaks Attributed to Fresh Leafy Vegetables, United States, 1973–2012. *Epidemiol. Infect.* **2015**, *143*, 3011–3021. [CrossRef]
16. Sirsat, S.; Neal, J. Microbial Profile of Soil-Free versus In-Soil Grown Lettuce and Intervention Methodologies to Combat Pathogen Surrogates and Spoilage Microorganisms on Lettuce. *Foods* **2013**, *2*, 488–498. [CrossRef]
17. Francis, G.A.; Gallone, A.; Nychas, G.J.; Sofos, J.N.; Colelli, G.; Amodio, M.L.; Spano, G. Factors Affecting Quality and Safety of Fresh-Cut Produce. *Crit. Rev. Food Sci. Nutr.* **2012**, *52*, 595–610. [CrossRef]
18. Bull, R.J.; Crook, J.; Whittaker, M.; Cotruvo, J.A. Therapeutic Dose as the Point of Departure in Assessing Potential Health Hazards from Drugs in Drinking Water and Recycled Municipal Wastewater. *Regul. Toxicol. Pharmacol.* **2011**, *60*, 1–19. [CrossRef]
19. Legay, C.; Rodriguez, M.J.; Sérodes, J.B.; Levallois, P. Estimation of Chlorination By-Products Presence in Drinking Water in Epidemiological Studies on Adverse Reproductive Outcomes: A Review. *Sci. Total Environ.* **2010**, *408*, 456–472. [CrossRef]
20. Kim, J.-G.; Yousef, A.E.; Dave, S. Application of Ozone for Enhancing the Microbiological Safety and Quality of Foods: A Review. *J. Food Prot.* **1999**, *62*, 1071–1087. [CrossRef]
21. Cullen, P.J.; Valdramidis, V.P.; Tiwari, B.K.; Patil, S.; Bourke, P.; O'Donnell, C.P. Ozone Processing for Food Preservation: An Overview on Fruit Juice Treatments. *Ozone Sci. Eng.* **2010**, *32*, 166–179. [CrossRef]

22. Khadre, M.A.; Yousef, A.E. Sporicidal Action of Ozone and Hydrogen Peroxide: A Comparative Study. *Int. J. Food Microbiol.* **2001**, *71*, 131–138. [CrossRef]
23. Zou, H.; Xu, L.; Xu, Z.; Xie, W.; Wang, Y.; Liao, X.; Kong, X. Effects of Ultra-High Temperature Treatment and Packages on Baked Purple Sweet Potato Nectar. *LWT* **2018**, *94*, 129–135. [CrossRef]
24. Gibson, K.E.; Almeida, G.; Jones, S.L.; Wright, K.; Lee, J.A. Inactivation of Bacteria on Fresh Produce by Batch Wash Ozone Sanitation. *Food Control* **2019**, *106*, 106747. [CrossRef]
25. Tzortzakis, N.; Chrysargyris, A. Postharvest Ozone Application for the Preservation of Fruits and Vegetables. *Food Rev. Int.* **2017**, *33*, 270–315. [CrossRef]
26. Miller, F.A.; Silva, C.L.M.; Brandão, T.R.S. A Review on Ozone-Based Treatments for Fruit and Vegetables Preservation. *Food Eng. Rev.* **2013**, *5*, 77–106. [CrossRef]
27. Horvitz, S.; Cantalejo, M.J. Application of Ozone for the Postharvest Treatment of Fruits and Vegetables. *Crit. Rev. Food Sci. Nutr.* **2014**, *54*, 312–339. [CrossRef]
28. Akata, I.; Torlak, E.; Erci, F. Efficacy of Gaseous Ozone for Reducing Microflora and Foodborne Pathogens on Button Mushroom. *Postharvest Biol. Technol.* **2015**, *109*, 40–44. [CrossRef]
29. Öztekin, S.; Zorlugenç, B.; Zorlugenç, F.K. Effects of Ozone Treatment on Microflora of Dried Figs. *J. Food Eng.* **2006**, *75*, 396–399. [CrossRef]
30. Nadas, A.; Olmo, M.; Garcia, J.M. Growth of Botrytis Cinerea and Strawberry Quality in Ozone-Enriched Atmospheres. *J Food Sci.* **2003**, *68*, 1798–1802. [CrossRef]
31. Boonkorn, P.; Gemma, H.; Sugaya, S.; Setha, S.; Uthaibutra, J.; Whangchai, K. Impact of High-Dose, Short Periods of Ozone Exposure on Green Mold and Antioxidant Enzyme Activity of Tangerine Fruit. *Postharvest Biol. Technol.* **2012**, *67*, 25–28. [CrossRef]
32. Sarig, P.; Zahavi, T.; Zutkhi, Y.; Yannai, S.; Lisker, N.; Ben-Arie, R. Ozone for Control of Post-Harvest Decay of Table Grapes Caused ByRhizopus Stolonifer. *Physiol. Mol. Plant Pathol.* **1996**, *48*, 403–415. [CrossRef]
33. Palou, L.; Crisosto, C.H.; Smilanick, J.L.; Adaskaveg, J.E.; Zoffoli, J.P. Effects of Continuous 0.3 Ppm Ozone Exposure on Decay Development and Physiological Responses of Peaches and Table Grapes in Cold Storage. *Postharvest Biol. Technol.* **2002**, *24*, 39–48. [CrossRef]
34. Tzortzakis, N.; Singleton, I.; Barnes, J. Impact of Low-Level Atmospheric Ozone-Enrichment on Black Spot and Anthracnose Rot of Tomato Fruit. *Postharvest Biol. Technol.* **2008**, *47*, 1–9. [CrossRef]
35. Wang, X.; Liu, R.; Yang, Y.; Zhang, M. Isolation, Purification and Identification of Antioxidants in an Aqueous Aged Garlic Extract. *Food Chem.* **2015**, *187*, 37–43. [CrossRef] [PubMed]
36. Minas, I.S.; Karaoglanidis, G.S.; Manganaris, G A.; Vasilakakis, M. Effect of Ozone Application during Cold Storage of Kiwifruit on the Development of Stem-End Rot Caused by Botrytis Cinerea. *Postharvest Biol. Technol.* **2010**, *58*, 203–210 [CrossRef]
37. Moscoso-Ramírez, P.A.; Palou, L. Effect of Ethylene Degreening on the Development of Postharvest Penicillium Molds and Fruit Quality of Early Season Citrus Fruit. *Postharvest Biol. Technol.* **2014**, *91*, 1–8. [CrossRef]
38. Diao, W.-R.; Zhang, L.-L.; Feng, S.-S.; Xu, J.-G. Chemical Composition, Antibacterial Activity, and Mechanism of Action of the Essential Oil from Amomum Kravanh. *J. Food Prot.* **2014**, *77*, 1740–1746. [CrossRef]
39. Yin, X.; Luo, Y.; Fan, H.; Feng, L.; Shen, H. Effect of Freeze-Chilled Treatment on Flavor of Grass Carp (*Ctenopharyngodon Idellus*) Fillets and Soups During Short-Term Storage. *J. Aquat. Food Prod. Technol.* **2016**, *25*, 777–787. [CrossRef]
40. Xu, Q.; Xing, Y.; Che, Z.; Guan, T.; Zhang, L.; Bai, Y.; Gong, L. Effect of Chitosan Coating and Oil Fumigation on the Microbiological and Quality Safety of Fresh-Cut Pear: Microbiological Safety of Fresh-Cut Pear. *J. Food Saf.* **2013**, *33*, 179–189. [CrossRef]
41. Xing, Y.; Xu, Q.; Li, X.; Che, Z.; Yun, J. Antifungal Activities of Clove Oil Against Rhizopus Nigricans, Aspergillus Flavus and Penicillium Citrinum in Vitro and in Wounded Fruit Test: Antifungal Activities of Clove Oil. *J. Food Saf.* **2012**, *32*, 84–93. [CrossRef]
42. Xing, Y.; Xu, Q.; Yang, S.; Chen, C.; Tang, Y.; Sun, S.; Zhang, L.; Che, Z.; Li, X. Preservation Mechanism of Chitosan-Based Coating with Cinnamon Oil for Fruits Storage Based on Sensor Data. *Sensors* **2016**, *16*, 1111. [CrossRef]
43. Salve, A.R.; Pegu, K.; Arya, S.S. Comparative Assessment of High-Intensity Ultrasound and Hydrodynamic Cavitation Processing on Physico-Chemical Properties and Microbial Inactivation of Peanut Milk. *Ultrason. Sonochem.* **2019**, *59*, 104728. [CrossRef] [PubMed]
44. Martiñon, M.E.; Moreira, R.G.; Castell-Perez, M.E.; Gomes, C. Development of a Multilayered Antimicrobial Edible Coating for Shelf-Life Extension of Fresh-Cut Cantaloupe (*Cucumis Melo* L.) Stored at 4 °C. *LWT—Food Sci. Technol.* **2014**, *56*, 341–350. [CrossRef]
45. Ong, M.K.; Ali, A. Antifungal Action of Ozone against Colletotrichum Gloeosporioides and Control of Papaya Anthracnose. *Postharvest Biol. Technol.* **2015**, *100*, 113–119. [CrossRef]
46. Antony-Babu, S.; Singleton, I. Effects of Ozone Exposure on the Xerophilic Fungus, Eurotium Amstelodami IS-SAB-01, Isolated from Naan Bread. *Int. J. Food Microbiol.* **2011**, *144*, 331–336. [CrossRef]
47. Ozkan, R.; Smilanick, J.L.; Karabulut, O.A. Toxicity of Ozone Gas to Conidia of *Penicillium Digitatum*, *Penicillium Italicum*, and Botrytis Cinerea and Control of Gray Mold on Table Grapes. *Postharvest Biol. Technol.* **2011**, *60*, 47–51. [CrossRef]
48. Palou, L.; Smilanick, J.L.; Crisosto, C.H.; Mansour, M. Effect of Gaseous Ozone Exposure on the Development of Green and Blue Molds on Cold Stored Citrus Fruit. *Plant Dis.* **2001**, *85*, 632–638. [CrossRef]

49. Tzortzakis, N.; Singleton, I.; Barnes, J. Deployment of Low-Level Ozone-Enrichment for the Preservation of Chilled Fresh Produce. *Postharvest Biol. Technol.* **2007**, *43*, 261–270. [CrossRef]
50. Yeoh, W.K.; Ali, A.; Forney, C.F. Effects of Ozone on Major Antioxidants and Microbial Populations of Fresh-Cut Papaya. *Postharvest Biol. Technol.* **2014**, *89*, 56–58. [CrossRef]
51. Aguayo, E.; Escalona, V.H.; Artés, F. Effect of Cyclic Exposure to Ozone Gas on Physicochemical, Sensorial and Microbial Quality of Whole and Sliced Tomatoes. *Postharvest Biol. Technol.* **2006**, *39*, 169–177. [CrossRef]
52. Zhou, Q. Discussion on the Ozone Disinfection and Sterilization Technology Applied in the Food Industry. *China Food Saf. Mag.* **2023**, *9*, 155–157. [CrossRef]

Article

Preservative Effects of Flaxseed Gum-Sodium Alginate Active Coatings Containing Carvacrol on Quality of Turbot (*Scophthalmus maximus*) during Cold Storage

Xinrui Yang [1], Shiyuan Fang [1], Yao Xie [1], Jun Mei [1,2,3,4,*] and Jing Xie [1,2,3,4,*]

1   College of Food Science & Technology, Shanghai Ocean University, Shanghai 201306, China; m210311052@st.shou.edu.cn (Y.X.)
2   Shanghai Professional Technology Service Platform on Cold Chain Equipment Performance and Energy Saving Evaluation, Shanghai Ocean University, Shanghai 201306, China
3   National Experimental Teaching Demonstration Center for Food Science and Engineering, Shanghai Ocean University, Shanghai 201306, China
4   Shanghai Engineering Research Center of Aquatic Product Processing & Preservation, Shanghai Ocean University, Shanghai 201306, China
*   Correspondence: jmei@shou.edu.cn (J.M.); jxie@shou.edu.cn (J.X.)

Abstract: In this article, the effect of active coatings of flaxseed gum (FG) and sodium alginate (SA) containing carvacrol (CA) on the quality of turbot (*Scophthalmus maximus*) after storage at 4 °C for 18 days was evaluated. The experimental results showed that FG/SA-CA could effectively inhibit the growth of microorganisms. At the same time, FG/SA-CA reduced the value of odorous-related compounds including thiobarbituric acid reactive substances (TBARS), total volatile base nitrogen (TVB-N), and K values. The FG/SA-CA significantly delayed the oxidation of myofibrillar protein (MP) through controlling the development of carbonyl groups and maintaining a high content of sulfhydryl groups. Thus, FG/SA-CA inhibits the growth of spoilage microorganisms, maintains the structure of the protein, and extends the refrigerated shelf life of turbot.

Keywords: *Scophthalmus maximus*; carvacrol; active coating; cold storage; quality

Citation: Yang, X.; Fang, S.; Xie, Y.; Mei, J.; Xie, J. Preservative Effects of Flaxseed Gum-Sodium Alginate Active Coatings Containing Carvacrol on Quality of Turbot (*Scophthalmus maximus*) during Cold Storage. *Coatings* **2024**, *14*, 338. https:// doi.org/10.3390/coatings14030338

Academic Editor: Elena Torrieri

Received: 11 January 2024
Revised: 23 February 2024
Accepted: 26 February 2024
Published: 12 March 2024

1. Introduction

Turbot (*Scophthalmus maximus*) is highly favored by customers due to its abundant nutritious content and good taste [1]. Different pre-treatment methods play a crucial role in quality changes during storage of turbot. However, turbot is perishable during cold storage owing to spoilage microorganisms. It is imperative to discover methods to impede the deterioration of turbot during cold storage. Scientists have looked to plant materials for the development of food preservatives over the past few years. Essential oils are volatile secondary metabolites of plants in the form of saturated or unsaturated hydrocarbons, alcohols, esters, aldehydes, ketones, ethers, and terpenes. Carvacrol (5-isopropyl-2-methyl phenol) is a monoterpenoid phenolic compound and considered to be a natural antibacterial and antioxidant agent. Carvacrol, as a broad-spectrum fungicide, is effective in preventing microbial infestation of fish during storage. It has been found as an indispensable and major acting component in both oregano and thyme oils [2]. The antibacterial and antioxidant properties of carvacrol can benefit foods without negatively affecting their senses. Chaparro et al. [3] found that combined carvacrol and chitosan on tilapia fillets inhibited the microbial growth during the refrigeration process, but reduced the evaluation of fish texture. Alves et al. [4] prepared carvacrol microcapsules that prolonged the storage duration of frozen salmon by 4–7 days and decreased the population of thermophilic and cryophilic bacteria during the preservation process in cold storage. However, carvacrol is sensitive to temperature, oxygen, humidity, and light, and the volatility and hydrophobicity of carvacrol prevents it from maximizing its effects [5]. Therefore, carvacrol presents difficulty for direct use in food preservation.

Edible packaging materials refer to natural substances that can be digested and absorbed by human beings as the base material, such as protein, polysaccharide, plant cellulose and fat, etc. The force between the molecules of these substances enables the generation of a with a porous network structure of packaging materials. Active packaging technology is a means of maintaining the quality of food products by changing the storage environment in food packaging, such as by adding various gas absorbers and releasers, eliminating oxygen, carbon dioxide, and other gaseous liquids inside the package, controlling temperature and humidity, or by adding bacteriostatic agents [6]. The primary goals include prolonging the food's quality assurance time, intensifying the food's scent and flavor, and enhancing the overall cleanliness and safety of the food [7]. Active packaging can be categorized as absorption, release, and coating based on the method of controlling the factors affecting preservation. Active coatings can prevent oxygen penetration, water dissipation, and inhibit microbial growth [8]. There is evidence that embedding natural preservatives into active coatings could extend the quality guarantee period of fish [9–11]. Polysaccharides are particularly excellent ingredients for making active coatings. For example, sodium alginate (SA) has been frequently used as a material for reactive or thin reactive coatings due to its excellent film-forming properties [12]. Pei et al. [13] successfully formulated gum tragacanth-SA-based active coatings containing epigallocatechin gallate and applied it to the preservation of *Larimichthys crocea*. However, the relatively poor water resistance and antioxidant activity of SA active coatings limit their application in food packaging. Flaxseed gum (FG) is a type of natural functional colloid with gel, emulsification, and thickening properties [14]. It can improve the hardness and ductility of the active coating [15]. The preservatives used on the surface of food are less effective against foodborne pathogenic microorganisms because preservatives quickly spread into food and denature with food composition [16]. Active coatings facilitate a gradual and uninterrupted long-term transfer from packaging materials to food surfaces.

Therefore, based on the study of Fang et al. [17], this research focuses on the effect of different concentrations of carvacrol FG-SA active coating on the refrigerated freshness of turbot.

2. Materials and Methods

2.1. Processing of Carvacrol Active Coatings

When making carvacrol emulsion, improvements were made on the basis of Fang et al. [18] to improve the stability of the emulsion. Carvacrol, at concentrations of 0.05%, 0.10%, and 0.20%, was mechanically mixed with 7.5 g β-cyclodextrin (βCD) and 50 g Tween 80 in a beaker to achieve a homogenous dispersion. Subsequently, 400 mL of ultra-pure water was added and agitated for a duration of 8 h at ambient temperature. The carvacrol/βCD emulsions contained carvacrol at concentrations of 0.75, 1.50, and 3.00 μL/mL. FG (0.5% *w/v*), glycerol (1.5% *v/w*) and SA (viscosity 200 ± 20 mPa, 1.5% *w/v*) were added to the prepared carvacrol/βCD emulsions at a temperature of 45 °C. Afterwards, this was followed by the addition of ultrapure water to the mixture, which resulted in the final emulsion volume being equal to 1 L. Next, the emulsions were agitated for 2 h and then the mixture was made uniform by subjecting it to ultrasonic homogenization using an Ultrasonic operating at a frequency of 20 KHz and a power of 600 W for 20 min. This process resulted in the formation of coating solutions that were consistent throughout. Finally, the FG/SA solutions with carvacrol concentrations of 0.05%, 0.10%, and 0.20% were designated as FG/SA-5CA, FG/SA-10CA, and FG/SA-20CA, respectively.

2.2. Preparation of Turbot and Immersion Sample Treatment

Seventy-two turbots (500 ± 10 g) were purchased from Luchaogang Market in Shanghai, China. Afterwards, market workers carefully removed the head, internal organs, skin, and bones, leaving the dorsal muscle of the turbot. They were then quickly transported on ice to the lab within 1 h. Subsequently, the specimens were meticulously cleansed using a 0.85% NaCl solution. The 72 turbots were randomly divided into 4 groups, with each group consisting of 18 turbots: (i) CK (uncoated); (ii) FG/SA-5CA (coated with FG/SA-5CA active coating solution); (iii) FG/SA-10CA (coated with FG/SA-10CA active coating solution) and (iv) FG/SA-20CA (coated with FG/SA-20CA active coating solution). Turbot samples from each batch were submerged in a coating solution at a ratio of 1:3 (w/v) for 10 min at 4 °C. A sterile biochemical incubator was used to generate the coating, and the samples were held at 4 °C for 60 min. Fish samples were removed from storage for analytical purposes on days 0, 3, 6, 9, 12, 15, and 18.

2.3. Bacteria Analysis

A representative 5 g sample of turbot dorsal muscle from each group was homogenized in 45 mL of saline solution, followed by a series of gradient dilutions. Then, 100 μL of bacterial solution (*Psychrophilic*, *Pseudomonas* spp., H_2S-producing, and lactic acid bacteria) were added to the muscle/saline medium. *Psychrophilic* bacteria were cultured on plate agar medium. *Pseudomonas* spp. were cultured on cetrimide agar medium. The H_2S-producing bacteria were cultured on iron agar medium and appear as black colonies. Lactic acid bacteria were cultured on MRS medium. *Psychrophilic* bacteria were counted after 7 days of incubation at 4 °C, and the remaining bacterial strains were cultured at 30 °C for 48 h before counting. The counts of all species of bacteria were calculated as lg CFU/g [19].

2.4. Total Volatile Basic Nitrogen (TVB-N) Analysis

Representative 5 g samples of turbot dorsal muscle from each of the four experimental groups were taken for determination of TVB-N values using a Kjeltec analyzer (FOSS 8400, Hilleroed, Denmark). The results were expressed as mg N/100 g.

2.5. K Value Analysis

A representative 5 g sample of turbot dorsal muscle from each experimental group was added to 10 mL 10% perchloric acid (PCA) and homogenized. The supernatant was collected after centrifugation at 4 °C at 8000× g for 10 min. The precipitate was mixed with 10 mL 5% PCA solution, and then centrifuged again at 4 °C at 8000× g for 10 min, repeated twice, to extract the supernatant. All supernatants were combined before adding 15 mL of ultra-pure water. The pH of the supernatant was adjusted to 6.5 with 10 mol/L KOH solution, 1 mol/L KOH solution, 5% PCA solution, and 10% PCA solution. After standing for 30 min, the supernatant was transferred to a 50 mL volumetric flask and diluted to 50 mL with ultrapure water. It was filtered with a 0.22 μm filter membrane and then injected into a sample bottle for the determination of each K value. Using the method of Cen et al. [20], ATP-related compounds were measured by HPLC (Waters e2695, Milford, CT, USA) and calculated as follows:

$$K \text{ value} = \frac{Hx + HxR}{ADP + AMP + Hx + HxR + ATP + IMP} \times 100\%$$

In the formula, HxR denotes inosine; Hx denotes hypoxanthine; ATP denotes adenosine triphosphate; ADP denotes adenosine diphosphate; AMP denotes adenosine 5′-monophosphate; and IMP denotes adenosine monophosphate.

2.6. Thiobarbituric Acid Reactive Substances (TBARS) Analysis

TBARS values were determined using the method devised by Chen et al. [21]. A representative 5 g sample of smashed dorsal muscle of turbot from each group was added to 97.5 mL of distilled water and 2.5 mL of 4 M HCl. Then, a 5 mL aliquot of the fraction was mixed with 5 mL of thiobarbituric reaction reagent (consisting of 0.02 M TBA and 90% glacial acetic acid) and heated in boiling water for 35 min. After cooling, the absorbance of the pink solution was quantified using a spectrophotometer at 532 nm and 600 nm. TBARS values are expressed as mg MDA/100 g. The formula is as follows:

$$\text{TBARS values}(\text{mg MDA}/100\text{ g}) = \frac{A_{532} - A_{600}}{155} \times 726$$

In the formula, A_{532} indicates the absorbance value of the solution at 532 nm; A_{600} indicates the absorbance value of the solution at 600 nm.

2.7. Myofibrillar Protein (MP) Extraction

The MP solutions were prepared with reference to the method described by Tan et al. [22], with some modifications. A representative 2 g sample of the dorsal muscle of turbot was selected and homogenized with 20 mL of 20 mmol/L cold buffer, with a buffer concentration of 20 mmol/L and 0.05 mol/L NaCl. The precipitate was obtained by subjecting it to centrifugation at 4 °C and 10,000× g for 15 min, which was repeated twice. The collected precipitates were combined with 20 mL of a cold sodium phosphate buffer solution (20 mmol/L, comprising 0.6 mol/L NaCl) and subjected to extraction at 4 °C for 2 h. Finally, the supernatants were collected by repeated centrifugation.

2.8. Measurement of Sulfydryl Groups (SH) and Carbonyl Groups

The index was determined according to the method of Xu et al. [23]. The SH contents measured as μmol/g of protein were determined using the extinction coefficient of 2-nitro-5-thiobenzoate (NTB). The carbonyl content was determined by 2,4-dinitrophenylhydrazine derivatization. The results are given as μmol/g.

2.9. Measurement of Ca^{2+}-ATPase Activity

A representative 5 g sample of turbot dorsal muscle from each experimental group was added to 18 mL of 0.85% NaCl saline solution and homogenized in an ice bath. The solution underwent centrifugation at a force of 2500× g of 15 min at 4 °C. The homogenized liquid was mixed with 0.85% NaCl solution in a ratio of 1:10. The quantitative determination of Ca^{2+}-ATPase was performed using a micro assay kit provided by the Nanjing Jianjian Bioengineering Institute (Nanjing, China). The measurement of the absorption spectra was conducted at a wavelength of 636 nm. The ultimate outcome was expressed in μmol/mg protein.

2.10. Surface Hydrophobicity of Myofibrillar Protein (MP)

Dilutions of MP solution were prepared to provide final concentrations of 0.2, 0.4, 0.8, and 1.0 mg/mL. A 4 mL sample of diluted MP solution was combined with 20 μL of ANS (8 mmol/L) and kept in darkness for 10 min. A fluorescence spectrophotometer (F-7100, Hitachi, Tokyo, Japan), with a slit width of 10 nm, an emission wavelength of 470 nm, and a detection wavelength of 390 nm was used to measure the relative fluorescence intensity (RFI) of the mixes [22].

2.11. Statistical Analysis

The experiments were carried out in triplicate. The results are shown as mean ± standard deviation. The data were analyzed by SPSS 22.0 analysis of variance (ANOVA). Graphing the data was done using Origin 2018.

3. Results and Discussion

3.1. Microbiological Analysis

The growth of five different microorganisms in turbot during cold storage is illustrated in Figure 1. The TVC of turbot was 3.3 lg CFU/g at the beginning of day 0, reflecting that the turbot was of good quality. Figure 1 shows an increase in TVC for all samples. However, the TVCs of the samples treated by other concentrations of CA were significantly higher than that of FG/SA-20CA during the whole storage process ($p < 0.05$). The microbiologically unacceptable limit is 7 lg CFU/g [24]. The CK, FG/SA-5CA, FG/SA-10CA, and FG/SA-20CA groups exceeded this limit on the 6th, 9th, 12th and 18th day, respectively.

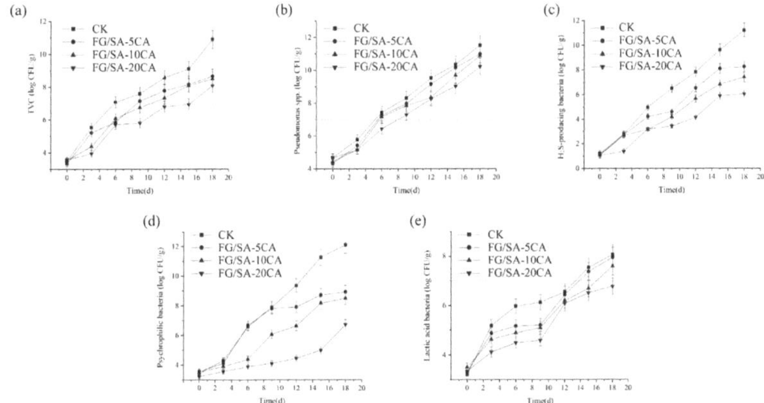

Figure 1. The growth trends of bacteria on turbot muscle with or without various concentrations of active coatings of FG/SA-CA during cold storage: total viable counts (**a**); *Pseudomonas* spp. counts (**b**); H_2S-producing bacteria counts (**c**); *Psychrophilic* bacteria counts (**d**); lactic acid bacteria counts (**e**).

The FG/SA-CA treatments showed a remarkable suppressive effect on bacteria ($p < 0.05$), and the inhibitory effect was more obvious at high concentrations. A lipid component of the bacterial cell wall may be penetrated by carvacrol, which then allows it to invade the bacterium. This changes the protein structure of the bacteria, causing it to die [25]. FG/SA-CA treatment showed similar inhibitory effects on the production of *Pseudomonas* spp., *Psychrophiles* bacteria, H_2S-producing bacteria, and lactic acid bacteria in turbot. The numbers of *Psychrophiles* in the FG/SA-10CA and FG/SA-CA samples were lower than that in CK, demonstrating that carvacrol effectively slowed down the growing process of *Psychrophilic* bacteria.

Pseudomonas spp. is a kind of bacteria that lives and flourishes in the presence of oxygen. It is one of the causes of spoilage of turbot during refrigerated storage [26]. The addition of carvacrol to the active coating decreased the number of *Pseudomonas* spp., suggesting that the colonization of aerobic spoilage bacteria is the primary factor in the deterioration of turbot that occurs during cold storage. The H_2S-producing bacteria are also a type of special spoilage bacteria in turbot during cold storage [27]. Bacterial counts for H_2S producing bacteria prior to the start of storage were 1.1 lg CFU/g. During refrigeration, turbot treated with carvacrol reduced the growth of H_2S-producing bacteria. Lactic acid bacteria generated organic acids in turbot during cold storage [28]. *Psychrophilic* bacteria produce metabolic chemicals such as volatile sulfides, ketones, biogenic amines and aldehydes that deteriorate the odor, texture and flavor of turbot. Following the conclusion of the storage period, the number of bacteria in the FG/SA-5CA, FG/SA-10CA and FG/SA-20CA treated groups decreased by 2.3%, 11.3%, and 13.7% compared to that of CK, respectively. Therefore, FG/SA-CA is effective in inhibiting the growth of these bacteria, and FG/SA-20CA has the strongest inhibitory effect. For this reason, FG/SA-20CA had the best effect of delaying the spoilage of turbot fillets. This is similar to the findings of Li et al. [29], who found that microencapsulated eugenol emulsion treatment of perch was effective in inhibiting microbial growth during refrigeration.

3.2. TVB-N Analysis

TVB-N indicates the content of volatile basic compounds and is one of the main indexes to determine the freshness in turbot [30]. Many basic volatile compounds are formed during the decomposition of proteinaceous and nonproteinaceous nitrogenous compounds. During cold storage, the spoilage bacteria degraded the proteins in the fish muscle and accumulated rapidly. As shown in Figure 2a, the turbot sample had a TVB-N value of 11 mg N/100 g on day 0. This indicates that the turbot samples were in a fresh state before treatment. In the above analysis, turbot also had a low initial TVC value, both of which testified to its freshness. During cold storage, TVB-N values increased due to bacterial growth and multiplication. The allowable threshold of TVB-N is 25 mg N/100 g [26] for marine fish. The turbot in the CK sample exceeded the acceptable limits for TVB-N values after 12 days of storage. Notably, all turbot treated with FG/SA-CA had lower TVB-N values. In particular, these values in the FG/SA-10CA and FG/SA-20CA groups remained below the limit throughout the cold storage. These two active coatings could therefore lead to reduced spoilage. This indicates that FG/SA-C inhibits microbial growth and slows down protein breakdown and amine production. Additionally, the TVB-N value of turbot treated with a high concentration of carvacrol was lower. The FG/SA-5CA group enabled a longer shelf life of about 4 d, and the FG/SA-10CA and FG/SA-20CA groups extended the shelf life by about 6 d. This is analogous to the findings of Li et al. [29], who found that TVB-N values were reduced when sea bass were treated with high concentrations of eugenol.

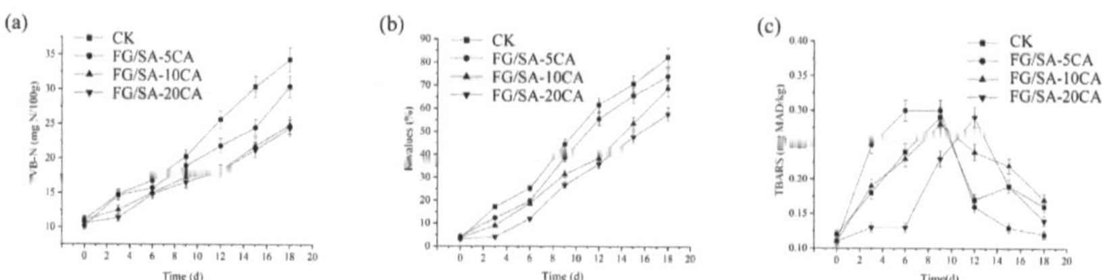

Figure 2. Changing trends in TVB-N (**a**), K values (**b**), and TBARS value (**c**) of turbot during cold storage.

3.3. K Values Analysis

The K values calculated from ATP catabolism are indispensable indicators for determining the spoilage of turbot [31]. The level of rejection by the K value is 60% [32]. Figure 2b illustrates the variations of ATP-related compounds during the cold storage period in turbot. As the duration of storage increases, the K values of turbot increased from 3.21% (d 0) to 80.25% (day 18) for the CK sample. By the end of cold storage, K values did not exceed 60% on day 18 in the FSG/SA-20CA group, but ranged from 60 to 80% in the FSG/SA-10CA and FSG/SA-5CA groups. The higher concentrations of FSG/SA-CA inhibited bacterial activity and nucleotide degradation, further delaying the increase in K values with storage time, which supports the conclusion of the TVB-N experiments. This result is also consistent with the colony count trends. It was demonstrated that FG/SA-CA could prevent the degradation of ATP controlled by enzymes and microorganisms, and could maintain good quality turbot during cold storage. This is analogous to the findings of Bazargani et al. [33], who investigated the effect of adding resveratrol to a sodium alginate coating on rainbow trout fillets during cold storage. They found that this treatment can effectively slow the increase of K values.

3.4. TBARS Value Analysis

TBARS represents the level of lipid oxidation. Elevated lipid oxidation levels result in the heightened buildup of lipid peroxides and the subsequent production of related secondary metabolites [34]. While turbot body fat content is low, the proportion of unsaturated fatty acids is high, and fat oxidation and acid produce odor and reduce food quality. Malondialdehyde (MDA) is one of the important oxidation products of polyunsaturated fatty acids [30]. Accumulation of MDA content reflects the increase in TBARS content during cold storage. The MDA content of the turbot sample on day 0 was 0.12 mg MDA/100 g (Figure 2c). The TBARS value exhibited an initial rise followed by a subsequent drop in all samples. On the 6th day, the FG/SA-CA groups possessed lower TBARS values in comparison to the CK group, indicating that the FG/SA-CA inhibited fatty acid oxidation. The antioxidant effect of carvacrol caused a reduction in lipid oxidation and a decrease in the oxygen permeability of fish lipids [35]. Kostaki et al. [36] suggested that since MDA had the potential to interact with other components in fish such as furfural, alkyl aldehyde, alkenal, ketones, and carbohydrates, thus creating secondary metabolites, the MDA content in fish decreases with storage. The other reason could be attributed to the interaction between MDA and other volatile compounds [37]. However, the TBARS values were consistently low and irregular, so the results obtained did not reflect the oxidation of turbot lipids satisfactorily.

3.5. Changes of Residual Groups in Amino Acid Side-Chains

3.5.1. Total SH Counts Analysis

It is possible to determine the oxidation of proteins during fish preservation by analyzing the levels of sulfhydryl (SH) and carbon groups. The SH groups are amongst most dynamic functional groups of protein [38]. They are easily oxidized to disulfide bonds and other oxidation products as storage time increases. It is the usual practice to take into account variations from the SH groups when determining the oxidation levels of proteins found in marine commodities [39]. As depicted in Figure 3a, the total SH counts were considerably reduced across all experimental groups throughout the storage period ($p < 0.05$). The initial SH values in the CK, FG/SA-5CA, FG/SA-10CA-CA, and FG/SA-20CA groups were approximately 41.96, 41.46, 42.32, and 41.75, respectively. The initial SH counts of CK, FG/SA-5CA, FG/SA-10CA, and FG/SA-20CA groups on the 18th day decreased by 42.12%, 47.93%, 35.52%, and 33.50%, respectively. It is noteworthy that the SH values of the CK group were significantly lower than those of the FG/SA-CA-treated groups at the same storage time ($p < 0.05$), and were particularly significant for the FG/SA-20CA group ($p < 0.05$). In addition, previous research has shown that several SH groups formed as a result of increased interactions within and between proteins caused by environmental and extrinsic energy changes [40]. The results of SH content were very similar to those of Pei et al. [41], who found that epigallocatechin gallate had a protective effect on the side chain groups of myofibrillar protein.

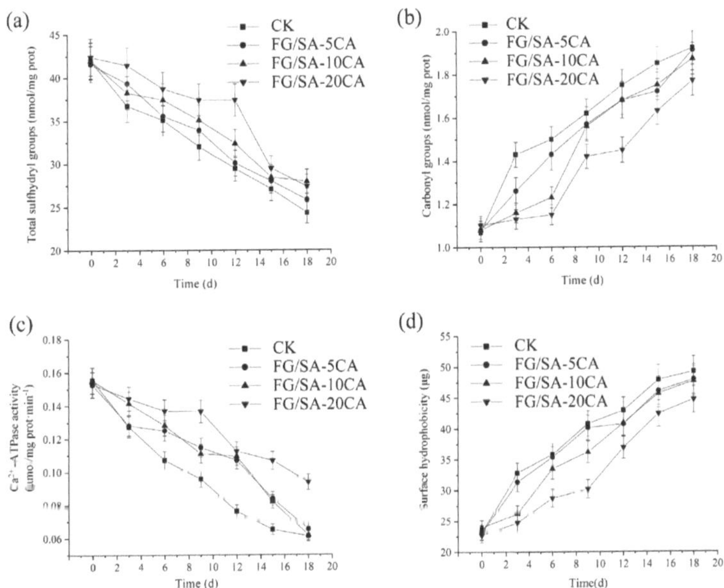

Figure 3. Changing trends in total sulfhydryl counts (**a**), carbonyl counts (**b**), Ca^{2+}-ATPase activity (**c**) and surface hydrophobicity (**d**) of turbot during cold storage.

3.5.2. Carbonyl Counts Analysis

The carbonyl groups are the primary chemical byproducts of MP oxidation. The oxidation of MPs causes backbone splitting and cross-linkage, resulting in a number of amino acid residues to turn into carbonyl groups [42]. As depicted in Figure 3b, the carbonyl content of all substances rose as the storage duration increased, suggesting that MP oxidation occurred during cold storage. The carbonyl groups in the CK group increased rapidly from 1.07 nmol/mg protein at day 0 to 1.92 nmol/mg protein at the end of storage. However, in the FG/SA-20CA group, the carbonyl groups were measured as 1.77 nmol/mg protein at the end of storage, showing that the FSG/SA-CA active coatings protected MPs from oxidation. Chamba et al. [43] previously reported an analogous outcome; discovering that there were no major differences in the carbonyl groups between FG/SA-5CA and FG/SA-10CA ($p > 0.05$). Similarly, Zhao et al. [42] found that the effect of grape seed extract on the carbonyl group was not concentration dependent. It can be inferred that concentration of the antioxidants being used has an inverse relationship with the carbonyl content of the sample.

3.6. Ca^{2+}-ATPase Activity Analysis

The activities of Ca^{2+}-ATPase could reflect the oxidative decomposition of MP [44]. Figure 3c shows the trend of Ca^{2+}-ATPase activity in each experimental group. Ca^{2+}-ATPase activity on day 0 was approximately 0.152 μmol/mg protein·min^{-1} (Figure 3c). At the end of the storage period, the Ca^{2+}-ATPase activity of CK, FG/SA-5CA, FG/SA-10CA, and FG/SA-20CA groups increased by 74.54, 73.63, 69.54, and 60.45%, respectively. The present study showed that carvacrol was protective against Ca^{2+}-ATPase activities, probably due to the inhibition of oxidative denaturation of myosin by the active coatings with the addition of carvacrol, similar to the results of the carbonyl groups. Through their research, Reza et al. [45] found that even minute alterations to the structure of the MP led to a reduction in the activity of Ca^{2+}-ATPase. It is possible that the function of Ca^{2+}-ATPase decreases as a result of the aggregation of myosin and the oxidation of the head SH bond. The inhibition of Ca^{2+}-ATPase function was associated with the modification of protein structure resulting from the creation and interconnection of disulfide bonds within or between polypeptides [46]. He et al. [47] found that eugenol had a certain inhibitory effect on proteases, which could slow down alteration in Ca^{2+}-ATPase activity.

3.7. Surface Hydrophobicity Analysis

The degree of protein denaturation could also be measured by surface hydrophobicity. Exposing hydrophobic groups results in an elevation of protein hydrophobicity, hence increasing the extent of MP denaturation [48]. The protein surface hydrophobicity increased with storage time (Figure 3d). This may be due to the slow dissociation rate of MP molecules during storage, resulting in the exposure of hydrophobic amino acid residues and the change of hydrophilic and hydrophobic groups [49]. On day 0, there was no significant difference in surface hydrophobicity between coated and uncoated proteins ($p > 0.05$). The surface hydrophobicity of MP in each group increased significantly with the prolongation of the storage term ($p < 0.05$). In contrast, the growth was slower in the carvacrol-treated group, probably due to the strong antioxidant properties of carvacrol that inhibited protein degradation [18]. This is analogous to the findings of Hu et al. [50], who treated beef with tea polyphenols, and found that the trend of increasing surface hydrophobicity was slowed down during cold storage. Therefore, turbot that has been treated with carvacrol has the potential to inhibit the exposure of some hydrophobic groups.

4. Conclusions

The results of the study showed that the FG/SA-CA active coating inhibits bacterial growth and reduces TVB-N, TBARS, and K levels, thus maintaining the high quality of turbot during cold storage and extending the quality guarantee period of food. The application of FG/SA-CA treatment can successfully suppress the oxidation of fish proteins and the formation of alkaline compounds, such as amines, in the fish. The shelf life of turbot is greatly extended as a result of this treatment, which also helps to maintain the fish's distinctive freshness. The preservation effect was augmented as the carvacrol level in the film increased. In comparison with the CK group, the FG/SA-CA groups exhibited obvious protective functions against protein oxidation, preventing the addition of carbonyl counts and the reduction in SH counts and Ca^{2+}-ATPase activity. Furthermore, the composite coating treatment suppressed the enhancement of surface hydrophobicity and stabilized the protein structure. Overall, FG/SA-20CA indicated optimal preservation performance and prolonged the shelf life for turbot during cold storage.

Author Contributions: Conceptualization, X.Y. and Y.X.; methodology, X.Y. and Y.X.; software, X.Y.; validation, S.F.; formal analysis, X.Y. and Y.X.; investigation, X.Y.; resources, J.M.; data curation, X.Y.; writing—original draft preparation, X.Y.; writing—review and editing, J.M. and J.X.; visualization, J.M. and J.X.; supervision, J.M. and J.X.; project administration, J.M. and J.X.; funding acquisition, J.X. All authors have read and agreed to the published version of the manuscript.

Funding: This work was supported by China Agriculture Research System (CARS-47), Shanghai Municipal Science and technology project to enhance the capabilities of the platform (20DZ2292200, 19DZ2284000).

Institutional Review Board Statement: Not applicable.

Informed Consent Statement: Not applicable.

Data Availability Statement: The data presented in this study are available in article. Data available in a publicly accessible repository.

Conflicts of Interest: The authors declare no conflicts of interest.

References

1. Jie, C.; Meijie, G.; Weiqiang, Q.; Jun, M.; Jing, X. Effect of tea polyphenol-trehalose complex coating solutions on physiological stress and flesh quality of marine cultured turbot (*Scophthalmus maximus*) during waterless transport. *J. Aquat. Anim. Health* **2024**. [CrossRef]
2. Ertan, K.; Celebioglu, A.; Chowdhury, R.; Sumnu, G.; Sahin, S.; Altier, C.; Uyar, T. Carvacrol/cyclodextrin inclusion complex loaded gelatin/pullulan nanofibers for active food packaging applications. *Food Hydrocoll.* **2023**, *142*, 108864. [CrossRef]
3. Chaparro-Hernández, S.; Ruíz-Cruz, S.; Márquez-Ríos, E.; Ocaño-Higuera, V.M.; Valenzuela-López, C.C.; Ornelas-Paz, J.D.; Del-Toro-Sánchez, C.L. Effect of chitosan-carvacrol edible coatings on the quality and shelf life of tilapia (*Oreochromis niloticus*) fillets stored in ice. *Food Sci. Technol.* **2015**, *35*, 734–741. [CrossRef]
4. Alves, V.; Rico, B.P.M.; Cruz, R.M.S.; Vicente, A.A.; Khmelinski, I.; Vieira, M.C. Preparation and characterization of a chitosan film with grape seed extract-carvacrol microcapsules and its effect on the shelf-life of refrigerated Salmon (*Salmo salar*). *LWT-Food Sci. Technol.* **2018**, *89*, 525–534. [CrossRef]
5. Guo, Q.; Li, S.; Du, G.A.; Chen, H.; Yan, X.H.; Chang, S.D.; Yue, T.L.; Yuan, Y.H. Formulation and characterization of microcapsules encapsulating carvacrol using complex coacervation crosslinked with tannic acid. *LWT-Food Sci. Technol.* **2022**, *165*, 113683. [CrossRef]
6. Domínguez, R.; Barba, F.J.; Gómez, B.; Putnik, P.; Kovacevic, D.B.; Pateiro, M.; Santos, E.M.; Lorenzo, J.M. Active packaging films with natural antioxidants to be used in meat industry: A review. *Food Res. Int.* **2018**, *113*, 93–101. [CrossRef]
7. Prasad, P.; Kochhar, A. Active Packaging in Food Industry: A Review. *IOSR J. Environ. Sci. Toxicol. Food Technol.* **2014**, *8*, 01–07. [CrossRef]
8. Remya, S.; Sivaraman, G.K.; Joseph, T.C.; Parmar, E.; Sreelakshmi, K.R.; Mohan, C.O.; Ravishankar, C.N. Influence of corn starch based bio-active edible coating containing fumaric acid on the lipid quality and microbial shelf life of silver pomfret fish steaks stored at 4 °C. *J. Food Sci. Technol.* **2022**, *59*, 3387–3398. [CrossRef]
9. Yang, H.; Li, Q.; Xu, Z.; Ge, Y.; Zhang, D.; Li, J.; Sun, T. Preparation of three-layer flaxseed gum/chitosan/flaxseed gum composite coatings with sustained-release properties and their excellent protective effect on myofibril protein of rainbow trout. *Int. J. Biol. Macromol.* **2022**, *194*, 510–520. [CrossRef]
10. Zhang, M.; Tao, N.; Li, L.; Xu, C.; Deng, S.; Wang, Y. Non-migrating active antibacterial packaging and its application in grass carp fillets. *Food Packag. Shelf Life* **2022**, *31*, 100786. [CrossRef]
11. Li, Y.; Zhang, L.T.; Zhuang, S.; Li, D.P.; Hong, H.; Lametsch, R.; Tan, Y.Q.; Luo, Y.K. Shelf life extension of chilled blunt snout bream fillets using coating based on chia seed gum and *Oleum ocimi gratissimi*. *Food Biosci.* **2023**, *54*, 102853. [CrossRef]

12. Xu, Z.; Pei, J.; Mei, J.; Yu, H.; Chu, S.; Xie, J. Effect of gum tragacanth-sodium alginate coatings incorporated with epigallocatechin gallate on the quality and shelf life of large yellow croaker (*Larimichthys crocea*) during superchilling storage. *Food Qual. Saf.* **2023**, *8*, fyad039. [CrossRef]

13. Pei, J.X.; Mei, J.; Yu, H.J.; Qiu, W.Q.; Xie, J. Effect of Gum Tragacanth-Sodium Alginate Active Coatings Incorporated with Epigallocatechin Gallate and Lysozyme on the Quality of Large Yellow Croaker at Superchilling Condition. *Front. Nutr.* **2022**, *8*, 812741. [CrossRef]

14. Ren, X.J.; Meng, X.; Zhang, Z.; Du, H.Y.; Li, T.P.; Wang, N. Effects of Ultrasound-Assisted Extraction on Structure and Rheological Properties of Flaxseed Gum. *Gels* **2023**, *9*, 318. [CrossRef]

15. Liu, J.; Shim, Y.Y.; Tse, T.J.; Wang, Y.; Reaney, M.J.T. Flaxseed gum a versatile natural hydrocolloid for food and non-food applications. *Trends Food Sci. Technol.* **2018**, *75*, 146–157. [CrossRef]

16. Presenza, L.; Teixeira, B.F.; Galvao, J.A.; Vieira, T. Technological strategies for the use of plant-derived compounds in the preservation of fish products. *Food Chem.* **2023**, *419*, 136069. [CrossRef]

17. Fang, S.Y.; Qiu, W.Q.; Mei, J.; Xie, J. Effect of Sonication on the Properties of Flaxseed Gum Films Incorporated with Carvacrol. *Int. J. Mol. Sci.* **2020**, *21*, 1637. [CrossRef]

18. Fang, S.Y.; Zhou, Q.Q.; Hu, Y.; Liu, F.; Mei, J.; Xie, J. Antimicrobial Carvacrol Incorporated in Flaxseed Gum-Sodium Alginate Active Films to Improve the Quality Attributes of Chinese Sea bass (*Lateolabrax maculatus*) during Cold Storage. *Molecules* **2019**, *24*, 3292. [CrossRef]

19. Saelens, G.; Houf, K. The involvement of *Pseudoterranova decipiens* fish infestation on the shelf-life of fresh Atlantic cod (*Gadus morhua*) fillet. *Int. J. Food Microbiol.* **2024**, *410*, 110426. [CrossRef]

20. Cen, S.J.; Fang, Q.; Tong, L.; Yang, W.G.; Zhang, J.J.; Lou, Q.M.; Huang, T. Effects of chitosan-sodium alginate-nisin preservatives on the quality and spoilage microbiota of *Penaeus vannamei* shrimp during cold storage. *Int. J. Food Microbiol.* **2021**, *349*, 109227. [CrossRef] [PubMed]

21. Chen, J.W.; Li, Y.X.; Wang, Y.F.; Yakubu, S.; Tang, H.B.; Li, T. Active polylactic acid/tilapia fish gelatin-sodium alginate bilayer films: Application in preservation of Japanese sea bass (*Lateolabrax japonicus*). *Food Packag. Shelf Life* **2022**, *33*, 100915. [CrossRef]

22. Tan, M.T.; Ding, Z.Y.; Mei, J.; Xie, J. Effect of cellobiose on the myofibrillar protein denaturation induced by pH changes during freeze-thaw cycles. *Food Chem.* **2022**, *373*, 131511. [CrossRef]

23. Xu, Z.; Zhao, X.; Yang, W.; Mei, J.; Xie, J. Effect of magnetic nano-particles combined with multi-frequency ultrasound-assisted thawing on the quality and myofibrillar protein-related properties of salmon (*Salmo salar*). *Food Chem.* **2024**, *445*, 138701. [CrossRef]

24. Remya, S.; Mohan, C.O.; Venkateshwarlu, G.; Sivaraman, G.K.; Nagarajarao, C.; Ravishankar, C.N. Combined effect of O_2 scavenger and antimicrobial film on shelf life of fresh cobia (*Rachycentron canadum*) fish steaks stored at 2 °C. *Food Control.* **2017**, *71*, 71–78. [CrossRef]

25. Laorenza, Y.; Harnkarnsujarit, N. Carvacrol, citral and α-terpineol essential oil incorporated biodegradable films for functional active packaging of Pacific white shrimp. *Food Chem.* **2021**, *363*, 130252. [CrossRef]

26. Wenru, L.; Jun, M.; Jing, X. Effect of locust bean gum-sodium alginate coatings incorporated with daphnetin emulsions on the quality of *Scophthalmus maximus* at refrigerated condition. *Int. J. Biol. Macromol.* **2021**, *170*, 129–139. [CrossRef]

27. Cheng, J.H.; Sun, D.W.; Wei, Q.Y. Enhancing Visible and Near-Infrared Hyperspectral Imaging Prediction of TVB-N Level for Fish Fillet Freshness Evaluation by Filtering Optimal Variables. *Food Anal. Methods* **2017**, *10*, 1888–1898. [CrossRef]

28. Cai, L.Y.; Cao, A.L.; Li, T.T.; Wu, X.S.; Xu, Y.X.; Li, J.R. Effect of the Fumigating with Essential Oils on the Microbiological Characteristics and Quality Changes of Refrigerated Turbot (*Scophthalmus maximus*) Fillets. *Food Bioprocess Technol.* **2015**, *8*, 844–853. [CrossRef]

29. Li, P.Y.; Peng, Y.F.; Mei, J.; Xie, J. Effects of microencapsulated eugenol emulsions on microbiological, chemical and organoleptic qualities of farmed Japanese sea bass (*Lateolabrax japonicus*) during cold storage. *LWT-Food Sci. Technol.* **2020**, *118*, 108831. [CrossRef]

30. Bian, C.H.; Yu, H.J.; Yang, K.; Mei, J.; Xie, J. Effects of single-, dual-, and multi-frequency ultrasound-assisted freezing on the muscle quality and myofibrillar protein structure in large yellow croaker (*Larimichthys crocea*). *Food Chem.-X* **2022**, *15*, 100362. [CrossRef] [PubMed]

31. Esua, O.J.; Sun, D.W.; Cheng, J.H.; Li, J.L. Evaluation of storage quality of vacuum-packaged silver Pomfret (*Pampus argenteus*) treated with combined ultrasound and plasma functionalized liquids hurdle technology. *Food Chem.* **2022**, *391*. [CrossRef]

32. Lan, W.; Che, X.; Xu, Q.; Wang, T.; Du, R.; Xie, J.; Hou, M.; Lei, H. Sensory and chemical assessment of silver pomfret (*Pampus argenteus*) treated with *Ginkgo biloba* leaf extract treatment during storage in ice. *Aquac. Fish.* **2018**, *3*, 30–37. [CrossRef]

33. Bazargani-Gilani, B.; Pajohi-Alamoti, M. The effects of incorporated resveratrol in edible coating based on sodium alginate on the refrigerated trout (*Oncorhynchus mykiss*) fillets' sensorial and physicochemical features. *Food Sci. Biotechnol.* **2020**, *29*, 207–216. [CrossRef]

34. Le, T.; Takahashi, K.; Okazaki, E.; Osako, K. Mitigation of lipid oxidation in tuna oil using gelatin pouches derived from horse mackerel (*Trachurus japonicus*) scales and incorporating phenolic compounds. *LWT-Food Sci. Technol.* **2020**, *128*, 109533. [CrossRef]

35. Jonusaite, K.; Venskutonis, P.R.; Martínez-Hernández, G.B.; Taboada-Rodríguez, A.; Nieto, G.; López-Gómez, A.; Marín-Iniesta, F. Antioxidant and Antimicrobial Effect of Plant Essential Oils and *Sambucus nigra* Extract in Salmon Burgers. *Foods* **2021**, *10*, 776. [CrossRef]

36. Kostaki, M.; Giatrakou, V.; Savvaidis, I.N.; Kontominas, M.G. Combined effect of MAP and thyme essential oil on the microbiological, chemical and sensory attributes of organically aquacultured sea bass (*Dicentrarchus labrax*) fillets. *Food Microbiol.* **2009**, *26*, 475–482. [CrossRef]

37. Piccini, J.L.; Evans, D.R.; Quaranta, H.O. Comparison of TBA number of irradiated fish with sensory quality. *Food Chem.* **1986**, *19*, 163–171. [CrossRef]

38. Gao, W.H.; Wu, X.R.; Ye, R.S.; Zeng, X.A.; Brennan, M.A.; Brennan, C.S.; Ma, J. Analysis of protein denaturation, and chemical visualisation, of frozen grass carp surimi containing soluble soybean polysaccharides. *Int. J. Food Sci. Technol.* **2022**, *57*, 5504–5513. [CrossRef]

39. Zhang, L.T.; Li, Q.; Bao, Y.L.; Tan, Y.Q.; Lametsch, R.; Hong, H.; Luo, Y.K. Recent advances on characterization of protein oxidation in aquatic products: A comprehensive review. *Crit. Rev. Food Sci. Nutr.* **2022**, *64*, 1572–1591. [CrossRef] [PubMed]

40. Ko, W.C.; Yu, C.C.; Hsu, K.C. Changes in conformation and sulfhydryl groups of tilapia actomyosin by thermal treatment. *LWT-Food Sci. Technol.* **2007**, *40*, 1316–1320. [CrossRef]

41. Pei, J.X.; Mei, J.; Wu, G.; Yu, H.J.; Xie, J. Gum tragacanth-sodium alginate active coatings containing epigallocatechin gallate reduce hydrogen peroxide content and inhibit lipid and protein oxidations of large yellow croaker (*Larimichthys crocea*) during superchilling storage. *Food Chem.* **2022**, *397*, 133792. [CrossRef]

42. Zhao, X.; Zhou, Y.G.; Zhao, L.; Chen, L.; He, Y.; Yang, H.S. Vacuum impregnation of fish gelatin combined with grape seed extract inhibits protein oxidation and degradation of chilled tilapia fillets. *Food Chem.* **2019**, *294*, 316–325. [CrossRef]

43. Chamba, M.V.M.; Hua, Y.F.; Katiyo, W. Oxidation and Structural Modification of Full-Fat and Defatted Flour Based Soy Protein Isolates Induced by Natural and Synthetic Extraction Chemicals. *Food Biophys.* **2014**, *9*, 193–202. [CrossRef]

44. Yang, F.; Jia, S.N.; Liu, J.X.; Gao, P.; Yu, D.W.; Jiang, Q.X.; Xu, Y.S.; Yu, P.P.; Xia, W.S.; Zhan, X.B. The relationship between degradation of myofibrillar structural proteins and texture of superchilled grass carp (*Ctenopharyngodon idella*) fillet. *Food Chem.* **2019**, *301*, 125278. [CrossRef]

45. Reza, M.S.; Bapary, M.A.J.; Ahasan, C.T.; Islam, M.N.; Kamal, M. Shelf life of several marine fish species of Bangladesh during ice storage. *Int. J. Food Sci. Technol.* **2009**, *44*, 1485–1494. [CrossRef]

46. Kong, B.H.; Guo, Y.Y.; Xia, X.F.; Liu, Q.; Li, Y.Q.; Chen, H.S. Cryoprotectants Reduce Protein Oxidation and Structure Deterioration Induced by Freeze-Thaw Cycles in Common Carp (*Cyprinus carpio*) Surimi. *Food Biophys.* **2013**, *8*, 104–111. [CrossRef]

47. Cui, H.Y.; Zhang, C.H.; Li, C.Z.; Lin, L. Antibacterial mechanism of oregano essential oil. *Ind. Crops Prod.* **2019**, *139*, 111498. [CrossRef]

48. Chelh, I.; Gatellier, P.; Santé-Lhoutellier, V. Technical note: A simplified procedure for myofibril hydrophobicity determination. *Meat Sci.* **2006**, *74*, 681–683. [CrossRef] [PubMed]

49. Benjakul, S.; Visessanguan, W.; Thongkaew, C.; Tanaka, M. Effect of frozen storage on chemical and gel-forming properties of fish commonly used for surimi production in Thailand. *Food Hydrocoll.* **2005**, *19*, 197–207. [CrossRef]

50. Hu, Y.P.; Gao, Y.F.; Solangi, I.; Liu, S.C.; Zhu, J. Effects of tea polyphenols on the conformational, functional, and morphological characteristics of beef myofibrillar proteins. *LWT-Food Sci. Technol.* **2022**, *154*, 112596. [CrossRef]

MDPI AG
Grosspeteranlage 5
4052 Basel
Switzerland
Tel.: +41 61 683 77 34

Coatings Editorial Office
E-mail: coatings@mdpi.com
www.mdpi.com/journal/coatings